Reducing salt in foods

Related titles:

Optimising sweet taste in foods
(ISBN-13: 978-1-84569-008-3; ISBN-10: 1-84569-008-7)
Consumer acceptance or rejection of a food can often be due to its taste. Sweet taste is especially attractive to the consumer and it is important to optimise this quality in food products. A wide range of compounds can be used to sweeten foods and with today's interest in diet and health, calorific sweeteners are often replaced with non-nutritive alternatives. Producing a high quality food product using alternatives to sugar, though, is not straightforward, as non-nutritive and low-calorie sweeteners do not have the same taste profiles and functional characteristics as sugar. With contributions by distinguished authors, this new book reviews factors affecting sweet taste perception, the types of sweet-tasting compound and their use in food products.

Flavour in food
(ISBN-13: 978-1-85573-960-4; ISBN-10: 1-85573-960-7)
The flavour of a food is one of its most important qualities. Edited by two leading authorities in the field, and with a distinguished international team of contributors, this important collection summarises the wealth of recent research on how flavour develops in food and is then perceived by the consumer. The first part of the book reviews ways of measuring flavour. Part II looks at the ways flavour is retained and released in food. It considers the way flavour is retained in particular food matrices, how flavour is released during the process of eating, and the range of influences governing how flavour is perceived by the consumer. *Flavour in food* guides the reader through a complex subject and provides the essential foundation in both understanding and controlling food flavour.

Handbook of herbs and spices (in three volumes)
(Vol 1: ISBN-13: 978-1-85573-562-0; ISBN-10: 1-85573-562-8)
(Vol 2: ISBN-13: 978-1-85573-721-1; ISBN-10: 1-85573-721-3)
(Vol 3: ISBN-13: 978-1-84569-017-5; ISBN-10: 1-84569-017-6)
Herbs and spices are among the most versatile and widely used ingredients in food processing. As well as their traditional role in flavouring and colouring foods, they have been increasingly used as natural preservatives and for their potential health-promoting properties, for example as antioxidants. Edited by a leading authority in the field, and with an international team of contributors, the *Handbook of herbs and spices* provides an essential reference for manufacturers wishing to make the most of these important ingredients. Chapters cover key issues from definition and classification to chemical structure, cultivation and post-harvest processing, uses in food processing, functional properties, regulatory issues, quality indices and methods of analysis.

Details of these books and a complete list of Woodhead titles can be obtained by:

- visiting our web site at www.woodheadpublishing.com
- contacting Customer Services (email: sales@woodhead-publishing.com; fax: +44 (0) 1223 893694; tel.: +44 (0) 1223 891358 ext. 130; address: Woodhead Publishing Limited, Abington Hall, Abington, Cambridge CB21 6AH, England)

Reducing salt in foods

Practical strategies

Edited by
David Kilcast and Fiona Angus

CRC Press
Boca Raton Boston New York Washington, DC

WOODHEAD PUBLISHING LIMITED
Cambridge England

Published by Woodhead Publishing Limited, Abington Hall, Abington,
Cambridge CB21 6AH, England
www.woodheadpublishing.com

Published in North America by CRC Press LLC, 6000 Broken Sound Parkway, NW,
Suite 300, Boca Raton, FL 33487, USA

First published 2007, Woodhead Publishing Limited and CRC Press LLC
© 2007, Woodhead Publishing Limited
The authors have asserted their moral rights.

British Library Cataloguing in Publication Data
A catalogue record for this book is available from the British Library.

Library of Congress Cataloging-in-Publication Data
A catalog record for this book is available from the Library of Congress.

Woodhead Publishing Limited ISBN-13: 978-1-84569-018-2 (book)
Woodhead Publishing Limited ISBN-10: 1-84569-018-4 (book)
Woodhead Publishing Limited ISBN-13: 978-1-84569-304-6 (e-book)
Woodhead Publishing Limited ISBN-10: 1-84569-304-3 (e-book)
CRC Press ISBN-13: 978-0-8493-9145-3
CRC Press ISBN-10: 0-8493-9145-8
CRC Press order number: WP9145

The publishers' policy is to use permanent paper from mills that operate a sustainable
forestry policy, and which has been manufactured from pulp which is processed using
acid-free and elementary chlorine-free practices. Furthermore, the publishers ensure that
the text paper and cover board used have met acceptable environmental accreditation
standards.

Project managed by Macfarlane Production Services, Dunstable, Bedfordshire, England
(e-mail: macfarl@aol.com)
Typeset by Godiva Publishing Services Ltd, Coventry, West Midlands, England
Printed by TJ International Limited, Padstow, Cornwall, England

Contents

Contributor contact details

(* = main contact)

Editors

Dr David Kilcast and Fiona Angus
Leatherhead Food International
Randalls Road
Leatherhead
Surrey KT22 7RY
UK

E-mail:
   DKilcast@LeatherheadFood.com
   FAngus@LeatherheadFood.com

Chapter 1

F. Angus
Leatherhead Food International
Randalls Road
Leatherhead
Surrey KT22 7RY
UK

E-mail:
   FAngus@LeatherheadFood.com

Chapter 2

Dr F. J. He* and G.A. MacGregor
Blood Pressure Unit
Cardiac & Vascular Sciences
St George's University of London
Cranmer Terrace
London SW17 0RE
UK

E-mail: fhe@sgul.ac.uk

Chapter 3

Dr J.D. Fernstrom
UPMC/Western Psychiatric Institute
   & Clinic
Room 1620
3811 O'Hara Street
Pittsburgh, PA 15213
USA

E-mail: fernstromjd@upmc.edu

Chapter 4

Dr S. McCaughey
Research Associate
Monell Chemical Senses Center
3500 Market St.
Philadelphia, PA 19104
USA

E-mail: McCaughey@monell.org

Chapter 5

Dr J. Purdy* and Dr G. Armstrong
Research and Policy Officer
Northern Ireland Public Health
    Alliance
Philip House
123 York Street
Belfast B15 1AB
UK

E-mail: j.purdy@cieh.org
    Ga.armstrong@ulster.ac.uk

Chapter 6

Dr C. Walsh
Department of Human Nutrition
University of the Free State
PO Box 339
Bloemfontein 9300
South Africa

E-mail: gnmvcw.md@mail.uovs.ac.za

Chapter 7

G. Bussell* and M. Hunt
Nutrition Consultant
Food and Drink Federation
6 Catherine Street
London WC2B 5JJ
UK

E-mail: gaynor@gaynorbussell.com

Chapter 8

D. Man
Department of Applied Science
Faculty of Engineering, Science & the
    Built Environment
London South Bank University
103 Borough Road
London SE1 0AA
UK

E-mail: mandc@lsbu.ac.uk

Chapter 9

Dr G. Betts
CCFRA
Station Road
Chipping Campden
Gloucester GL55 6LD
UK

E-mail: g.betts@campden.co.uk

Chapter 10

D. Kilcast* and C. den Ridder
Leatherhead Food International
Randalls Road
Leatherhead
Surrey KT22 7RY
UK

E-mail:
    DKilcast@LeatherheadFood.com

Chapter 11

R. McGregor
2005 Eastpark Boulevard
Cranbury, NJ 08512
USA

E-mail: rmcgregor@linguagen.com

Chapter 12

E. Desmond
AllinAll Ingredients
33 Lavery Avenue
Park West
Dublin 12
Ireland

E-mail: research@alllinall.ie

Chapter 13

S. Pedro* and M. L. Nunes
Av. Brasília 1449-006
Lisboa
Portugal

E-mail: spedro@ipimar.pt

Chapter 14

S.P. Cauvain
BakeTran
97 Guinions Road
High Wycombe
Bucks HP13 7NU
UK

E-mail: spc@baketran.demon.co.uk

Chapter 15

Professor P. Ainsworth* and Dr A.
   Plunkett
School of Food, Consumer, Tourism
   and Hospitality Management
Manchester Metropolitan University
Old Hall Lane
Manchester M14 6HR
UK

E-mail: p.ainsworth@mmu.ac.uk
   a.plunkett@mmu.ac.uk

Chapter 16

T. Guinee* and B.T. O'Kennedy
Moorepark Food Research Centre
Teagasc
Fermoy
Co. Cork
Ireland

E-mail: tim.guinee@teagasc.ie

Chapter 17

Dr T. Robinson
Company Nutritionist
H J Heinz Company Limited
Kitt Green
Wigan WN5 0JL
UK

E-mail:
   tristan.robinson@uk.hjheinz.com

Introduction

The question of the level of salt in our diets, and the impact on health, is not new. For many years concerns have been expressed that, at least in the Western world, consumption levels of salt (or, more correctly, sodium) are well above those needed for nutritional purposes, and that these are having an adverse effect on health, in particular the incidence of cardiovascular disease. The problem facing health professionals, governments and food manufacturers has been to identify the steps that can be taken to reduce salt intake whilst maintaining the essential functions of salt: its preservative effects, its contribution to flavour and its function in maintaining processability.

Of these three major functions, the easiest target in salt reduction programmes is through its contribution to flavour, via its function as a basic taste. This is in direct contrast to the primary historical function as a food preservative. The argument that high levels of salt, especially in processed foods, are used by the industry to boost the flavour of otherwise insipid foods has been used with increasing force in recent years. Whilst highly publicised surveys that have revealed high levels of salt in some foods that cannot be justified on grounds of either preservation, flavour or processability have given this view some credence, this represents a naïve view that ignores the multiple functions of salt. However, the message that has been broadcast loudly in the media is that salt is largely unnecessary and can be reduced across all food products. This has been underlined by well-publicised pressure from health professionals and governments through media channels.

The ability of the industry to reduce salt is made particularly difficult not only by the multifunctional nature of salt, but also by the practical difficulties in finding an alternative that delivers the same functions. If comparisons are made with the parallel pressures to reduce sugar consumption, then one major obstacle lies in the severely limited range of optional materials in contrast to the wide range of alternative bulk and intense sweeteners available for sugar reduction. In addition, the importance of sugar in delivering quality has driven an enormous amount of research into understanding the basis of sweetness perception, and into understanding and countering the sensory defects that are present to some degree in many sugar substitutes. Saltiness, in comparison, has been seen to originate through a comparatively simple mechanism, and has been subjected to much less research. As a consequence, the food manufacturing industry now finds itself under immense pressure to reduce salt levels, but without the necessary tools to achieve this whilst maintaining the required functions.

In view of the many obstacles, many sectors of the industry have been remarkably successful in achieving substantial reductions in salt content. This has been seen even in the case of major brands, in which the sensory consequences of the reduction can readily be detected, but which have had little effect on consumer purchase, flying in the face of the dictum that consumers will not tolerate any change to their favourite brands. This success has relied on an important principle, that we do not have an innate liking for salt taste, and that we will adapt, eventually, to lower levels of perceived saltiness. This principle has been demonstrated in the achievements of companies in 'reduction by stealth', but there is evidence that naïve assumptions have been made, even by professionals associated with the food business, that this can be applied generally. This is not true, and several manufacturers are now facing reduced sales in some product areas as a direct consequence of reducing salt content too far, too quickly, or both, and ignoring the complexities of salt reduction.

This volume brings together the major issues associated with salt reduction, and offers insights into how reductions might be achieved, but also into the limitations associated with salt reduction programmes. Part I outlines the nutritional drivers for salt reduction, the actions being taken by governments, particularly in the UK, and the labelling implications. This part also explains the sensory basis of saltiness perception, and gives views on salt reduction from a consumer viewpoint. Part II describes different strategies that can be taken or are being investigated for reducing salt. These chapters cover the technological, microbiological and sensory functions of salt, and approaches as to how these can be maintained. Part III describes approaches that have been taken to reduce salt in specific foods: meat and poultry, seafood, bread, snack foods, cheese and dairy products and also canned foods.

It is hoped that this volume will provide most sectors of the industry with guidance for approaching salt reduction logically, and will identify the opportunities that are open to them, and also the pitfalls that might present difficulties if over-hasty decisions are made. In addition, the book contains indications of future directions, especially accruing from the progress being made in understanding the biological mechanisms of taste perception, and in the way in which salty stimuli reach the sense organs.

David Kilcast and Fiona Angus
Leatherhead

Part I

Dietary salt, health and the consumer

1

Dietary salt intake: sources and targets for reduction

F. Angus, Leatherhead Food International, UK

1.1 Introduction

There has been considerable discussion regarding sodium intakes for a number of years, since it was established that intakes are high compared with physiological needs. The concerns have been largely focused on the link between sodium and hypertension, the evidence for this is well reviewed in other chapters.

1.2 Intakes of sodium

Intakes of sodium vary widely in the UK. The greatest contributor to sodium intakes is salt but sodium is also obtained from other food ingredients and OTC medicines. Data from the UK Dietary and Nutritional Survey of British Adults indicated that mean daily salt intakes (estimated from total urinary sodium) were 11 g among men and 8.1 g among women (Hoare *et al.*, 2004). This is of concern, not only since the levels are far in excess of those required but that intakes have risen from the previous survey in 1986/7 (Gregory *et al.*, 1990). Whilst salt is still added by consumers during cooking and at the table, this is thought to form a small part of total salt intake. Processed foods are thought to account for as much as 80% of total salt intake. In the NDNS survey, a number of key food groups were found to be the major contributors to daily intake from food among British adults. These included cereals and cereal products providing a third of sodium from food, meat and meat products contributing 26% and milk and milk products 8%. Figure 1.1 shows the percentage contribution of food types to average daily intake of sodium. This explains why the UK Government

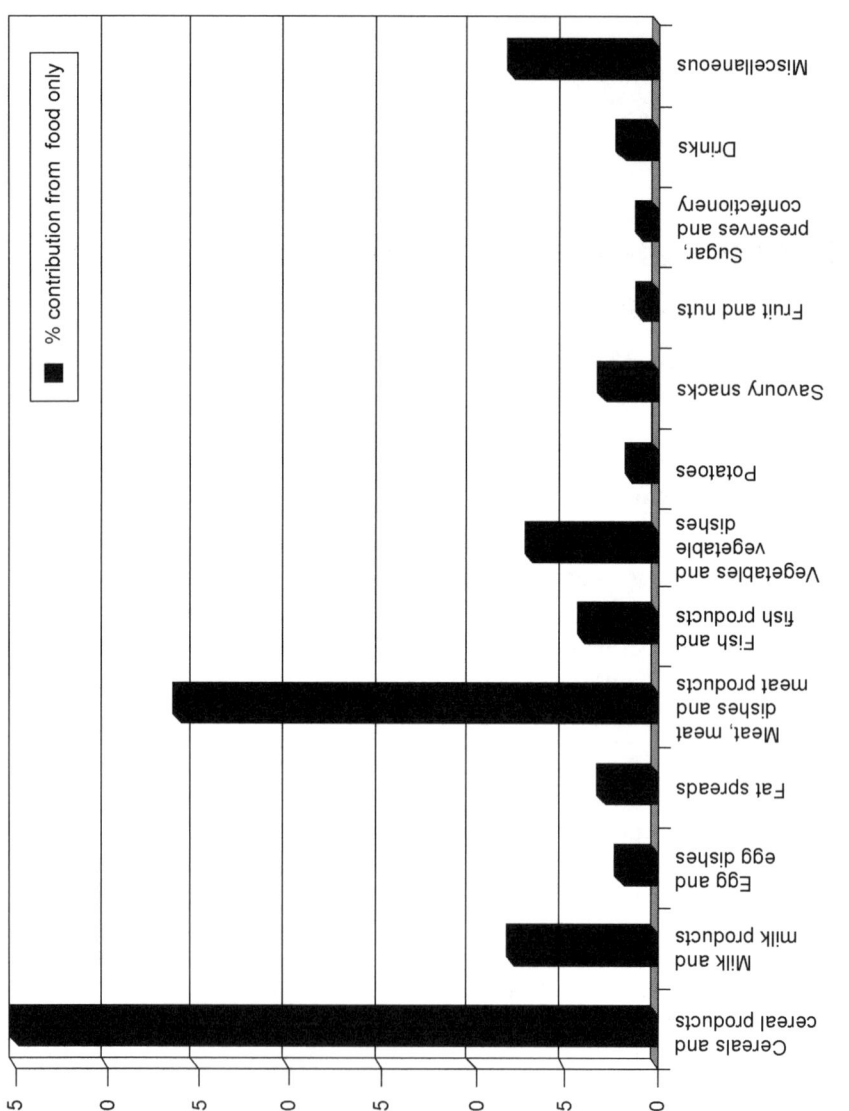

Fig. 1.1 Percentage contribution of food types to average daily intake of sodium: all respondents.

has focused on the food industry to reduce salt levels in processed foods, as well as increasing awareness among consumers about the need to reduce salt intakes.

In other European countries, intake of sodium varies. The Intersalt study considered mean sodium excretion in 12 countries in Western Europe and found that it varied from 3.1–4.1 g per day, equivalent to 7.9–10.5 g salt. Estimates of salt intake were 8–9 g in countries such as Belgium, Denmark and The Netherlands and 9–11 g in Finland, Italy and Portugal (Intersalt Cooperative Research Group, 1988).

In the United States, dated collected some years ago suggested that mean sodium intakes were 4.1 g for men and around 2.7 g for women. This is equivalent to salt intakes of 10.25 g for men and 6.75 g for women (USDA, 1994).

1.3   Recommendations for intake

In 1991, the UK government issued guideline intakes through COMA for a variety of nutrients including sodium (DoH, 1991). It advised that sodium intakes were 'needlessly high' and cautioned against any rise, setting a reference nutrient intake (RNI) for sodium of 1600 mg/day for people 11–50+ years (Table 1.1).

In 1994, these COMA recommendations were re-evaluated in its report on the Nutritional Aspects of Cardiovascular Disease (COMA, 1994). In this report, the committee maintained the RNI for sodium of 1600 mg a day for adults and advised a target reduction on average intake of salt by the population from 9 g to 6 g a day. This was endorsed by the Scientific Advisory Committee on Nutrition (SACN) in its report on Salt and Health (SACN, 2003). SACN also acknowledged that benefits would be gained from a reduction in average salt consumption among children. The recommendations for children are shown in Table 1.2 and represent what the committee considered to be achievable targets, rather than optimal levels.

Table 1.1   Reference nutrient intakes for sodium

Age	Reference nutrient intake mg/d
0–3 months	210
4–6 months	280
7–9 months	320
10–12 months	350
1–3 years	500
4–6 years	700
7–10 years	1200
11–14 years	1600
15–18 years	1600
19–50 years	1600
50+ years	1600

Source: COMA (1991)

Table 1.2   Daily target average salt intakes for infants and children

Age	Target average salt intake g/d
0–6 months	<1
7–12 months	1
1–3 years	2
4–6 years	3
7–10 years	5
11–14 years	6

Source: SACN (2003)

Several UK food manufacturers and retailers have developed Guideline Daily Amounts (GDA) to assist consumers understand COMA recommendations (IGD, 2005). The GDA for salt was developed in line with SACN recommendations and a level of 6 g for men and women was agreed. A GDA of sodium was developed since EU labelling law requires that sodium be calculated for labelling purposes. The GDA for sodium of 2.4 g was based on the RNI for sodium multiplied by 1.5.

In addition, a GDA for sodium and salt were developed for children (IGD, 2005). For this, the recommendations of SACN were adopted and age bandings retained. A GDA for children 5–10 years was developed from a mean of the values for children 4–6 years and 7–10 years (Table 1.3). This GDA was 1.4 g for sodium and 4 g for salt.

Salt intake recommendations are similar in Europe and the United States, although there has been less focus to date outside the UK on reducing salt intakes. In the USA, the Dietary Guidelines for Americans 2005, recommended that Americans should consume no more than 2300 mg sodium per day, equivalent to 5.75 g salt. Individuals with hypertension, blacks and middle-aged and older adults were advised to consumer less than 1500 mg sodium a day or 3.75 g salt per day (Department of Health and Human Sciences, 2005).

Table 1.3   Sodium and salt GDAs for children

Children	4–6y	7–10y	11–14y	15–18y
Sodium (g)	1.1	1.8	2.4	2.4
Salt (g)	3	5	6	6
Children	5–10y			
Sodium (g)	1.4			
Salt (g)	4			

Source: IGD (2005)

1.4   Targets for reduction

The Food Standards Agency (FSA) began working with the UK food industry to try and reduce the salt levels in processed foods in 2004. To assist industry as to the type of foods in which reductions were required and the level of reductions needed to reduce intakes, the government developed proposals for target levels of salt and these were published for consultation in August 2005. The feasibility of these was discussed and commented on by industry and other stakeholders and the final salt reduction targets were finally published in March 2006. The targets stand until 2008, at which time the FSA will review industry progress towards these targets and what further action is needed.

Table 1.4 lists the FSA targets for salt and sodium by food category. The full table is available on the Food Standards Agency website: (http://www.food.gov.uk/multimedia/pdfs/salttargetsapril06.pdf). The table gives the target levels originally published in the initial salt model, a target for 2010 and a final target for sodium and salt for each product.

There are specific technical issues with salt reduction in certain products and these are discussed in more detail in subsequent chapters. In a number of cases, the FSA final salt target was above that originally proposed in the salt model and comments are provided on the complete table to support these revisions. In summary, a number of products were affected and some of the key issues identified by the FSA, relating to specific products have been summarised below.

1.4.1   Bacon and ham
The salt target was based on the process of injection cured bacon and ham being managed. Dry and immersion cured bacon were excluded from the targets since this process made it more difficult to control salt levels. It was likely that producers would have to change processing techniques in order to meet the target.

1.4.2   Cooked sausages and sausage meats
The target salt levels set for cooked sausages were higher than other sausages to allow for fat and moisture loss during cooking.

1.4.3   Meat pies
Technical difficulties with reduction of salt in pastry meant that these products had been set higher salt targets until further research work was undertaken on them.

1.4.4   Pre-packed bread
The bread industry has made significant reductions in salt to date and was reported to prefer a staged approach to the target. The FSA has said it will review the target in 2008.

Table 1.4 Food Standards Agency salt reduction targets for industry 2006

Main product category	Sub-category	FSA salt model illustrative average value		FSA proposed targets to be achieved by 2010 (g salt, sodium per 100 g food, as sold)	Final targets (g salt or sodium per 100 g)
		mg sodium per 100 g	g salt per 100 g		
Meat products	Bacon	750	1.9	3.5 g salt or 1.4 g sodium (maximum)	3.5 g salt or 1.4 g sodium per 100 g (average)
	Ham/other cured meats	750	1.9	2.5 g salt or 1 g sodium (maximum)	2.5 g salt or 1 g sodium (average)
	Sausages	550	1.4	1.4 g salt or 550 mg sodium (maximum)	1.4 g salt or 550 mg sodium (maximum)
	Sausages – Cooked sausages and sausage meat products	550	1.4	1.4 g salt or 550 mg sodium (maximum)	1.8 g salt or 700 mg sodium (maximum)
	Meat pies – delicatessan, pork pies and sausage rolls	300	0.8	1.3 g salt or 500 mg sodium (maximum)	1.5 g salt or 600 mg sodium (maximum)
	Meat pies – Cornish and meat-based pastries	300	0.8	1.1 g salt or 450 mg sodium (maximum)	1.3 g salt or 500 mg sodium (maximum)
	Meat pies – other meat-based pastry products, including pies and slices	300	0.8	1.1 g salt or 450 mg sodium (maximum)	1.1 g salt or 450 mg sodium (maximum)
	Cooked uncured meat	450	1.8	1.8 g salt or 700 mg sodium (maximum	1.5 g salt or 600 mg sodium (maximum)
	Burgers and grillsteaks	300	0.8	1.0 g salt or 400 mg sodium (maximum	1.0 g salt or 400 mg sodium (maximum)
	Burgers and grillsteaks – speciality and topped burgers and grillsteaks	300	0.8	1.0 g salt or 400 mg sodium (maximum	1.3 g salt or 500 mg sodium (maximum

Category	Food				
	Poultry (coated)	450	1.1	1.0g salt or 400mg sodium (maximum)	1.0g salt or 400mg sodium (maximum)
	Canned frankfurters, hot dogs and burgers	550 (sausages); 300mg (burgers)	1.4 (sausages); 0.8 (burgers)	1.4g salt or 550mg sodium (maximum)	1.4g salt or 550mg sodium (maximum)
Bread	Pre-packed bread and rolls	350	0.9	1g salt or 400mg sodium (maximum)	1.1g salt or 430mg sodium (average)
	Bread and rolls with additions	350	0.9	1.2g salt or 470mg sodium (maximum)	1.3g salt or 500mg sodium (average)
	Morning goods	350	0.9	1g salt or 400mg sodium (maximum)	1.3g salt or 500mg sodium (average)
Breakfast cereals	Breakfast cereals	300	0.8	0.8g salt or 300mg sodium (average)	0.8g salt or 300mg sodium (average)
Cheese	Cheddar and other similar 'hard' cheeses Mild cheddar and other similar	500	1.3	1.7g salt or 670mg sodium (average)	1.7g salt or 670mg sodium (average)
	Mature cheddar and other similar	500	1.3	1.95g salt or 750mg sodium (average)	1.95g salt or 750mg sodium (average)
	Fresh cheeses Soft white cheese	500	1.3	0.5g salt or 200mg sodium (average)	0.8g salt or 320mg sodium (maximum)
	Cottage cheese, plain	500	1.3	0.5g salt or 200mg sodium (average)	0.54g salt or 215mg sodium (average)
	Cottage cheese, flavoured	500	1.3	0.5g salt or 200mg sodium (average)	0.64g salt or 250mg sodium (average)
	Mozzarella	500	1.3	1.8g salt or 750mg sodium (average)	1.8g salt or 750mg sodium (average)
	Blue cheese	500	1.3	1.9g salt or 750mg sodium (average)	No target to be set at present

Table 1.4 Continued

Main product category	Sub-category	FSA salt model illustrative average value		FSA proposed targets to be achieved by 2010 (g salt, sodium per 100 g food, as sold)	Final targets (g salt or sodium per 100 g)
		mg sodium per 100 g	g salt per 100 g		
	Processed cheese				
	Cheese spreads	500	1.3	2.0g salt or 800mg sodium (average)	2.0g salt or 800mg sodium (average)
	Other processed cheese	500	1.3	2.9g salt or 1170mg sodium (average)	2.9g salt or 1170mg sodium (average)
Butter	Butter	400	1.0	3.8g salt or 1.5g sodium (average)	3.0g salt or 1200mg sodium (average)
	Welsh and regional butter	400	1.0	1.4g salt or 550mg sodium (average)	1.7g salt or 670mg sodium (average)
	Salted butter	400	1.0	1.4g salt or 550mg sodium (average)	1.7g salt or 670mg sodium (average)
	Lightly salted butter	400	1.0	1.2g salt or 470mg sodium (average)	1.2g salt or 470mg sodium (average)
	Unsalted butter	400	1.0	0.1g salt or 40mg sodium (average)	0.1g salt or 40mg sodium (average)
Fat spreads	Margarines/other spreads	400	1.0	1.4g salt or 550mg sodium (average)	1.5g salt or 600mg sodium (average)
Baked beans	Baked beans in tomato sauce without accompaniments	350	0.9	0.75g salt or 300mg sodium (maximum)	0.8g salt or 300mg sodium (maximum)
	Baked beans in tomato sauce with accompaniments	350	0.9	0.75g salt or 300mg sodium (maximum)	1.0g salt or 400mg sodium (maximum)

Category	Item				
Ready meals and meal centres – meat, fish and vegetable based	Chinese/Thai/Indian – ready meals	250	0.6	0.8g salt or 300mg sodium (average)	0.8g salt or 300mg sodium (average)
	Chinese/Thai/Indian – meal centres	250	0.6	1.0g salt or 400mg sodium (average)	1.0g salt or 400mg sodium (average)
	Italian/Traditional/other – ready meals	250	0.6	0.6g salt or 250mg sodium (average)	0.6g salt or 250mg sodium (average)
	Italian/Traditional/other – meal centres	250	0.6	0.8g salt or 300mg sodium (average)	0.8g salt or 300mg sodium (average)
Soups	Dried soups (as consumed)	200	0.5	0.7g salt or 280mg sodium (average)	0.6g salt or 250mg sodium (average)
	Wet soups	200	0.5	0.6g salt or 250mg sodium (average)	0.6g salt or 250mg sodium (average)
Pizza	Pizza with higher salt toppings	300	0.8	1.2mg or 470mg sodium (average)	1.2mg or 470mg sodium (average)
	Pizza without higher salt toppings	300	0.8	1.0g salt or 400mg sodium (maximum)	1.0g salt or 400mg sodium (maximum)
Crisps and savoury snacks	Standard potato crisps	550	14	1.5g salt or 600mg sodium (maximum)	1.5g salt or 600mg sodium (average)
	Extruded snacks	550	1.4	2.8g salt or 1.1g sodium (maximum)	2.8g salt or 1.1g sodium (average)
	Pelleted snacks	550	1.4	3.2g salt or 1.3g sodium (maximum)	3.4g salt or 1.4g sodium (average)
	Salt and vinegar snacks	550	1.4	3.1g salt or 1.2g sodium (maximum)	3.1g salt or 1.2g sodium (average)
Buns, cakes, pastries and fruit pies	Buns	200	0.5	0.5g salt or 200mg sodium (maximum)	0.5g salt or 200mg sodium (average)
	Cakes	200	0.5	0.6g salt or 250mg sodium (maximum)	0.6g salt or 240 mg sodium (average)

Table 1.4 Continued

Main product category	Sub-category	FSA salt model illustrative average value		FSA proposed targets to be achieved by 2010 (g salt, sodium per 100 g food, as sold)	Final targets (g salt or sodium per 100 g)
		mg sodium per 100 g	g salt per 100 g		
	Pastries	200	0.5	0.5g salt or 200mg sodium (maximum)	0.5g salt or 185mg sodium (average)
	Fruit pies	200	0.5	0.5g salt or 200mg sodium (maximum)	0.4g salt or 130mg sodium (average)
Bought sandwiches	With high salt fillings	350	0.9	1.3g salt or 500mg sodium (average)	1.3g salt or 500mg sodium (average
	Without high salt fillings	350	0.9	1g salt or 400mg sodium (average)	1g salt or 400mg sodium (average)
Table sauces	Tomato ketchup	600	1.5	1.8g salt or 700mg sodium (maximum)	2.4g salt or 1g sodium (maximum)
	Brown sauce	600	1.5	1.5g salt or 600mg sodium (maximum)	1.5g salt or 600mg sodium (maximum)
	Salad cream	600	1.5	1.8g salt or 700mg sodium (maximum)	1.8g salt or 700mg sodium (maximum)
	Mayonnaise (not reduced calorie)	600	1.5	1.5g salt or 600mg sodium (maximum)	1.5g salt or 600mg sodium (maximum)
	Mayonnaise (reduced fat/ calorie)	600	1.5	2.5g salt or 1g sodium (maximum)	2.5g salt or 1g sodium (maximum)
	Salad dressing	600	1.5	2.5g salt or 1g sodium (maximum)	2.5g salt or 1g sodium (maximum)

Cook-in and pasta sauces	All cook in and pasta sauces (except pesto)	250	1.2g salt or 470mg sodium (average)	1.1g salt or 430mg sodium (average)
	Pesto and other thick sauces	250	3g salt or 1.2mg sodium (average)	3g salt or 1.2mg sodium (average)
Biscuits	Sweet biscuits unfilled	250	0.7g salt or 280mg sodium (maximum)	1.1g salt or 416mg sodium (average)
	Sweet biscuits filled	250	0.7g salt or 280mg sodium (maximum)	0.5g salt or 205mg sodium (average)
	Savoury biscuits unfilled	250	1.3g salt or 500mg sodium (maximum)	2.2g salt or 860mg sodium (average)
	Savoury biscuits filled	250	1.3g salt or 500mg sodium (maximum)	1.9g salt or 740mg sodium (average)
Pasta	Pasta excluding ready meals	78	0.5g salt or 200mg sodium (maximum)	0.5g salt or 200mg sodium (maximum)
Rice	Rice unflavoured (as consumed)	87	0.2g salt or 87mg sodium (maximum)	0.2g salt or 87mg sodium (maximum)
	Rice flavoured (as consumed)	87	0.2g salt or 87mg sodium (maximum)	0.8g salt or 300mg sodium (average)
Other cereals	Other cereals	300	0.8g salt or 300mg sodium (maximum)	0.8g salt or 300mg sodium (maximum)
Processed pudding products	Dessert mixes as consumed	80	0.2g salt or 80mg sodium (maximum)	0.5g salt or 200mg sodium (maximum)
	Cheesecake	80	0.2g salt or 80mg sodium (maximum)	0.5g salt or 200mg sodium (maximum)
	Sponge-based processed puddings	80	0.2g salt or 80mg sodium (maximum)	1.0g salt or 400mg sodium (maximum)
	All other processed puddings	80	0.2g salt or 80mg sodium (maximum)	0.3g salt or 120mg sodium (maximum)

Table 1.4 Continued

Main product category	Sub-category	FSA salt model illustrative average value		FSA proposed targets to be achieved by 2010 (g salt, sodium per 100 g food, as sold)	Final targets (g salt or sodium per 100 g)
		mg sodium per 100 g	g salt per 100 g		
Quiches	Quiches	250	0.6	0.6g salt or 250mg sodium (maximum)	0.8g salt or 300mg sodium (maximum)
Other processed egg products	Other processed egg products	300	0.8	0.8g salt or 300mg sodium (maximum)	0.8g salt or 300mg sodium (maximum)
	Scotch eggs			0.8g salt or 300mg sodium (maximum)	1.0g salt or 400mg sodium (maximum)
Canned fish	Canned tuna	300	0.8	0.8g salt or 300mg sodium (maximum)	1.0g salt or 400mg sodium (average)
	Canned salmon	300	0.8	0.8g salt or 300mg sodium (maximum)	1.2g salt or 470mg sodium (average)
	Other canned fish	300	0.8	0.8g salt or 300mg sodium (maximum)	1.5g salt or 600mg sodium (average)
Canned vegetables	Canned vegetables	50	0.13	0.13g salt or 50mg sodium (maximum)	0.13g salt or 50mg sodium (maximum)
	Processed/marrowfat/mushy peas	50	0.13	0.13g salt or 50mg sodium (maximum)	0.5g salt or 200mg sodium (maximum)
	Processed vegetable-based products	260	0.66	0.7g salt or 280mg sodium (maximum)	0.7g salt or 280mg sodium (maximum)
	Dehydrated instant mashed potato	100	0.25	0.25g salt or 100mg sodium (maximum)	0.25g salt or 100mg sodium (maximum)

	Other processed potato products	100	0.25g salt or 100mg sodium (maximum)	0.5g salt or 195mg sodium (maximum)
Beverages, dry weight	Dried beverages, as consumed	50	0.13g salt or 50mg sodium (maximum)	0.25g salt or 100mg sodium (maximum)
Take away meat based	Take away meat based	250	0.6g salt or 250mg sodium (maximum)	0.6g salt or 250mg sodium (maximum)
Take away fish based	Take away fish based	200	0.5g salt or 200mg sodium (maximum)	0.5g salt or 200mg sodium (maximum)
Take away vegetable and potato based	Take away vegetable and potato based	200	0.5g salt or 200mg sodium (maximum)	0.5g salt or 200mg sodium (maximum)

1.4.5   Morning goods
The salt target for morning goods was increased to take into account the raising agents used in scones.

1.4.6   Cheese
Dairy UK has specified that its members will work to its own salt targets of 300 mg/100 g sodium for plain white soft cheese and 350 mg/100 g sodium for flavoured soft white cheeses. No target was set for blue cheese due to technical and safety issues. Similarly technical issues meant that a target for processed cheeses will not be set until 2008.

1.4.7   Butter
The dairy industry has said that there are natural variations in salt content of butter that make meeting the FSA target difficult. The industry has been asked to consider ways of reducing process variation.

1.4.8   Chinese/Thai and Indian ready meals and centres
Acceptance was made that some authentic dishes may need to be higher in salt but industry was asked to look at innovative ways of reducing sodium content in these meals.

1.4.9   Crisps and snacks
Higher target levels have been set for extruded snacks and pelleted snacks due to processing difficulties encountered if the salt level is too low. The situation will be reviewed by the FSA in 2008.

1.4.10   Bought sandwiches
Higher salt levels for bought sandwiches were accepted based on the difficulty of meeting the target for a wide variety of fillings. The FSA said it will review the target in 2008.

1.4.11   Sauces and dressings
Higher salt levels have been accepted due to stability and food safety issues. Reduced fat mayonnaise requires higher levels of salt to maintain stability but manufacturers have been asked to keep levels as low as possible.

1.4.12   Biscuits
Salt levels have already been reduced in these products and the industry has commented that it believes only small further reductions are possible.

1.4.13 Pasta

Flavoured noodle products cannot currently meet the salt target. The FSA said that these products would be monitored.

1.4.14 Processed pudding products

Sodium in these products comes from sources other than salt.

1.4.15 Canned fish

The target levels of salt for canned fish were higher to reflect natural variation in salt levels in fish and standard processing techniques.

1.4.16 Canned vegetables

Technical issues limited the level of salt reduction possible in processed peas (see Table 1.4).

In conclusion, there are now clear targets for industry to meet with regard to salt in foods. If these are achieved by industry, then the UK government will be well on the way to achieving the 6 g a day target intake for the average population. Significant reductions in salt have already been made in several key food categories but more work is needed. It is clear that further reductions will take time to achieve, however, and for some products, new and innovative technologies are still needed to get there. Outside the UK, focus on salt intakes is increasing but other governments have yet to engage with industry on salt in the same way as has developed in the UK over the last few years.

1.5 References

COMA (1994). 'Nutritional Aspects of Cardiovascular Disease'. Report of the Cardio-vascular Review Group of the Committee on Medical Aspects of Food Policy. Report No.46. London: HMSO.

DEPARTMENT OF HEALTH (1991). 'Dietary Reference Values for Food Energy and Nutrients for the United Kingdom'. Report of Health and Social Subjects No. 41. London: HMSO.

DEPARTMENT OF HEALTH AND HUMAN SCIENCES (2005). 'Dietary Guidelines for Americans 2005)'. USDA.

GREGORY J. et al. (1990). The Dietary and Nutrition Survey of British Adults. London: HMSO.

HOARE J. et al. (2004). The National Diet & Nutrition Survey: adults aged 19 to 64 years. London: TSO.

INTERSALT COOPERATIVE RESEARCH GROUP (1988). 'Intersalt: an international study of electrolyte excretion and blood pressure. Results for 24-hour urinary sodium and potassium excretion'. British Medical Journal, 297, 319–28.

IGD (2005). Guideline Daily Amounts. Report of the IGD/PIC Industry Nutrition Strategy Group. IGD.

SACN (2003). Salt and Health. www.sacn.gov.uk/pdfs/sacn_salt_final.pdf.

USDA (1994). Continuing Survey of Food Intakes by Individuals. USDA.

2

Dietary salt, high blood pressure and other harmful effects on health

F. J. He and G. A. MacGregor, St George's University of London, UK

2.1 Introduction

Cardiovascular disease (strokes, heart attacks and heart failure) is the leading cause of death and disability worldwide. Increasing blood pressure throughout the range is the major cause of strokes and heart failure. It is also a very important cause of coronary heart disease (Lewington *et al.* 2002). Worldwide, 26.4% of the adult population had high blood pressure (greater than 140/90 mmHg) and the estimated total number of adults with high blood pressure was 972 million in 2000. This number is predicted to increase to a total of 1.56 billion in 2025 (Kearney *et al.* 2005).

Salt (sodium chloride) is the primary cause of raised blood pressure and is largely responsible for the rise in blood pressure that occurs in almost all adults as they grow older. Evidence that relates salt intake to blood pressure comes from six different lines of evidence – epidemiology (Elliott *et al.* 1996), migration (Poulter *et al.* 1990), intervention (Forte *et al.* 1989), treatment (He and MacGregor 2002), animal (Denton *et al.* 1995) and genetic studies (Lifton 1996). Evidence from these different sources all suggests that salt intake is important in blood pressure regulation and a reduction in salt intake would lead to a reduction in population blood pressure, a reduction in the rise in blood pressure with age and a reduction in blood pressure in those with high blood pressure whether on or off antihypertensive treatment.

Salt intake in nearly all countries around the world is between 9 and 12 g/day (INTERSALT 1988; Henderson *et al.* 2003). The World Health Organisation has set a worldwide target of reducing salt to 5 g/day or less for all adults (WHO 2003). The UK and US recommendations are 6 g/day or less (Whelton *et al.*

2002; SACN 2003). A reduction in salt intake can be achieved by a public campaign about the dangers of salt leading to less use of table and cooking salt combined with a gradual and sustained reduction in the salt concentration of all processed and catering foods. More than three-quarters of salt that is consumed in most developed countries is added by the food industry, i.e. processed, canteen, restaurant, fast and takeaway foods. Only 15% is added in cooking or at the table and 5% is naturally present in foods. Clearly, therefore, a major reduction must be made in the amount of salt added to food by the food industry in order to achieve the target of 5 to 6 g/day.

If these targets were achieved, the benefits would be very large. For instance, in the UK, approximately 35,000 stroke and heart attack deaths would be saved each year (He and MacGregor 2003), as well as many people dying from heart failure. Approximately, a further 35,000 nonfatal strokes and heart attacks as well as many people suffering from heart failure would be prevented each year. A reduction in salt intake is, therefore, one of the most important strategies for improving public health and preventing people dying or suffering unnecessarily from strokes, heart attacks and heart failure.

There is also increasing evidence that our current high salt intake has other harmful effects on health, which may be independent of and additive to the effect of salt on blood pressure, e.g. a direct effect on stroke (Perry and Beevers 1992), left ventricular hypertrophy (Kupari *et al.* 1994; Schmieder and Messerli 2000), progression of renal disease and albuminuria (Heeg *et al.* 1989; Cianciaruso *et al.* 1998; Swift *et al.* 2005), stomach cancer (Joossens *et al.* 1996; Tsugane *et al.* 2004), and bone demineralization (Devine *et al.* 1995).

In this chapter, we will review the evidence that relates salt to raised blood pressure as well as the possible mechanisms whereby salt increases blood pressure. We will also briefly discuss other harmful effects of salt on health.

2.2   Definition of hypertension

The relationship between blood pressure and cardiovascular disease displays a continuous graded relationship, and there is no evidence of any threshold level of blood pressure below which lower levels of blood pressure are not associated with lower risks of cardiovascular disease (Lewington *et al.* 2002). Thus, any classification of people into dichotomous categories ('normotensive' and 'hypertensive') is inherently arbitrary. Nevertheless, it is useful to provide a classification of blood pressure for the purpose of identifying high-risk individuals and providing guidelines for treatment with tablets and follow-up.

Geoffrey Rose suggested that 'the operational definition of hypertension is the level at which the benefits ... of action exceed those of inaction'. The criteria for the classification of hypertension have changed over the past 40 years as more recent studies have shown benefit at lower levels of blood pressure (Vasan *et al.* 2001). The recent Seventh Joint National Committee (JNC VII) (Chobanian *et al.* 2003) defined individuals with blood pressure less than 120/80 mmHg as 'normal' and those with blood pressure ≥140 mmHg systolic or ≥90 mmHg diastolic as 'hypertension', whereas for those with blood pressure ranging from 120 to 139 mmHg systolic and/or 80 to 89 mmHg diastolic, the JNC VII report has introduced a new term 'prehypertension'. This new designation is intended to identify those individuals in whom early intervention by adoption of healthy lifestyles could reduce blood pressure, decrease the rate of progression of blood pressure to hypertensive levels with age, or prevent hypertension entirely.

Hypertension is extremely common in Western countries. For instance, in England just under 40% of the entire adult population have hypertension (systolic ≥140 mmHg and/or diastolic ≥90 mmHg) (Primatesta *et al.* 2001). The prevalence of hypertension increases with age, e.g. at the age of 50–59 years, approximately 50% have high blood pressure, and at the age of 60–79 years, 70% have high blood pressure. Many treatment trials have demonstrated a clear benefit of lowering blood pressure in hypertensive individuals (Staessen *et al.* 2001).

2.3   Benefits of lowering blood pressure in the 'normal' range

In the general population, blood pressure is distributed in a roughly normal or Gaussian manner in a bell-shaped curve with a slight skew towards higher readings. Although the risk of cardiovascular mortality increases progressively with increase in blood pressure, for the population at large, the greatest number of strokes, heart attacks, and heart failure attributable to blood pressure occur in the upper range of normal (i.e. systolic between 130 and 140 mmHg and diastolic between 80 and 90 mmHg) because there are so many individuals who have blood pressure at these levels in the population (MacMahon 1996). Therefore, a population-based approach aimed at achieving a downward shift in the distribution of blood pressure in the whole population, even by a small amount, will have a large impact on reducing the number of strokes, heart attacks and heart failure.

2.4   Salt and blood pressure

2.4.1   History of salt

Salt is a chemical that consists of sodium and chloride and, as something added to food, is not a normal constituent of the human or any mammalian diet. For several million years the evolutionary ancestors of humans ate a diet that contained less than 0.25 g/day of salt (Eaton and Konner 1985). Sufficient sodium exists in natural foods to ensure that mammals could develop away from the sea. However, the chemical salt has played an important role in the development of civilization (MacGregor and de Wardener 1998). It was first found to have the magical property of preserving foods, probably by the Chinese around 5000 years ago, when they found that meat or fish could be preserved for a long time when they were soaked in saline solutions. This ability to preserve food allowed the development of settled communities. Salt became one of the most traded commodities in the world, as well as one of the most taxed. Salt was initially expensive to produce and was regarded as a luxury, but with the mining of salt it became much more plentiful and was added to fresh food as this food tasted bland compared to the highly salted preserved foods that most people were used to eating. It was also found that when salt was added to food that was going putrid, the bitter flavours were removed and the food became edible. Salt was seen as almost magical and became a symbol of purity in most religions. In the late nineteenth century deep freezers and refrigerators were invented and salt lost its importance as a preservative. Since that time salt intake has been gradually falling. However, with the increased consumption of processed, canteen, restaurant and fast food, which contains large amounts of hidden salt, salt intake is now increasing. More than three-quarters of salt intake in most developed countries now comes from salt added to processed foods (James *et al.*, 1987; Nestle 2002).

Evolving on a low salt diet of no more than 0.25 g/day, humans are genetically programmed to this amount of salt, and have exquisite mechanisms to conserve sodium within the body, i.e. they are able to reduce sodium excretion in the urine and sweat to almost zero. The recent change, in evolutionary terms, to the current very high salt intake of 9 to 12 g/day (40 times more than previously consumed) presents a major challenge to the physiological systems in the body to excrete these very large amounts of salt through the kidney into the urine. Thus, there has been little time for physiological systems to adapt. The consequence of this is that our current high salt intake increases blood pressure, cardiovascular disease, renal disease, and the risk of bone demineralization (de Wardener and MacGregor 2002).

2.4.2 Evidence on salt and blood pressure

The earliest comment that relates dietary salt to blood pressure was recorded in the ancient Chinese medical literature – the Yellow Emperor's classic on internal medicine, Huang Ti Nei Ching Su Wein, 2698–2598 BC. It is stated that 'If too much salt is used for food, the pulse hardens ...'. However, the first meaningful scientific evidence for a link between salt intake and blood pressure only emerged in the early 1900s. There is now overwhelming evidence for a causal relationship between salt intake and blood pressure. The evidence comes from the following sources:

- epidemiological studies
- migration studies
- population-based intervention studies
- treatment trials
- animal studies
- genetic studies.

Epidemiological studies

In 1960, Dahl reported a strong relationship between average salt intake and prevalence of hypertension in an ecological study of five geographically diverse populations (Dahl 1960). Subsequently, several other authors have confirmed Dahl's findings. The limitations of these across-population studies are that the data were from several different studies that used unstandardized methods, and few of the studies collected data on potential confounding factors and the multiple social and environmental differences among populations around the world may affect the salt–blood pressure relationship.

Within-population studies have been hampered by a number of methodological challenges including measurement difficulties caused by large variations in day-to-day salt intake, and a wide range of blood pressure values at any level of salt intake caused by the multifactorial nature of environmental and genetic influence on blood pressure. As such, a large number of individuals would be required to demonstrate a significant association between habitual salt intake and blood pressure. The large international study – INTERSALT (INTERSALT 1988), which involved 10,079 individuals from 52 centres around the world using standardized methods of measuring blood pressure and 24 h urinary sodium demonstrates that salt intake is an important factor in determining population blood pressure level and the rise in blood pressure with age (Fig. 2.1). It was estimated that an increase of 6 g/day in salt intake is related to a rise of 10/6 mmHg in blood pressure over 30 years (e.g., from age of 25 to 55 years), which represents a large increase in population blood pressure.

Many of the INTERSALT investigators have collaborated in another major nutrition–blood pressure study, the INTERMAP study, which shed additional light on the salt–blood pressure relationship. One article from the INTERMAP group reports nutrient intakes in four countries: China, Japan, UK, and USA (Zhou *et al.* 2003). The results confirm findings from INTERSALT that China and Japan, where both the prevalence of hypertension and stroke mortality are very high, have a higher salt intake, lower potassium intake, and therefore, a higher salt/potassium ratio, compared with the UK and the USA. Another INTERMAP article (Stamler *et al.* 2003) demonstrates that most of the adverse effect of a low education level which is known to be inversely related to blood pressure, is attributable to dietary variables, including a higher salt intake in those with a lower education level.

In the INTERSALT study, only four communities had salt intake less than 3 g/day. A number of other studies have studied the nonacculturated tribes which have a low salt intake (less than 3 g/day). Individuals in these tribes have lower levels of blood pressure and more importantly, their blood pressure does not rise with age. The most striking example is the Yanomamo Indians on the border between Venezuela and Brazil (Oliver *et al.* 1975; Mancilha-Carvalho *et al.* 1989). They have a salt intake of 0.05 g/day. The average blood pressure for adults is only 96/61 mmHg, and there is no rise in blood pressure with age and no evidence of cardiovascular disease. Whilst there may be other factors that also account for the lower blood pressure, several studies have clearly

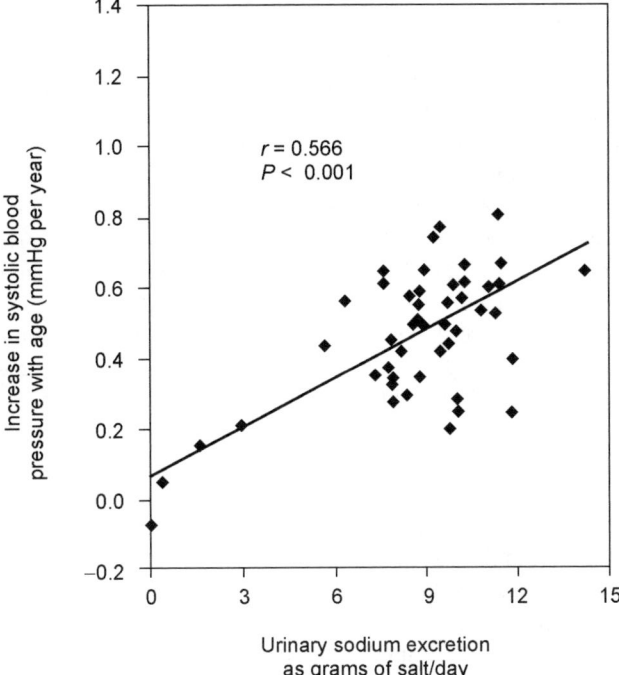

Fig. 2.1   The relationship between salt intake and the slope of the rise in systolic blood pressure with age in 52 centres in the INTERSALT study. Adapted from INTERSALT (1988).

demonstrated the profound importance of salt. A study in the Pacific Islands where one undeveloped community used salt water in their food and the other did not, showed that the community using salt had higher blood pressure (Page *et al.* 1974). Another study of two rural communities in Nigeria, one of which had access to salt from a salt lake and the other did not, showed differences in salt intake and differences in blood pressure, and yet in all other aspects of lifestyle and diet the two communities were similar. The Qash'qai, an undeveloped tribe living in Iran who have access to salt deposits on the ground, develop high blood pressure and a rise in blood pressure with age similar to that which occurs in Western communities, but they live a lifestyle similar to nonacculturated societies (Page *et al.* 1981).

Migration studies
A number of studies have shown that migration from isolated low-salt societies to Westernized environment with a high salt intake is associated with an increase in blood pressure, a rise in blood pressure with age, and a higher prevalence of hypertension. For instance, a well controlled migration study of a rural tribe in Kenya showed that on migration to an urban environment, there was an increase in salt intake and a reduction in potassium intake, and blood

pressure rose after a few months, compared to those in the control group who remained in the rural environment (Poulter *et al.* 1990).

Another example of migration study is that of the Yi people, an ethnic minority living in southwestern China (He *et al.* 1991a,b). Blood pressure rose very little with increasing age (0.13/0.23 mmHg/year) in the Yi farmers who lived in their natural remote mountainous environment and consumed a low salt diet. In contrast, Yi migrants and Han people who lived in urban areas consumed a high salt diet and experienced a much greater increase in blood pressure with increasing age (0.33/0.33 mmHg/year) (He *et al.* 1991a). In a sample of 417 recent migrants (Yi) or native (Han) men living in the urban areas, there was a significant positive relationship between salt intake and blood pressure (He *et al.* 1991b). These findings suggest that changes in lifestyle, including higher salt intake, contribute to the higher blood pressure among Yi migrants.

Population-based intervention studies
In the late 1950s the Japanese became aware that certain parts of Japan, particularly the north, had a high salt consumption and deaths from stroke were amongst the highest in the world. It was then found that the number of strokes in different parts of Japan was directly related to the levels of salt intake. In view of these findings, there was a government campaign to reduce salt intake, which was successful in reducing salt intake over the following decade from an average of 13.5 g/day to 12.1 g/day. However, in the north of Japan the salt intake fell from 18 g/day to 14 g/day. This resulted in a gradual fall in blood pressure both in adults and children, and an 80% reduction in stroke mortality (Tanaka *et al.* 1982; Kimura 1983; Iso *et al.* 1999).

Several population-based well-controlled intervention studies have been carried out. One was conducted in two similar villages in Portugal (Forte *et al.* 1989). Each village had approximately 800 inhabitants who had salt intakes of 21 g/day and the prevalence of hypertension and stroke mortality were very high. During the two-year intervention period, there was a vigorous, widespread health education effort to reduce salt intake, especially from those foods that had previously been identified as the major sources of salt in the intervention village. Whereas, in the control village, no dietary advice was given. The intervention was successful in achieving a difference of approximately 50% in salt intake between the two villages. This caused a significant difference in blood pressure at one year and a more pronounced difference at two years (a difference of 13/6 mmHg in blood pressure between the two villages) (Fig. 2.2).

Another population-based intervention study was carried out in Tianjin, China, where the salt intake, as well as the prevalence of hypertension and stroke mortality are also very high. The intervention was based on examinations of independent cross-sectional population samples in 1989 (1719 persons) and 1992 (2304 persons) (Tian *et al.* 1995). During the intervention period, there was a small reduction in salt intake in the intervention area, whereas in the control area there was a small increase in salt intake, so that the net difference in the change in salt intake between the intervention and control area was 2.4 g/day in

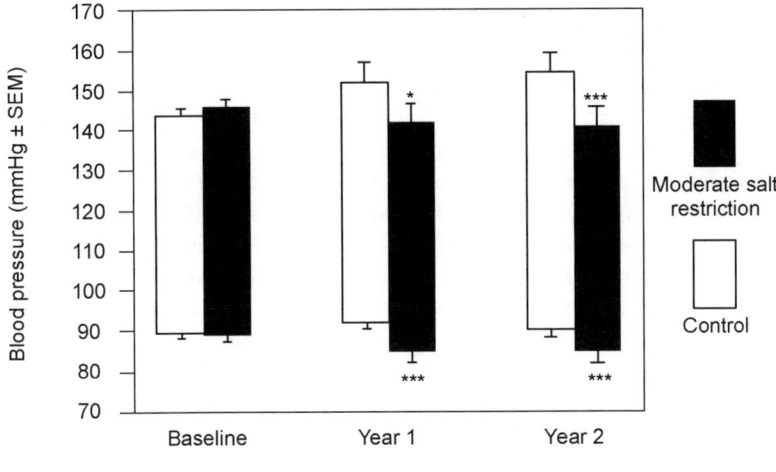

Fig. 2.2 Blood pressure changes with time in two Portuguese villages, one of which had salt intake reduced, the other had similar measurements of blood pressure but no advice on diet. Note the significant differences in blood pressure at 1 year and continuing differences at 2 years. Adapted from Forte *et al.* (1989).

men and 0.9 g/day in women. This was associated with a difference in systolic blood pressure of 5 mmHg in men and 4 mmHg in women.

The above two population-based intervention studies clearly demonstrate that a reduction in population salt intake lowers population blood pressure. A reduction in population blood pressure, even by a small amount, will have a large impact on reducing cardiovascular morbidity and mortality.

Two other intervention studies (one in Belgium, and the other in North Karelia) are often quoted as being negative (Tuomilehto *et al.* 1984; Staessen *et al.* 1988). However, neither were successful in reducing salt intake and it is not surprising that there was no difference in blood pressure in these studies between the community that was instructed on reducing salt, but failed to do so, and the community that was not given such instructions.

Treatment trials

Ambard and Beaujard (1904) were the first to show that reducing salt intake lowered blood pressure. These results were confirmed over the next 30 years by several workers, but it was not until Kempner (1948) revived the idea of a large reduction in salt intake that salt restriction became widely used in the treatment of hypertension. The first double-blind placebo-controlled trial of a more modest reduction in salt intake was performed in the 1980s in a group of untreated patients with essential hypertension (MacGregor *et al.* 1982a). It clearly demonstrated that a reduction in salt intake from 10 to 5 g/day for one month caused a significant fall in blood pressure, which was equivalent to that seen with a diuretic. Since then, a large number of salt reduction trials have been carried out not only in hypertensive individuals, but also in normotensive subjects.

Several meta-analyses of salt reduction trials have been performed (Law *et al.* 1991; Midgley *et al.* 1996; Cutler *et al.* 1997; Graudal *et al.* 1998; Hooper *et al.* 2002). In two meta-analyses (Midgley *et al.* 1996; Graudal *et al.* 1998), it was claimed that the results showed that salt reduction had no or very little effect on blood pressure in individuals with normal blood pressure. The authors concluded that a reduction in population salt intake is not warranted. Furthermore, these papers were used as the basis of a commentary in Science (Taubes 1998) casting doubt on the link between salt intake and blood pressure, and have also been used to oppose public health recommendations for a reduction in salt intake (Swales 2000). However, detailed examination of these two meta-analyses (Midgley *et al.* 1996; Graudal *et al.* 1998) shows that they are flawed. Both meta-analyses included trials of very short duration of salt restriction, many for only 5 days. On average, the median duration of salt reduction in individuals with normal blood pressure was only 8 days in one meta-analysis and 14 days in the other. Furthermore, around half of these trials compared the effects of acute salt loading to abrupt and severe salt restriction, e.g. from 20 to less than 1 g/day of salt for only a few days. It is known that these acute and large changes in salt intake cause an increase in sympathetic activity, plasma renin activity and angiotensin II concentration (He *et al.* 2001a), which would counteract the effects on blood pressure. It is also known that most blood pressure-lowering drugs do not exert their maximal effect within 5 days; this is particularly true with diuretics which are likely to work by a similar mechanism to that of a reduction in salt intake. For these reasons it is inappropriate to include the acute salt restriction trials in a meta-analysis that attempts to apply them to public health recommendations for a longer-term modest reduction in salt intake.

A recent meta-analysis by Hooper *et al.* (2002) is an important attempt to look at whether long-term salt reduction (i.e. more than six months) in randomized trials causes a fall in blood pressure. However, most trials included in this meta-analysis only achieved a very small reduction in salt intake and, on average, salt intake was only reduced by 2 g/day. It is, therefore, not surprising that there was only a small, but still highly significant fall in blood pressure.

More recently, we carried out a meta-analysis of randomized trials of modest reductions in salt intake with duration of one month or longer (He and MacGregor 2002). Our meta-analysis demonstrates that a modest reduction in salt intake, as currently recommended, does have a significant effect on blood pressure not only in hypertensive individuals, but also in those with normal blood pressure (Fig. 2.3). Furthermore, our study demonstrates a dose-response to salt reduction, within the range of 12 to 3 g/day, the lower the salt intake achieved, the lower the blood pressure (He and MacGregor 2003).

Although the recent DASH (Dietary Approaches to Stop Hypertension) – sodium trial (Sacks *et al.* 2001) was included in the above meta-analysis, it is still worth mentioning. It is a well controlled feeding trial studying three levels of salt intake (8, 6 and 4 g/day) on two different diets (i.e. the normal American diet and

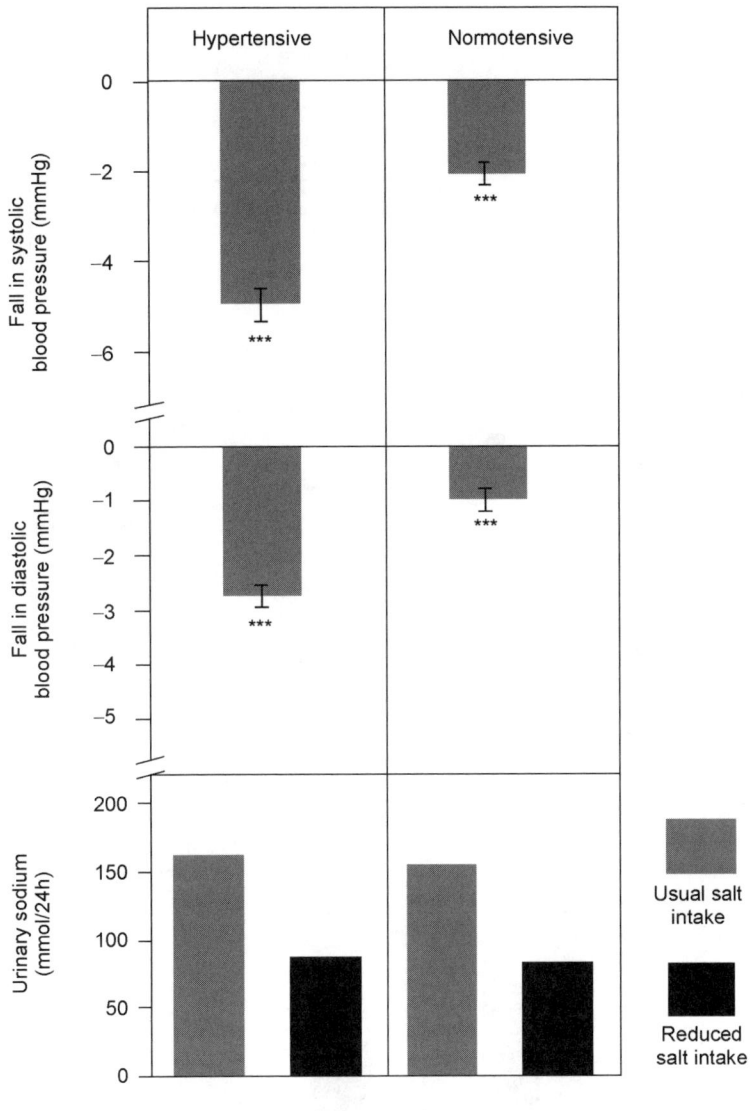

Fig. 2.3 Effect of modest salt reduction on blood pressure in hypertensive and normotensive individuals in a meta-analysis of 28 randomized controlled trials of one month or longer.

the DASH diet, which is rich in fruits, vegetables, and low-fat dairy products). This study demonstrates that salt reduction lowers blood pressure in both hypertensive and normotensive individuals, and there is a dose-response to salt reduction. Furthermore, salt reduction causes a further fall in blood pressure in individuals who consume the DASH diet. This is of importance in that it

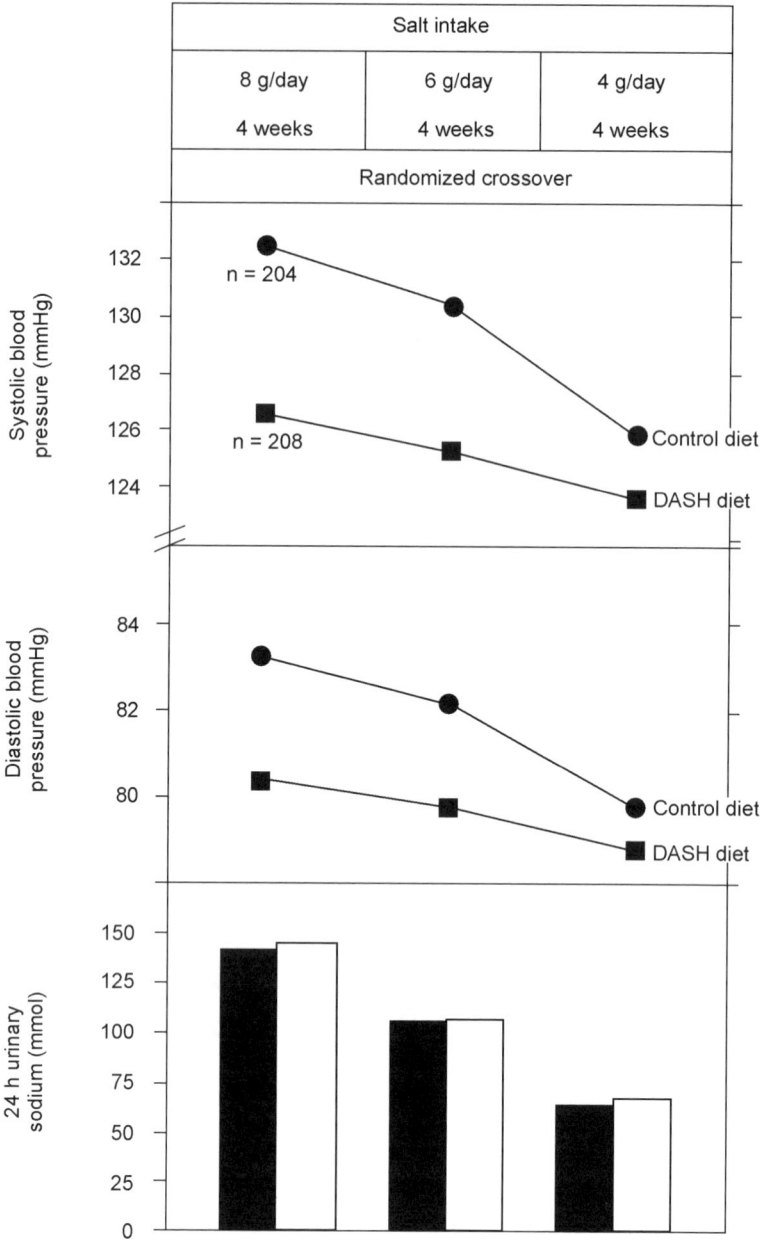

Fig. 2.4   Changes in blood pressure and 24 h urinary sodium excretion with the reduction in salt intake in all participants (hypertensives: $n = 169$; normotensives: $n = 243$) on the normal American diet (i.e. control diet) and on DASH diet. For 24 h urinary sodium solid bar represents control diet and open bar represents DASH-diet. Redrawn from Sacks *et al.* (2001).

demonstrates that not only would there be benefit from reducing salt intake on the diet we currently consume, but there would be additional benefits if salt reduction is combined with an increase in fruit and vegetable consumption (Fig. 2.4).

Treatment trials also show that, for a given reduction in salt intake, the falls in blood pressure are larger in individuals of African origin and in the older subjects (Sacks *et al.* 2001). The greater falls in blood pressure with salt reduction are related to lower levels of plasma renin activity and, thereby, angiotensin II, as well as the lower responsiveness of the renin-angiotensin system in these individuals (He *et al.* 1998).

A modest reduction in salt intake is also additive to antihypertensive drug treatments (MacGregor *et al.* 1987). Longer-term trials have shown that salt reduction enhances blood pressure control and reduces the need for antihypertensive drug therapy (Whelton *et al.* 1998).

Salt reduction in children
A number of studies have looked at the effect of reducing dietary salt intake on blood pressure in children. However, most of these studies either did not achieve any reduction in salt intake, or were underpowered to detect a small fall in blood pressure with reducing salt intake in children. One study (the Exeter-Andover project) was successful in reducing salt intake and did demonstrate a significant fall in blood pressure in adolescents (Ellison *et al.* 1989). The Exeter-Andover project was carried out in two boarding high schools with the intervention applied in each of the schools in alternate school years, with the second school serving as a control in each year. A total of 650 students (average age: 15 years old) participated in the study (341 in the intervention group and 309 in the control group). The intervention through changes in food purchasing and in preparation practices in the schools' kitchens achieved a reduction in salt intake of 15–20%. By the 24 weeks of the study the estimated net intervention effects were a decrease of 1.7/1.5 mmHg in blood pressure ($P < 0.01$) with adjustments for sex and baseline blood pressure.

A well-controlled double-blind study in just under 500 newborn babies showed that when salt intake was reduced by about half (intervention vs. control group) for six months, as judged by spot urinary sodium concentrations, there was a progressive difference in systolic blood pressure (Hofman *et al.* 1983). After six months intervention, the babies on the lower salt intake had a 2.1 mmHg lower systolic blood pressure ($P < 0.01$) (Fig. 2.5). The study was discontinued at six months. Fifteen years later, 35% of these babies were restudied. There remained a significant difference in blood pressure, when adjusted for confounding factors, between those babies who in the first six months of life had had a reduced salt intake compared to those who had not (Geleijnse *et al.* 1997). This study suggests that there is a programming effect of salt intake in early life, which fits with several studies in animals (Dahl *et al.* 1968).

Animal studies
There are numerous studies in the rat, dog, chicken, rabbit, baboon and chimpanzee, all of which have shown that when there is a prolonged increase in

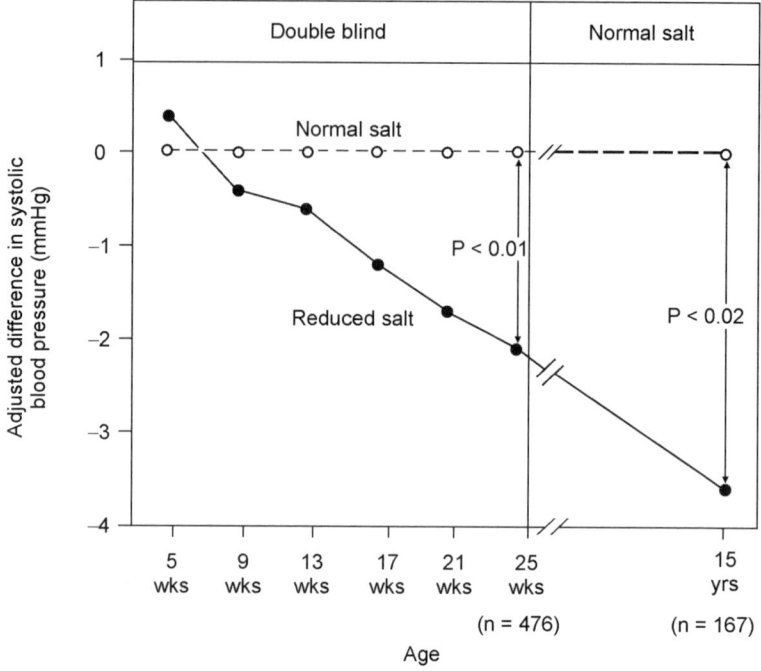

Fig. 2.5   Difference in systolic blood pressure in newborn babies, randomized to either a normal salt intake or a reduced salt intake over the first six months of life. At six months, the study was discontinued, with all participants resuming their usual salt intake. Fifteen years later, a subgroup of those in the study had blood pressure re-measured. Adapted from Hofman *et al.* (1983) and Geleijnse *et al.* (1997).

salt intake there is an increase in blood pressure. Furthermore, in all forms of experimental hypertension, whatever the animal model, a high salt intake is essential for blood pressure to rise.

A recent study was carried out in chimpanzees (98.8% genetic homology with man) (Denton *et al.* 1995). In a randomized parallel study, one group of chimpanzees was maintained on their normal diet of around half a gram of salt per day ($n = 12$), and the other had salt intake increased to 5, 10 and 15 g/day ($n = 10$). During the study there was no significant change in blood pressure in the control group. However, in the 10 animals assigned to the increased salt intake, mean systolic blood pressure was increased by 12 mmHg compared to the corresponding baseline level ($P < 0.05$) after the first 19 weeks of supplementary salt intake (5 g/day). Following the 39 weeks of supplementation with 10 g/day of salt (3 weeks) and 15 g/day of salt (36 weeks), mean systolic was increased by 26 mmHg ($P < 0.001$). Following a further 26 weeks of supplementation with 15 g/day of salt (a total of 84 weeks of supplementation with salt), mean systolic was increased by 33 mmHg ($P < 0.001$). Twenty weeks after the end of the salt supplementation period, the animals' average level of blood pressure returned to its baseline level (Fig. 2.6). This experiment provides

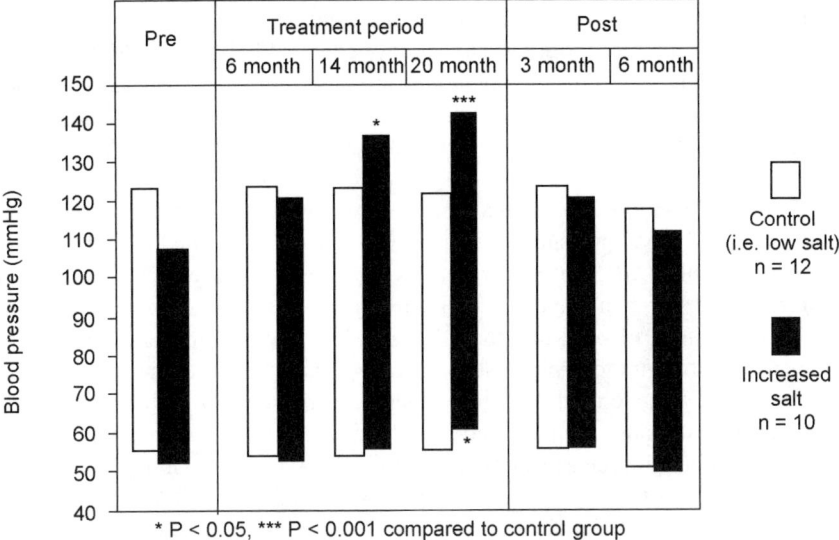

Fig. 2.6 Blood pressure in chimpanzees who either continued on their usual low-salt diet or were given an increased salt intake. At the end of the 20-month study, the salt supplements were stopped and blood pressure declined to that of the control group. Adapted from Denton *et al.* (1995).

direct evidence for a causal relationship between a high salt intake and a rise in blood pressure.

Human genetic studies

Attempts have been made to identify the genes whose function or dysfunction have shed light on the importance of salt in regulating blood pressure. It is generally accepted that several genes may contribute to the blood pressure levels which reflects a complex network of gene-gene and gene-environment interactions. However, in some individuals defects in a single gene cause marked abnormalities of blood pressure regulation. Genetic studies of these rare Mendelian forms of high and low blood pressure have shown an underlying common pathway: the kidney's ability to excrete or retain sodium (Lifton 1996; Lifton *et al.* 2001). The monogenic causes of high blood pressure reduce the kidney's ability to excrete sodium and cause high blood pressure if salt is consumed. The monogenic causes of low blood pressure result in the kidney being unable to hold on to sodium normally, thereby causing low blood pressure. These forms of low blood pressure are ameliorated by a high salt intake. Overall these genetic studies clearly indicate the vital importance of salt in regulating blood pressure in humans.

2.4.3 How far should salt intake be reduced?

Salt intake in many countries is between 9 and 12 g/day (INTERSALT 1988; Henderson *et al.* 2003). The current World Health Organisation recommendations for adults are to reduce salt intake to 5 g/day or less (WHO 2003) and the UK and US recommendations are to 6 g/day or less (Whelton *et al.* 2002; SACN 2003). However, these recommendations are based on what is feasible and not on what might have the maximum impact on blood pressure and cardiovascular disease. Recent evidence suggests that these levels, whilst they may be feasible, are too high.

Studies in experimental animals have shown a clear dose-response between salt intake and blood pressure, i.e. the higher the salt intake, the higher the blood pressure (Dahl *et al.* 1968). A recent study in chimpanzees which has been referred to earlier, demonstrated a dose-response when salt was increased from their usual intake of 0.5 g/day to 5, 10, and 15 g/day (Denton *et al.* 1995). In humans it is difficult to conduct such trials, particularly to keep individuals on a low salt diet long term due to the widespread presence of salt in nearly all processed, restaurant, canteen and fast foods. However, two well-controlled trials have studied three salt intakes (i.e. from 11.2, 6.4 to 2.9 g/day in one trial and from 8.3, 6.2 to 3.8 g/day in the other) (MacGregor *et al.* 1989; Sacks *et al.* 2001) and both showed a clear dose-response relationship, i.e. the lower the salt intake, the lower the blood pressure (Fig. 2.4).

A recent meta-analysis of randomized trials of modest salt reduction for one month or longer demonstrates a significant relationship between the reduction in 24-hour urinary sodium and the fall in blood pressure, indicating the greater the reduction in salt intake, the greater the fall in blood pressure (Fig. 2.7) (He and MacGregor 2002). A comparison of the dose-response found in the meta-analysis

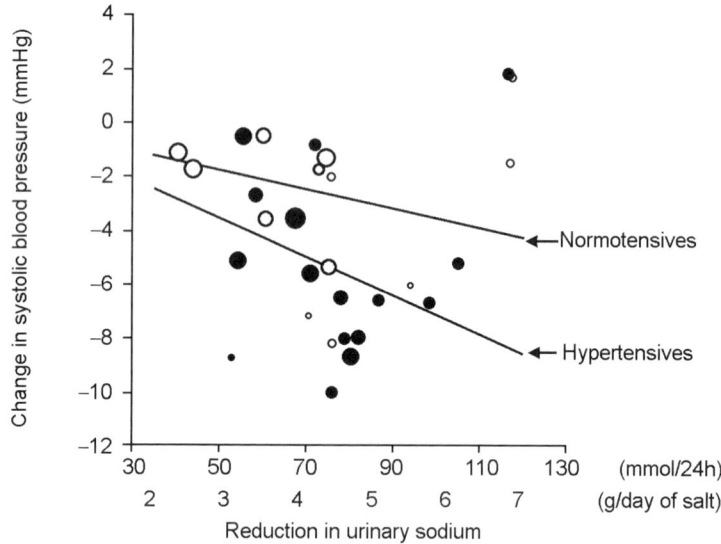

Fig. 2.7   Relationship between the net change in urinary sodium excretion and systolic blood pressure. The open circles represent normotensives and the solid circles represent hypertensives. The slope is weighted by the inverse of the variance of the net change in systolic blood pressure. The size of the circle is in proportion to the weight of the trial.

with the two studies that had three levels of salt intake showed a consistent dose-response to salt reduction within the range that was studied, i.e. 12 to 3 g/day, though the falls in blood pressure were greater in the better controlled studies (MacGregor *et al.* 1989; Sacks *et al.* 2001). A reduction of 6 g/day (e.g., from the current intake of 12 g/day to the recommended level of 6 g/day) predicts a fall in blood pressure of 7–11/4–6 mmHg in hypertensives and 4–7/2–4 mmHg in normotensives depending on study (He and MacGregor 2003). The effect would be much larger if salt intake were reduced further to 3 g/day.

From the evidence above, it is clear that the current public health recommendations to reduce salt intake from 9–12 g/day to 5–6 g/day will have a major effect on blood pressure, but are not ideal. A further reduction to 3 g/day will have a much greater effect and should, therefore, now become the long-term target for population salt intake worldwide.

2.5   Mechanisms by which salt raises blood pressure

The mechanisms whereby salt raises blood pressure are not fully understood. However, there is much evidence that individuals who develop high blood pressure have an underlying defect in the kidneys' ability to excrete sodium. The kidney cross-transplantation experiments clearly demonstrate the important role of the kidneys in blood pressure regulation (Dahl *et al.* 1972; 1974). When a kidney from a normotensive rat is inserted into a young bilaterally nephrectomized hypertensive rat, the blood pressure of the hypertensive rat does not rise, and conversely, when a kidney from a young hypertensive rat (before it has developed hypertension) is inserted into a bilaterally nephrectomized normotensive rat, the blood pressure of the normotensive rat rises. Similarly the high blood pressure of patients with essential hypertension who developed kidney failure, became normal (over a mean follow-up of 4.5 yr) when, following bilateral nephrectomy, they were transplanted with a kidney from a young normotensive donor (Curtis *et al.* 1983). These findings clearly indicate that whatever functional abnormalities may occur at other sites, the primary disturbance that initiates the rise in blood pressure resides in the kidneys.

The impaired ability of the kidneys to excrete sodium causes a tendency to retain salt and water, particularly on a high salt intake. This stimulates various compensatory mechanisms. The persistent presence of some of the compensatory mechanisms eventually cause the blood pressure to rise which also helps overcome the kidneys' difficulties in excreting sodium. Based on experiments in 70% nephrectomized dogs given large amounts of saline intravenously daily for two weeks Guyton (Guyton 1980) suggested that volume expansion raises the blood pressure by the autoregulatory effect on resistance vessels of the increase in blood flow which accompanies the associated persistent increase in cardiac output, even if this slight increase in cardiac output is usually unmeasurable (Safar *et al.* 1976). Others (Blaustein 1977) have proposed that among the multiple changes which counter the kidneys' impaired ability to excrete sodium and the resultant tendency to volume expansion, there is an increase in the plasma's capacity to inhibit Na-K-ATPase, which not only increases sodium excretion but also raises the blood pressure by inhibiting the sodium-calcium pump in vascular smooth muscle (Bagrov *et al.* 1996). Another hypothesis on the pressor mechanism induced by dietary salt suggests that the tendency for an increase in extracellular fluid volume is responsible for the documented increase in right and left (wedge) pressures in the auricles. It is proposed that the resultant increase in vagal afferent stimulation is responsible for the observed hypo-thalamic pressor changes (de Wardener 2001).

There is now increasing evidence that small increases in plasma sodium may be an important mechanism for the rise in blood pressure with increasing salt intake. We have recently looked at three types of studies of changing salt intake, (1) An acute and large reduction in salt intake from 20 to 1 g/day for 5 days in both hypertensives and normotensives was associated with a fall in plasma sodium of ≈ 3 mmol/L ($P < 0.001$); (2) Progressive increases in salt intake from 1 to 15 g/day by a daily amount of 3 g/day in normotensives, caused increases in plasma sodium ($P < 0.001$); (3) Longer-term modest reduction in salt intake in hypertensives was studied in double-blind randomized crossover studies; 1 month of usual salt intake (≈ 10 g/day) compared with reduced salt intake (≈ 5 g/day). There was a decrease in plasma sodium of 0.4 ± 0.2 mmol/L ($P < 0.05$) which was weakly but significantly correlated with the fall in systolic blood pressure ($r = 0.18, P < 0.05$) (He *et al.* 2005c). These studies demonstrate that an increase or decrease in dietary salt intake causes parallel changes in plasma sodium in both hypertensive and normotensive individuals.

Small changes in plasma sodium are the immediate drive to the changes in extracellular volume (Fig. 2.8). At the same time, small changes in plasma sodium may directly influence blood pressure, independent of, and additive to, the effect plasma sodium has on extracellular volume (de Wardener *et al.* 2004). Intraperitoneal dialysis with differing physiological salt solutions in rat has demonstrated that blood pressure rises or falls in direct relation to plasma sodium though extracellular volume changes occur in the opposite direction (Friedman *et al.* 1990). Tissue culture experiments varying the bath sodium within the physiological range have shown marked cellular changes in both arterial smooth muscle and cardiac myocytes (Gu *et al.* 1998). In addition, small changes in plasma sodium may also directly affect the hypothalamus's control of blood pressure through the local renin angiotensin system (de Wardener *et al.* 2004). In hypertension, a defect in a kidney's ability to excrete salt is likely to cause a greater increase in plasma sodium for a given salt intake. This may partially explain the raised blood pressure.

2.6   Salt and cardiovascular mortality

One of the difficulties of drawing conclusions about the importance of dietary or other lifestyle changes in cardiovascular disease is the gap in the evidence that

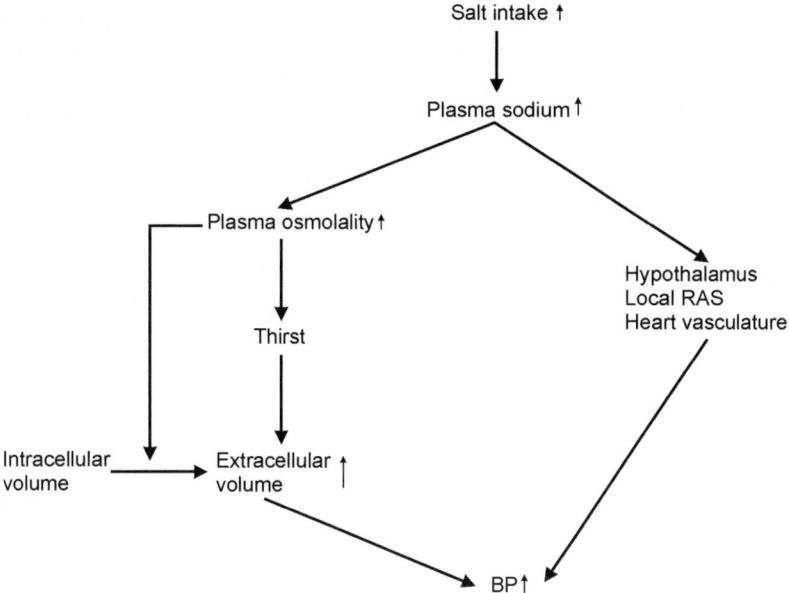

Fig. 2.8   Hypothesis on the possible links between salt intake, plasma sodium and blood pressure. BP: blood pressure. RAS: renin-angiotensin-system.

relates to mortality. One has to accept that outcome studies of changing diet in the population are extremely difficult and there is unlikely to ever be outcome evidence on mortality for dietary variables, e.g. fruit and vegetables or other lifestyle changes, e.g. stopping smoking, losing weight, or taking exercise. For instance, a study on salt would need to randomize subjects at the time of conception to a lower and higher salt intake and then follow up the two groups of offspring on a high and low salt intake for the rest of their lives. Such studies are impractical and would be unethical in the light of current knowledge.

Increasing blood pressure throughout the range is an important risk factor for cardiovascular disease. A reduction in salt intake lowers blood pressure, and would therefore be predicted to reduce cardiovascular disease. Based on falls in blood pressure from the meta-analysis of randomized salt reduction trials and the relationship between blood pressure and stroke and ischaemic heart disease from a recent meta-analysis of 1 million adults from 62 prospective studies, we estimated that a reduction of 3 g/day in salt intake would reduce strokes by 13% and ischaemic heart disease (IHD) by 10% (He and MacGregor 2003). In the UK the total number of stroke deaths is 60,666 a year and the total number of IHD deaths is 124,037 a year. Therefore, a reduction of 3 g/day in salt intake would prevent approximately 7800 stroke deaths and 11,500 IHD deaths a year. The effects on strokes and IHD would be almost doubled if salt intake were reduced by 6 g/day, and tripled with a 9 g/day reduction. A reduction of 9 g/day in salt intake (e.g., from 12 to 3 g/day) would reduce strokes by approximately a third and IHD by a quarter. In the UK this would prevent around 20,500 stroke deaths and 31,400 IHD deaths a year.

Approximately 50% of patients who suffer strokes or heart attacks survive, therefore, there would be a proportionate reduction in the numbers of these people. This would result in a reduction in disability and major cost savings both to individuals, their families and the Health Service. Furthermore, high blood pressure is an important risk factor for heart failure. A reduction in salt intake would therefore have a major effect on heart failure.

A reduction in salt intake not only lowers blood pressure, but also has other beneficial effects on the cardiovascular system, independent of and additive to the effect of salt reduction on blood pressure, e.g. a direct effect on stroke (Perry and Beevers 1992), left ventricular hypertrophy (Kupari *et al.* 1994; Schmieder and Messerli 2000), renal disease and proteinuria. Therefore, the true effect of salt reduction on the cardiovascular outcome may be larger than those estimated from blood pressure fall alone.

Several epidemiological studies have looked at the relationship between salt intake and cardiovascular disease. A recently published prospective study of 29,079 Japanese men and women living in Takayama City and Gifu showed a significant association between salt intake and death from ischaemic stroke (hazard ratio 3.22) and intracerebral haemorrhage (HR 3.85) in men and borderline associations (HR 1.70 and 2.10, respectively) in women (Nagata *et al.* 2004). More convincingly, because avoiding potential inaccuracy of estimating the habitual salt intake from 24-h dietary recall, a prospective Finnish cohort study conducted in 1173 men and 1263 women aged 25–64 demonstrates a significant association between salt intake (judged by 24-hour urinary sodium excretion) and coronary heart disease mortality, cardiovascular disease mortality and total mortality (Fig. 2.9) (Tuomilehto *et al.* 2001). The hazard ratios for coronary heart disease, cardiovascular disease, and all deaths associated with a 6 g increase in salt intake were 1.56, 1.36 and 1.22 respectively. In this study, the hazard ratio for cardiovascular disease mortality was 1.23 in men with normal body weight, while it was 1.44 for men who are overweight (body mass index $\geq 27 \, kg/m^2$). This suggests that a high salt intake increased the risk of subsequent cardiovascular disease in both normal weight and overweight people, and the effect is larger in those who are overweight.

Alderman *et al.* have also attempted to look at the relationship between salt intake and cardiovascular disease in two cohort studies. The first one was in hypertensive individuals who had renin profiling performed prior to entering a study of long-term follow-up on blood pressure-lowering drugs (Alderman *et al.* 1995). In order to perform the renin profiling, all subjects had their salt intake restricted for five days to stimulate renin release. This enabled the subjects to be sub-grouped into low, normal or high renin groups. Alderman found that the 24-hour urinary sodium excretion on the fifth day of a reduced salt intake was related to subsequent myocardial infarction and made the extraordinary claim that a lower salt intake led to more heart attacks. However, no measurement of salt intake had been carried out on the subjects' normal diet. Furthermore, no attempt was made to monitor salt intake during the follow-up period. Analysis of the 24-hour urinary sodium data also revealed severe methodological problems

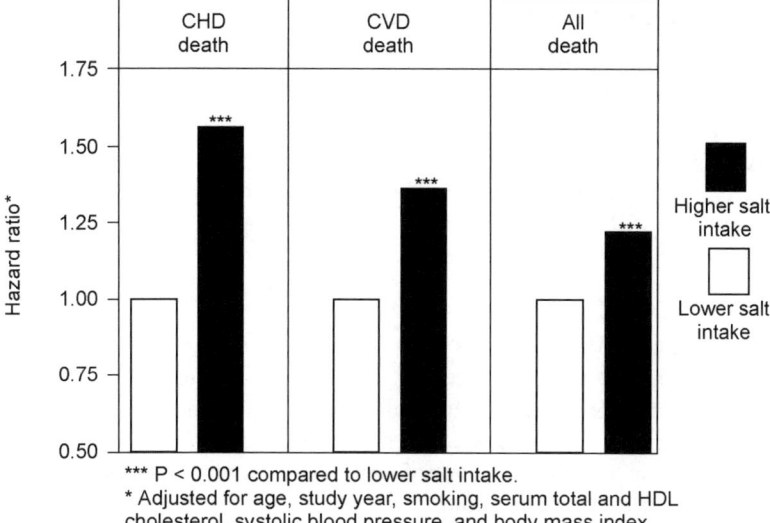

Fig. 2.9 The hazards ratios for coronary heart disease (CHD), cardiovascular disease (CVD), and all-cause mortality associated with a 6 g/day increase in salt intake as judged by 24-hour urinary sodium excretion. Adapted from Tuomilehto *et al.* (2001).

as the lowest salt quartile had a much lower 24-hour urinary creatinine excretion. This demonstrated that many of those who had been attributed to the lowest salt quartile on the fifth day of a reduced salt intake were there not because they had been more successful in reducing their salt intake, but had collected incomplete 24-hour urine samples (MacGregor 1996).

The second study by Alderman *et al.* involved the NHANES 1 – a dietary survey of US adults from the mid-1970s (Alderman *et al.* 1998). However, any analysis of salt intake from this study is difficult to judge as 24-hour urinary sodium excretion was not measured and dietary salt was assessed by dietary history with no account taken of discretionary salt (i.e. salt added at the table or in cooking), which at that time would have accounted for more than half of the salt intake. Alderman claimed that salt intake was inversely related to cardiovascular disease. However, examination of the data showed major discrepancies, e.g. subjects in the lower salt intake were on a calorie intake that was near starvation levels compared to the higher salt group and yet the lower calorie group weighed 4 kg more than those on the higher salt and calorie intake (de Wardener and MacGregor 1998; Engelman 1998; Karppanen and Mervaala 1998; de Wardener 1999).

In view of the serious flaws in these two studies by Alderman *et al.*, they cannot be used in any way to interpret the long-term effects of salt reduction. Indeed, a more appropriate analysis of the same NHANES 1 data showed a positive relationship between dietary salt intake and risk of stroke, coronary heart disease and heart failure in overweight individuals (body mass index $>27 \, \text{kg/m}^2$) (He *et al.* 1999).

2.7   Salt and other harmful effects

There is increasing evidence that salt has other harmful effects on human health (de Wardener and MacGregor 2002), which may be independent of and additive to the effect of salt on blood pressure, e.g. a direct effect on stroke (Perry and Beevers 1992), left ventricular hypertrophy (Kupari *et al.* 1994; Schmieder and Messerli 2000), progression of renal disease and albuminuria (Heeg *et al.* 1989; Cianciaruso *et al.* 1998; Swift *et al.* 2005), stomach cancer (Joossens *et al.* 1996; Tsugane *et al.* 2004), and bone demineralization (Devine *et al.* 1995).

2.7.1   Salt and water retention

When salt intake is increased, there is retention of sodium and thereby water, and this expands the extracellular fluid volume. This increase in extracellular volume is a trigger for various compensatory mechanisms to increase urinary sodium excretion but at the expense of continued retention of sodium and water. The increase in extracellular fluid exacerbates all forms of sodium and water retention, e.g. heart failure, and is a major cause of oedema in women, aggravating both cyclical and idiopathic oedema (MacGregor and de Wardener 1997).

2.7.2   Salt and stroke

Increasing blood pressure throughout the range is the major cause of stroke. A high salt intake causes a rise in blood pressure which will increase the risk of stroke. However, experimental studies in animals (Tobian and Hanlon 1990) and epidemiological studies in humans (Perry and Beevers 1992; Xie *et al.* 1992) have shown that a high salt diet may have a direct effect on stroke, independent of and additive to its effect on blood pressure. Perry and Beevers performed an ecological analysis of the relationship between urinary sodium excretion (data from INTERSALT study) and stroke mortality in Western Europe. They found a significant positive correlation between urinary sodium excretion and stroke mortality (Fig. 2.10) (Perry and Beevers 1992), and this relationship is much stronger than that found when urinary sodium is plotted against blood pressure.

2.7.3   Salt and left ventricular hypertrophy

Left ventricular hypertrophy is an important independent predictor of cardiovascular morbidity and mortality and is highly prevalent in individuals with raised blood pressure (Laufer *et al.* 1989; Levy *et al.* 1990). Several cross-sectional studies have shown a positive correlation between urinary sodium excretion and left ventricular mass in both hypertensives and normotensives (Schmieder *et al.* 1988; du Cailar *et al.* 1992; Kupari *et al.* 1994) (Fig. 2.11). More importantly, 24-hour urinary sodium has been shown to be an independent and more powerful determinant for left ventricular wall thickness than blood pressure (Schmieder *et al.* 1988). A reduction in salt intake has been shown to

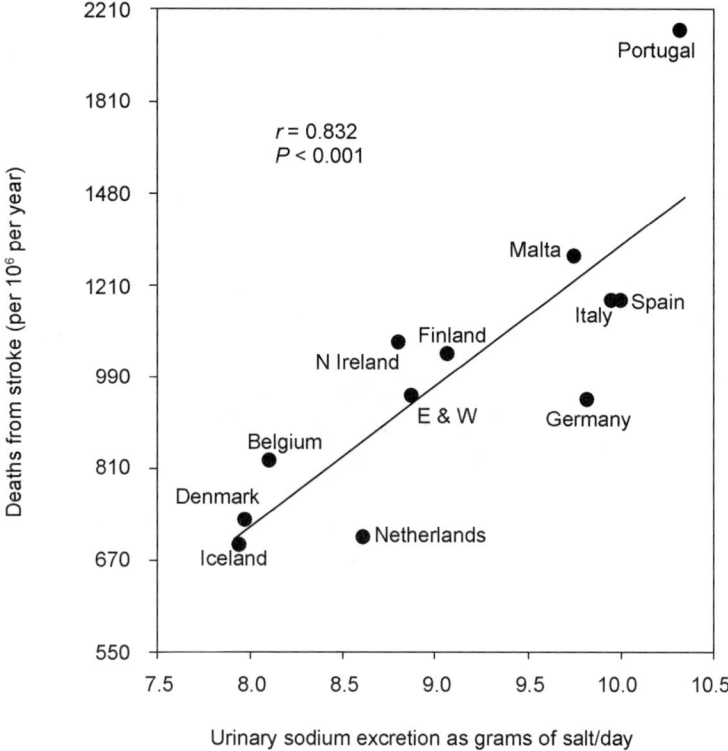

Fig. 2.10   Relationship between salt intake and deaths from strokes in 12 European countries. Adapted from Perry and Beevers (1992).

decrease left ventricular mass in patients with essential hypertension (Ferrara *et al.* 1984; Jula and Karanko 1994; Liebson *et al.* 1995).

2.7.4   Salt and blood vessels

Stiffness of conduit arteries, measured as an increase in pulse wave velocity or pulse pressure, is a strong independent predictor of cardiovascular risk (Blacher *et al.* 1999; Gasowski *et al.* 2002). Studies in both humans and experimental animals have shown that an increase in salt intake increases the stiffness of conduit arteries and the reactivity of the small resistance vessels and the wall thickness of both (Safar *et al.* 2000; Simon and Illyes 2001). In a study of two Chinese populations, the age-associated increase in pulse wave velocity was blunted in the population with a lower salt intake (Avolio *et al.* 1985). Another study in normotensive subjects showed that a low salt diet reduced arterial stiffness, independent of blood pressure (Avolio *et al.* 1986). A recent random-ized double-blind study shows that a modest reduction in dietary salt intake reduces pulse pressure both in individuals with isolated systolic hypertension and in those with both raised systolic and diastolic blood pressure, suggesting that salt

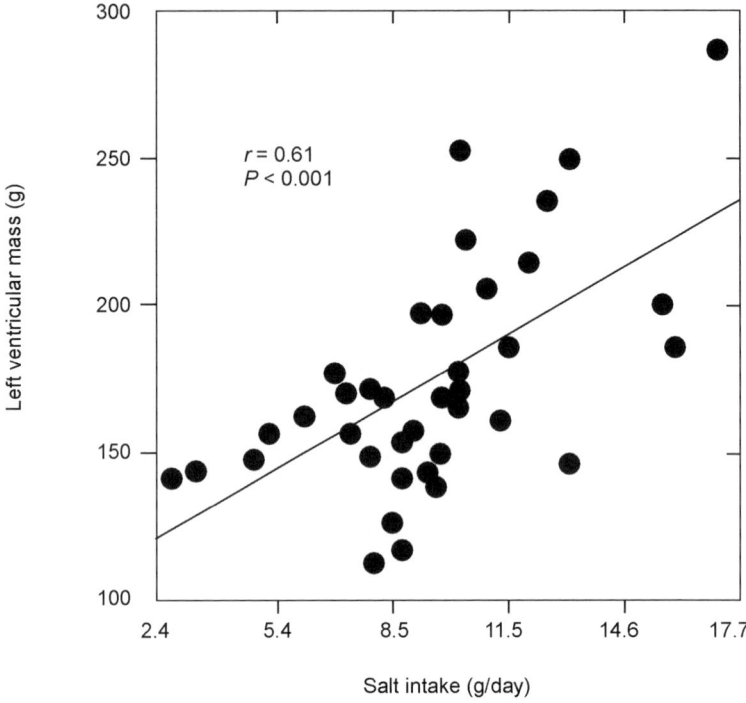

Fig. 2.11   Relationship between salt intake and left ventricular mass in individuals with systolic blood pressure >121 mmHg. Adapted from Kupari *et al.* (1994).

reduction improves arterial distensibility (He *et al.* 2005b). Another recently published paper by Gates *et al.* demonstrates that directly measured large elastic artery compliance is increased by dietary salt restriction in middle-aged and older men and women with stage 1 systolic hypertension (Gates *et al.* 2004).

2.7.5   Salt and albuminuria
Urinary albumin excretion has been shown to be an independent and continuous risk factor for both renal and cardiovascular disease (Grimm *et al.* 1997; Gerstein *et al.* 2001; Hillege *et al.* 2002). Epidemiological studies have shown a direct association between salt intake and urinary albumin excretion, i.e. the higher the salt intake, the higher the urinary albumin excretion (du Cailar *et al.* 2002; Verhave *et al.* 2004). A recent randomized double-blind trial in 40 hypertensive blacks demonstrates that a modest reduction in salt intake from approximately 10 to 5 g/day, as currently recommended, reduces urinary protein excretion significantly (Fig. 2.12) (Swift *et al.* 2005). Other studies have shown that the antiproteinuric effect of angiotensin-converting-enzyme (ACE) inhibitor is dependent on salt intake, i.e. a low salt intake enhances and a high salt intake abolishes the antiproteinuric effect of ACE inhibitor (Heeg *et al.* 1989; Cianciaruso *et al.* 1998).

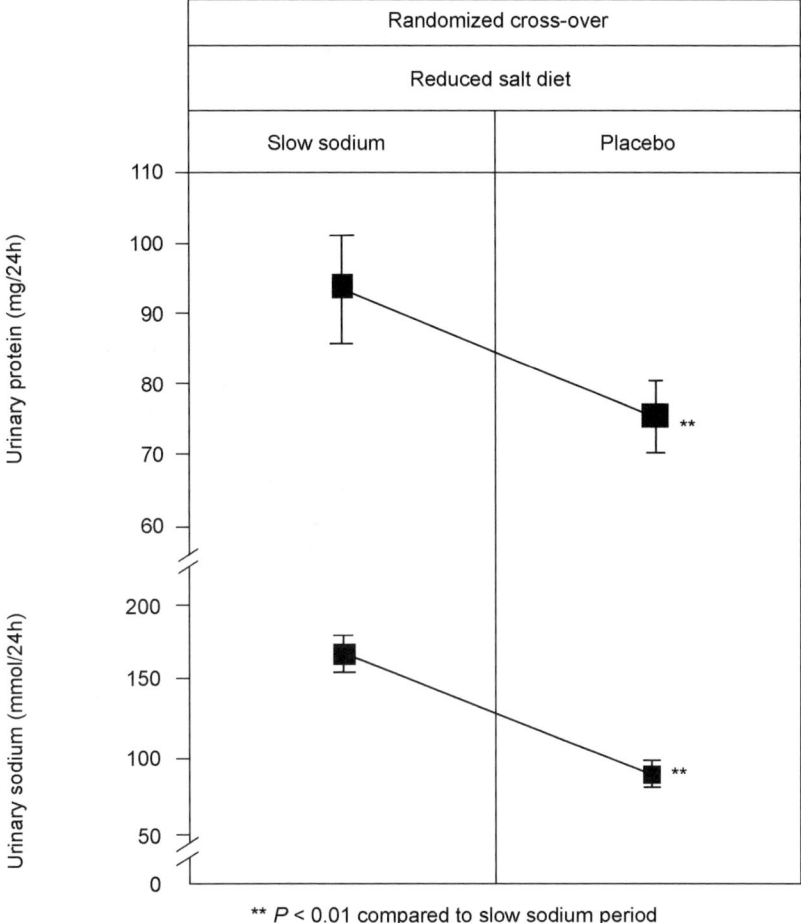

Fig. 2.12   Change in urinary sodium and protein excretion with a modest reduction in salt intake in 40 hypertensive blacks.

2.7.6   Salt and stomach cancer

Death from stomach cancer is the second commonest form of death from cancer worldwide. An ecological analysis showed a significant direct association between salt intake (as judged by 24-hour urinary sodium excretion) and deaths from stomach cancer among 39 populations from 24 countries (Joossens *et al.* 1996) (Fig. 2.13). A recent study from Japan confirms a close relationship between salt intake and stomach cancer within a single country (Tsugane *et al.* 2004). A number of studies have shown that H-pylori infection, which underlies the cause of both duodenal and gastric ulcers and stomach cancer, is also closely associated with salt intake in different countries in both women and men (Forman *et al.* 1991; Beevers *et al.* 2004; Wong *et al.* 2004). Foods that contain high concentrations of salt are irritating to the delicate lining of the stomach. It is

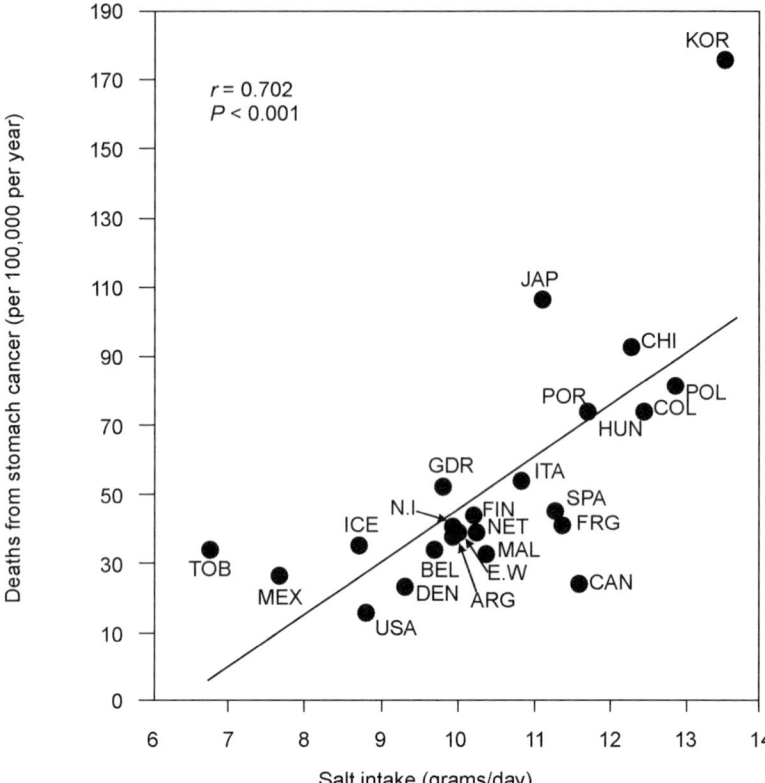

Fig. 2.13   Relationship between salt intake and deaths from stomach cancer. Adapted from Joossens *et al.* (1996).

possible that this makes H-pylori infection more likely or more severe and that the H-pylori infection then leads to stomach cancer. A modest reduction in salt intake may reduce H-pylori infection and therefore lead to stomach cancer prevention.

2.7.7   Salt and renal stones and bone mineral density

Salt intake is one of the major dietary determinants of urinary calcium excretion. Both epidemiological studies and randomized trials show that a reduction in salt intake causes a decrease in urinary calcium excretion (Matkovic *et al.* 1995; Cappuccio *et al.* 1997; 2000; Lin *et al.* 2003). As calcium is the main component of most urinary stones, salt intake is therefore an important cause of renal stones.

   Until recently, it was assumed that when salt intake was increased, the increase in calcium excretion was compensated for by an increase in intestinal calcium absorption. There is now evidence to suggest that, when salt intake is increased, there is a negative calcium balance with stimulation of mechanisms not only to increase intestinal absorption of calcium, but also to mobilize calcium from bone. A study in post-menopausal women showed that the loss of hip bone density over two years was related to the 24-hour urinary sodium excretion at entry to the study and was as strong as that relating to calcium intake (Devine *et al.* 1995). Diuretics, through a reduction in extracellular volume, reduce calcium excretion leading to a positive calcium balance, increase bone density and reduce bone fractures. Salt reduction is likely to do the same.

2.7.8   Salt and asthma

Epidemiological evidence suggests that the severity of asthma may be related to salt intake in different countries (Burney 1987). A double-blind study of modest salt reduction caused a decrease in the severity of asthma attacks and a reduction in the use of medication and an improvement in the measurement of airways resistance (Carey *et al.* 1993). A more recent double-blind study illustrates the mechanism whereby a higher salt intake exacerbates asthma (Mickleborough *et al.* 2005).

2.8   Other dietary and lifestyle factors in the development of hypertension

Much evidence suggests that potassium intake plays an important role in regulating blood pressure (He and MacGregor 1999; 2001). Obesity coupled with a lack of exercise are also important factors in the development of hypertension. Excess alcohol intake is also related to high blood pressure, but the effect appears to be transient and there is debate as to whether excess alcohol intake causes a sustained increase in blood pressure. Other minerals, e.g. calcium, magnesium, fat and protein intake have also been studied but so far the results are inconsistent.

2.8.1   Potassium

Addison was the first to suggest in 1928 that increasing potassium intake might lower blood pressure through its natriuretic effect (Addison 1988). Consequently potassium chloride was used in some patients with heart failure, but the effects on blood pressure were largely ignored. Recently, epidemiological and clinical studies in man and experimental studies in animals have shown that increasing potassium intake does lower blood pressure and that communities with a high potassium intake tend to have lower population blood pressures. The INTERSALT study confirmed that potassium intake, as judged by 24-hour urinary potassium excretion, was an important independent determinant of population blood pressure. A 30–45 mmol increase in potassium intake was associated with an average reduction in population systolic blood pressure of 2 to 3 mmHg (Dyer *et al.* 1994). Randomized trials have shown that increasing potassium intake lowers blood pressure both in individuals with high blood pressure and to a lesser, but still significant amount in those with normal blood pressure (Cappuccio and MacGregor 1991; Whelton *et al.* 1997).

In relation to potassium intake and blood pressure, there are two continuing areas of controversy. One is whether salt and potassium intake, which have opposite effects on blood pressure, have additive effects, i.e. when potassium intake is increased and salt intake is reduced. The INTERSALT study, where blood pressure was measured in 52 communities in the world, did show a direct relationship to salt intake as judged by 24-hour urinary sodium excretion and an inverse and independent relationship to potassium intake as judged by 24-hour urinary potassium excretion (Dyer *et al.* 1994). However, some small clinical trials indicated that increasing potassium intake, when salt intake was reduced, had less effect on blood pressure (Smith *et al.* 1985). The recent DASH-sodium study clearly demonstrates that salt restriction is additive to the DASH diet, but the effect is less than when a similar reduction in salt intake is made on a normal American diet, i.e. when the potassium intake is lower. Nevertheless, it clearly shows an additive effect of both increasing potassium intake and reducing salt intake (Sacks *et al.* 2001).

The other area of controversy is whether potassium in the form of chloride has a greater or lesser effect on blood pressure compared to other potassium salts. Most previous trials have looked at the effect of potassium chloride on blood pressure. However, potassium in fruits and vegetables is present with phosphate, sulphate, citrate, and many organic anions including proteins rather than as potassium chloride. A comparison of the DASH study (Appel *et al.* 1997) to double-blind studies of potassium chloride supplementation (MacGregor *et al.* 1982b) does seem to indicate that the fall in blood pressure with increasing fruits and vegetables is very similar to that found when it is done by increasing potassium chloride intake. We have recently carried out a randomized crossover trial comparing potassium chloride with potassium citrate (96 mmol/day, each for one week) in 14 hypertensive individuals. The results show that, for a similar increase in 24-hour urinary potassium excretion, potassium chloride and potassium citrate have a similar blood pressure-lowering effect (He *et al.* 2005a). This suggests that potassium does not need to be given in the form of chloride to lower blood pressure. Increasing the consumption of foods high in potassium is likely to have the same effect on blood pressure as potassium chloride.

2.8.2   Obesity, alcohol and physical exercise

Both epidemiological studies and randomized trials have shown that overweight, high alcohol consumption and physical inactivity are important contributing factors to raised blood pressure. Maintaining normal body weight (body mass index 18.5–24.9 kg/m^2), limiting alcohol consumption to no more than two drinks per day in most men and to no more than one drink per day in women and lighter-weight persons, and engaging in regular aerobic physical activity such as brisk walking (at least 30 minutes per day, most days of the week) have all been recommended for the prevention of hypertension (Chobanian *et al.* 2003).

2.9   Conclusions and perspectives

The totality of evidence that links salt intake to blood pressure is now overwhelming. Current recommendations are to reduce salt intake from 9–12 g/ day to 5–6 g/day. From the evidence reviewed in this chapter, it is clear that these reductions will have a major effect on blood pressure and cardiovascular disease, but reducing salt intake further to 3 g/day will have additional large effects. Therefore, the target of 5–6 g/day should be seen as an interim target and the long-term target for population salt intake worldwide should now be 3 g/day.

One important point is how to reduce salt intake to the target levels of 5–6 g/ day. In most developed countries 75–80% of salt intake now comes from salt added to processed foods. In our view, the best strategy would be to have the food industry gradually reduce the salt concentration of all processed foods, starting with a 10–25% reduction which is not detectable by human salt taste receptors (Girgis *et al.* 2003) and causes no problem in the food technology, and continuing a sustained reduction over the course of the next decade. This strategy has now been adopted in the UK both by the Department of Health and Food Standards Agency, and several leading supermarkets and food manufacturers have already started to implement such changes. Of all the dietary changes to try and prevent cardiovascular disease, a reduction in salt intake is the easiest change to make as it can be done without the consumers' knowing but it requires the co-operation of the food industry. Clearly it would be helped if individuals also reduced the amount of salt they add to their own cooking or to their food.

Some members of the food industry are reluctant to reduce the salt content of processed foods. This is because salt makes cheap, unpalatable food edible at no cost (Nestle 2002). If high salt foods are consistently consumed, the salt taste receptors are suppressed and habituation to high salt foods occurs, with greater demand for cheap but profitable high salt processed foods (Fig. 2.14). Salt also has two other important properties – one is in meat products where increasing the salt concentration in conjunction with other water-binding chemicals increases the amount of water that can be bound into the meat product and the weight of the product can be increased by up to 20% with water. The other important property is that salt is a major determinant of thirst and any reduction in salt intake will reduce fluid consumption with a subsequent reduction in soft drink and mineral water sales (He *et al.* 2001b). Some of the largest snack companies in the world are part of companies selling soft drinks. It is therefore not surprising that the salt industry and some members of the food industry are very reluctant to see any reduction in salt intake and have been largely responsible for trying to make salt such a controversial issue relative to other dietary changes. Their strategies are identical to the techniques used by the tobacco industry and the tobacco manufacturers' association. The commercial

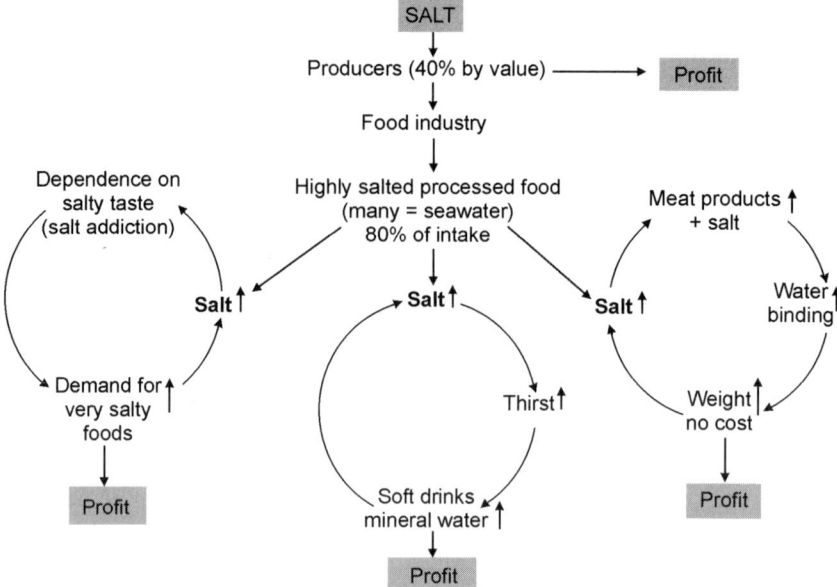

Fig. 2.14   The commercial importance of salt in processed food.

reasons for this opposition need to be acknowledged. However, they should not be allowed to stand in the way of a reduction in salt intake as this reduction will be of major benefit to the future health of the whole population, particularly if it is combined with other dietary and lifestyle changes, e.g. increasing fruit and vegetable consumption, reducing saturated fat intake, and stopping smoking. The bulk of the food industry has nothing to fear from gradually reducing the very high salt concentration of the many foods they produce. Indeed these foods are healthier. They lower blood pressure, reduce the risk of cardiovascular disease, renal disease, stomach cancer, and bone demineralization. The population will therefore live longer and there will be an increase in the number of consumers.

2.10   References

ADDISON W L (1988), 'The Canadian Medical Association Journal, vol. XVIII, 1928: The use of sodium chloride, potassium chloride, sodium bromide, and potassium bromide in cases of arterial hypertension which are amenable to potassium chloride', *Nutr Rev*, 46(8), 295–296.

ALDERMAN M H, MADHAVAN S, COHEN H, SEALEY J E, LARAGH J H (1995), 'Low urinary sodium is associated with greater risk of myocardial infarction among treated hypertensive men', *Hypertension*, 25(6), 1144–1152.

ALDERMAN M H, COHEN H, MADHAVAN S (1998), 'Dietary sodium intake and mortality: the National Health and Nutrition Examination Survey (NHANES I)', *Lancet*, 351(9105), 781–785.

AMBARD L, BEAUJARD E (1904), 'Cause de l'hypertension arterielle', *Arch Gen Med*, 1, 520–533.

APPEL L J, MOORE T J, OBARZANEK E, VOLLMER W M, SVETKEY L P, SACKS F M, BRAY G A, VOGT T M, CUTLER J A, WINDHAUSER M M, LIN P H, KARANJA N (1997), 'A clinical trial of the effects of dietary patterns on blood pressure. DASH Collaborative Research Group', *N Engl J Med*, 336(16), 1117–1124.

AVOLIO A P, DENG F Q, LI W Q, LUO Y F, HUANG Z D, XING L F, O'ROURKE M F (1985), 'Effects of aging on arterial distensibility in populations with high and low prevalence of hypertension: comparison between urban and rural communities in China', *Circulation*, 71(2), 202–210.

AVOLIO A P, CLYDE K M, BEARD T C, COOKE H M, HO K K, O'ROURKE M F (1986), 'Improved arterial distensibility in normotensive subjects on a low salt diet', *Arteriosclerosis*, 6(2), 166–169.

BAGROV A Y, FEDOROVA O V, DMITRIEVA R I, FRENCH A W, ANDERSON D E (1996), 'Plasma marinobufagenin-like and ouabain-like immunoreactivity during saline volume expansion in anesthetized dogs', *Cardiovasc Res*, 31(2), 296–305.

BEEVERS D G, LIP G Y, BLANN A D (2004), 'Salt intake and Helicobacter pylori infection', *J Hypertens*, 22(8), 1475–1477.

BLACHER J, ASMAR R, DJANE S, LONDON G M, SAFAR M E (1999), 'Aortic pulse wave velocity as a marker of cardiovascular risk in hypertensive patients', *Hypertension*, 33(5), 1111–1117.

BLAUSTEIN M P (1977), 'Sodium ions, calcium ions, blood pressure regulation, and hypertension: a reassessment and a hypothesis', *Am J Physiol*, 232(5), C165–173.

BURNEY P (1987), 'A diet rich in sodium may potentiate asthma. Epidemiologic evidence for a new hypothesis', *Chest*, 91(6 Suppl), 143S–148S.

CAPPUCCIO F P, MACGREGOR G A (1991), 'Does potassium supplementation lower blood pressure? A meta-analysis of published trials', *J Hypertens*, 9(5), 465–473.

CAPPUCCIO F P, MARKANDU N D, CARNEY C, SAGNELLA G A, MACGREGOR G A (1997), 'Double-blind randomised trial of modest salt restriction in older people', *Lancet*, 350(9081), 850–854.

CAPPUCCIO F P, KALAITZIDIS R, DUNECLIFT S, EASTWOOD J B (2000), 'Unravelling the links between calcium excretion, salt intake, hypertension, kidney stones and bone metabolism', *J Nephrol*, 13(3), 169–177.

CAREY O J, LOCKE C, COOKSON J B (1993), 'Effect of alterations of dietary sodium on the severity of asthma in men', *Thorax*, 48(7), 714–718.

CHOBANIAN A V, BAKRIS G L, BLACK H R, CUSHMAN W C, GREEN L A, IZZO J L, JR., JONES D W, MATERSON B J, OPARIL S, WRIGHT J T, JR., ROCCELLA E J (2003), 'Seventh report of the Joint National Committee on Prevention, Detection, Evaluation, and Treatment of High Blood Pressure', *Hypertension*, 42(6), 1206–1252.

CIANCIARUSO B, BELLIZZI V, MINUTOLO R, TAVERA A, CAPUANO A, CONTE G, DE NICOLA L (1998), 'Salt intake and renal outcome in patients with progressive renal disease', *Miner Electrolyte Metab*, 24(4), 296–301.

CURTIS J J, LUKE R G, DUSTAN H P, KASHGARIAN M, WHELCHEL J D, JONES P, DIETHELM A G (1983), 'Remission of essential hypertension after renal transplantation', *N Engl J Med*, 309(17), 1009–1015.

CUTLER J A, FOLLMANN D, ALLENDER P S (1997), 'Randomized trials of sodium reduction: an overview', *Am J Clin Nutr*, 65(2 Suppl), 643S–651S.

DAHL L (1960). Possible role of salt intake in the development of essential hypertension. In: Cottier P, Bock KD (eds). *Essential Hypertension – An International*

Symposium. Berlin, Springer-Verlag.

DAHL L K, KNUDSEN K D, HEINE M A, LEITL G J (1968), 'Effects of chronic excess salt ingestion. Modification of experimental hypertension in the rat by variations in the diet', *Circ Res*, 22(1), 11–18.

DAHL L K, HEINE M, THOMPSON K (1972), 'Genetic influence of renal homografts on the blood pressure of rats from different strains', *Proc Soc Exp Biol Med*, 140(3), 852–856.

DAHL L K, HEINE M, THOMPSON K (1974), 'Genetic influence of the kidneys on blood pressure. Evidence from chronic renal homografts in rats with opposite predispositions to hypertension', *Circ Res*, 40(4), 94–101.

DE WARDENER H, MacGREGOR G A (1998), 'Sodium intake and mortality', *Lancet*, 351(9114), 1508; author reply 1509–1510.

DE WARDENER H E (1999), 'Salt reduction and cardiovascular risk: the anatomy of a myth', *J Hum Hypertens*, 13(1), 1–4.

DE WARDENER H E (2001), 'The hypothalamus and hypertension', *Physiol Rev*, 81(4), 1599–1658.

DE WARDENER H E, MacGREGOR G A (2002), 'Harmful effects of dietary salt in addition to hypertension', *J Hum Hypertens*, 16(4), 213–223.

DE WARDENER H E, HE F J, MacGREGOR G A (2004), 'Plasma sodium and hypertension', *Kidney Int*, 66(6), 2454–2466.

DENTON D, WEISINGER R, MUNDY N I, WICKINGS E J, DIXSON A, MOISSON P, PINGARD A M, SHADE R, CAREY D, ARDAILLOU R, PAILLARD F, CHAPMAN J, THILLET J, MICHEL J (1995), 'The effect of increased salt intake on blood pressure of chimpanzees', *Nat Med*, 1(10), 1009–1016.

DEVINE A, CRIDDLE R A, DICK I M, KERR D A, PRINCE R L (1995), 'A longitudinal study of the effect of sodium and calcium intakes on regional bone density in postmenopausal women', *Am J Clin Nutr*, 62(4), 740–745.

DU CAILAR G, RIBSTEIN J, DAURES J P, MIMRAN A (1992), 'Sodium and left ventricular mass in untreated hypertensive and normotensive subjects', *Am J Physiol*, 263(1 Pt 2), H177–181.

DU CAILAR G, RIBSTEIN J, MIMRAN A (2002), 'Dietary sodium and target organ damage in essential hypertension', *Am J Hypertens*, 15(3), 222–229.

DYER A R, ELLIOTT P, SHIPLEY M (1994), 'Urinary electrolyte excretion in 24 hours and blood pressure in the INTERSALT Study. II. Estimates of electrolyte-blood pressure associations corrected for regression dilution bias. The INTERSALT Cooperative Research Group', *Am J Epidemiol*, 139(9), 940–951.

EATON S B, KONNER M (1985), 'Paleolithic nutrition. A consideration of its nature and current implications', *N Engl J Med*, 312(5), 283–289.

ELLIOTT P, STAMLER J, NICHOLS R, DYER A R, STAMLER R, KESTELOOT H, MARMOT M (1996), 'Intersalt revisited: further analyses of 24 hour sodium excretion and blood pressure within and across populations. Intersalt Cooperative Research Group', *BMJ*, 312(7041), 1249–1253.

ELLISON R C, CAPPER A L, STEPHENSON W P, GOLDBERG R J, HOSMER D W, JR., HUMPHREY K F, OCKENE J K, GAMBLE W J, WITSCHI J C, STARE F J (1989), 'Effects on blood pressure of a decrease in sodium use in institutional food preparation: the Exeter-Andover Project', *J Clin Epidemiol*, 42(3), 201–208.

ENGELMAN K (1998), 'Sodium intake and mortality', *Lancet*, 351(9114), 1508–1509.

FERRARA L A, DE SIMONE G, PASANISI F, MANCINI M (1984), 'Left ventricular mass reduction during salt depletion in arterial hypertension', *Hypertension*, 6(5), 755–759.

FORMAN D, NEWELL D G, FULLERTON F, YARNELL J W, STACEY A R, WALD N, SITAS F (1991), 'Association between infection with Helicobacter pylori and risk of gastric cancer: evidence from a prospective investigation', *BMJ*, 302(6788), 1302–1305.

FORTE J G, MIGUEL J M, MIGUEL M J, DE PADUA F, ROSE G (1989), 'Salt and blood pressure: a community trial', *J Hum Hypertens*, 3(3), 179–184.

FRIEDMAN S M, McINDOE R A, TANAKA M (1990), 'The relation of blood sodium concentration to blood pressure in the rat', *J Hypertens*, 8(1), 61–66.

GASOWSKI J, FAGARD R H, STAESSEN J A, GRODZICKI T, POCOCK S, BOUTITIE F, GUEYFFIER F, BOISSEL J P (2002), 'Pulsatile blood pressure component as predictor of mortality in hypertension: a meta-analysis of clinical trial control groups', *J Hypertens*, 20(1), 145–151.

GATES P E, TANAKA H, HIATT W R, SEALS D R (2004), 'Dietary sodium restriction rapidly improves large elastic artery compliance in older adults with systolic hypertension', *Hypertension*, 44(1), 35–41.

GELEIJNSE J M, HOFMAN A, WITTEMAN J C, HAZEBROEK A A, VALKENBURG H A, GROBBEE D E (1997), 'Long-term effects of neonatal sodium restriction on blood pressure', *Hypertension*, 29(4), 913–917.

GERSTEIN H C, MANN J F, YI Q, ZINMAN B, DINNEEN S F, HOOGWERF B, HALLE J P, YOUNG J, RASHKOW A, JOYCE C, NAWAZ S, YUSUF S (2001), 'Albuminuria and risk of cardio-vascular events, death, and heart failure in diabetic and nondiabetic individuals', *JAMA*, 286(4), 421–426.

GIRGIS S, NEAL B, PRESCOTT J, PRENDERGAST J, DUMBRELL S, TURNER C, WOODWARD M (2003), 'A one-quarter reduction in the salt content of bread can be made without detection', *Eur J Clin Nutr*, 57(4), 616–620.

GRAUDAL N A, GALLOE A M, GARRED P (1998), 'Effects of sodium restriction on blood pressure, renin, aldosterone, catecholamines, cholesterols, and triglyceride: a meta-analysis', *JAMA*, 279(17), 1383–1391.

GRIMM R H, JR., SVENDSEN K H, KASISKE B, KEANE W F, WAHI M M (1997), 'Proteinuria is a risk factor for mortality over 10 years of follow-up. MRFIT Research Group. Multiple Risk Factor Intervention Trial', *Kidney Int Suppl*, 63, S10–14.

GU J W, ANAND V, SHEK E W, MOORE M C, BRADY A L, KELLY W C, ADAIR T H (1998), 'Sodium induces hypertrophy of cultured myocardial myoblasts and vascular smooth muscle cells', *Hypertension*, 31(5), 1083–1087.

HE F J, MacGREGOR G A (1999), 'Potassium intake and blood pressure', *Am J Hypertens*, 12(8 Pt 1), 849–851.

HE F J, MacGREGOR G A (2001), 'Fortnightly review: Beneficial effects of potassium', *BMJ*, 323(7311), 497–501.

HE F J, MacGREGOR G A (2002), 'Effect of modest salt reduction on blood pressure: a meta-analysis of randomized trials. Implications for public health', *J Hum Hypertens*, 16(11), 761–770.

HE F J, MacGREGOR G A (2003), 'How far should salt intake be reduced?' *Hypertension*, 42(6), 1093–1099.

HE F J, MARKANDU N D, SAGNELLA G A, MacGREGOR G A (1998), 'Importance of the renin system in determining blood pressure fall with salt restriction in black and white hypertensives', *Hypertension*, 32(5), 820–824.

HE F J, MARKANDU N D, MacGREGOR G A (2001a), 'Importance of the renin system for determining blood pressure fall with acute salt restriction in hypertensive and normotensive whites', *Hypertension*, 38(3), 321–325.

HE F J, MARKANDU N D, SAGNELLA G A, MacGREGOR G A (2001b), 'Effect of salt intake on renal excretion of water in humans', *Hypertension*, 38(3), 317–320.

HE F J, MARKANDU N D, COLTART R, BARRON J, MacGREGOR G A (2005a), 'Effect of short-term supplementation of potassium chloride and potassium citrate on blood pressure in hypertensives', *Hypertension*, 45(4), 571–574.

HE F J, MARKANDU N D, MacGREGOR G A (2005b), 'Modest salt reduction lowers blood pressure in isolated systolic hypertension and combined hypertension', *Hypertension*, 46(1), 66–70.

HE F J, MARKANDU N D, SAGNELLA G A, DE WARDENER H E, MacGREGOR G A (2005c), 'Plasma sodium: ignored and underestimated', *Hypertension*, 45(1), 98–102.

HE J, KLAG M J, WHELTON P K, CHEN J Y, MO J P, QIAN M C, MO P S, HE G Q (1991a), 'Migration, blood pressure pattern, and hypertension: the Yi Migrant Study', *Am J Epidemiol*, 134(10), 1085–1101.

HE J, TELL G S, TANG Y C, MO P S, HE G Q (1991b), 'Relation of electrolytes to blood pressure in men. The Yi people study', *Hypertension*, 17(3), 378–385.

HE J, OGDEN L G, VUPPUTURI S, BAZZANO L A, LORIA C, WHELTON P K (1999), 'Dietary sodium intake and subsequent risk of cardiovascular disease in overweight adults', *JAMA*, 282(21), 2027–2034.

HEEG J E, DE JONG P E, VAN DER HEM G K, DE ZEEUW D (1989), 'Efficacy and variability of the antiproteinuric effect of ACE inhibition by lisinopril', *Kidney Int*, 36(2), 272–279.

HENDERSON L, IRVING K, GREGORY J, BATES C J, PRENTICE A, PERKS J, SWAN G, FARRON M (2003), *National Diet & Nutrition Survey: Adults aged 19 to 64*, Volume 3, page 127–136.

HILLEGE H L, FIDLER V, DIERCKS G F, VAN GILST W H, DE ZEEUW D, VAN VELDHUISEN D J, GANS R O, JANSSEN W M, GROBBEE D E, DE JONG P E (2002), 'Urinary albumin excretion predicts cardiovascular and noncardiovascular mortality in general population', *Circulation*, 106(14), 1777–1782.

HOFMAN A, HAZEBROEK A, VALKENBURG H A (1983), 'A randomized trial of sodium intake and blood pressure in newborn infants', *JAMA*, 250(3), 370–373.

HOOPER L, BARTLETT C, DAVEY SMITH G, EBRAHIM S (2002), 'Systematic review of long term effects of advice to reduce dietary salt in adults', *BMJ*, 325(7365), 628.

INTERSALT (1988), 'Intersalt: an international study of electrolyte excretion and blood pressure. Results for 24 hour urinary sodium and potassium excretion. Intersalt Cooperative Research Group', *BMJ*, 297(6644), 319–328.

ISO H, SHIMAMOTO T, YOKOTA K, OHKI M, SANKAI T, KUDO M, HARADA M, WAKABAYASHI Y, INAGAWA M, KITAMURA A, SATO S, IMANO H, IIDA M, KOMACHI Y (1999), '[Changes in 24-hour urinary excretion of sodium and potassium in a community-based heath education program on salt reduction]', *Nippon Koshu Eisei Zasshi*, 46(10), 894–903.

JAMES W P, RALPH A, SANCHEZ-CASTILLO C P (1987), 'The dominance of salt in manufactured food in the sodium intake of affluent societies', *Lancet*, 1(8530), 426–429.

JOOSSENS J V, HILL M J, ELLIOTT P, STAMLER R, LESAFFRE E, DYER A, NICHOLS R, KESTELOOT H (1996), 'Dietary salt, nitrate and stomach cancer mortality in 24 countries. European Cancer Prevention (ECP) and the INTERSALT Cooperative Research Group', *Int J Epidemiol*, 25(3), 494–504.

JULA A M, KARANKO H M (1994), 'Effects on left ventricular hypertrophy of long-term nonpharmacological treatment with sodium restriction in mild-to-moderate essential hypertension', *Circulation*, 89(3), 1023–1031.

KARPPANEN H, MERVAALA E (1998), 'Sodium intake and mortality', *Lancet*, 351(9114), 1509–1510.

KEARNEY P M, WHELTON M, REYNOLDS K, MUNTNER P, WHELTON P K, HE J (2005), 'Global burden of hypertension: analysis of worldwide data', *Lancet*, 365(9455), 217–223.

KEMPNER W (1948), 'Treatment of hypertensive vascular disease with rice diet', *Am J Med*, 4, 545–577.

KIMURA N (1983), 'Changing patterns of coronary heart disease, stroke, and nutrient intake in Japan', *Prev Med*, 12(1), 222–227.

KUPARI M, KOSKINEN P, VIROLAINEN J (1994), 'Correlates of left ventricular mass in a population sample aged 36 to 37 years. Focus on lifestyle and salt intake', *Circulation*, 89(3), 1041–1050.

LAUFER E, JENNINGS G L, KORNER P I, DEWAR E (1989), 'Prevalence of cardiac structural and functional abnormalities in untreated primary hypertension', *Hypertension*, 13(2), 151–162.

LAW M R, FROST C D, WALD N J (1991), 'By how much does dietary salt reduction lower blood pressure? III – Analysis of data from trials of salt reduction', *BMJ*, 302(6780), 819–824.

LEVY D, GARRISON R J, SAVAGE D D, KANNEL W B, CASTELLI W P (1990), 'Prognostic implications of echocardiographically determined left ventricular mass in the Framingham Heart Study', *N Engl J Med*, 322(22), 1561–1566.

LEWINGTON S, CLARKE R, QIZILBASH N, PETO R, COLLINS R (2002), 'Age-specific relevance of usual blood pressure to vascular mortality: a meta-analysis of individual data for one million adults in 61 prospective studies', *Lancet*, 360(9349), 1903–1913.

LIEBSON P R, GRANDITS G A, DIANZUMBA S, PRINEAS R J, GRIMM R H, JR., NEATON J D, STAMLER J (1995), 'Comparison of five antihypertensive monotherapies and placebo for change in left ventricular mass in patients receiving nutritional-hygienic therapy in the Treatment of Mild Hypertension Study (TOMHS)', *Circulation*, 91(3), 698–706.

LIFTON R P (1996), 'Molecular genetics of human blood pressure variation', *Science*, 272(5262), 676–680.

LIFTON R P, GHARAVI A G, GELLER D S (2001), 'Molecular mechanisms of human hypertension', *Cell*, 104(4), 545–556.

LIN P H, GINTY F, APPEL L J, AICKIN M, BOHANNON A, GARNERO P, BARCLAY D, SVETKEY L P (2003), 'The DASH diet and sodium reduction improve markers of bone turnover and calcium metabolism in adults', *J Nutr*, 133(10), 3130–3136.

MACGREGOR G (1996), 'Low urinary sodium and myocardial infarction', *Hypertension*, 27(1), 156.

MACGREGOR G A, DE WARDENER H E (1997), 'Idiopathic edema,' in G. C. Schrier RW, *Diseases of the Kidney*, Boston, Little Brown and Company, III, 2343–2352.

MACGREGOR G A, DE WARDENER H E (1998). *Salt, Diet and Health*, Cambridge: Cambridge University Press.

MACGREGOR G A, MARKANDU N D, BEST F E, ELDER D M, CAM J M, SAGNELLA G A, SQUIRES M (1982a), 'Double-blind randomised crossover trial of moderate sodium restriction in essential hypertension', *Lancet*, 1(8268), 351–355.

MACGREGOR G A, SMITH S J, MARKANDU N D, BANKS R A, SAGNELLA G A (1982b), 'Moderate potassium supplementation in essential hypertension', *Lancet*, 2(8298), 567–570.

MACGREGOR G A, MARKANDU N D, SINGER D R, CAPPUCCIO F P, SHORE A C, SAGNELLA G A (1987), 'Moderate sodium restriction with angiotensin converting enzyme inhibitor in essential hypertension: a double blind study', *Br Med J (Clin Res Ed)*, 294(6571), 531–534.

MacGREGOR G A, MARKANDU N D, SAGNELLA G A, SINGER D R, CAPPUCCIO F P (1989), 'Double-blind study of three sodium intakes and long-term effects of sodium restriction in essential hypertension', *Lancet*, 2(8674), 1244–1247.

MacMAHON S (1996), 'Blood pressure and the prevention of stroke', *J Hypertens Suppl*, 14(6), S39–46.

MANCILHA-CARVALHO J J, DE OLIVEIRA R, ESPOSITO R J (1989), 'Blood pressure and electrolyte excretion in the Yanomamo Indians, an isolated population', *J Hum Hypertens*, 3(5), 309–314.

MATKOVIC V, ILICH J Z, ANDON M B, HSIEH L C, TZAGOURNIS M A, LAGGER B J, GOEL P K (1995), 'Urinary calcium, sodium, and bone mass of young females', *Am J Clin Nutr*, 62(2), 417–425.

MICKLEBOROUGH T D, LINDLEY M R, RAY S (2005), 'Dietary salt, airway inflammation, and diffusion capacity in exercise-induced asthma', *Med Sci Sports Exerc*, 37(6), 904–914.

MIDGLEY J P, MATTHEW A G, GREENWOOD C M, LOGAN A G (1996), 'Effect of reduced dietary sodium on blood pressure: a meta-analysis of randomized controlled trials', *JAMA*, 275(20), 1590–1597.

NAGATA C, TAKATSUKA N, SHIMIZU N, SHIMIZU H (2004), 'Sodium intake and risk of death from stroke in Japanese men and women', *Stroke*, 35(7), 1543–1547.

NESTLE M (2002). *Food Politics – How the Food Industry Influences Nutrition and Health*. London, University of California Press.

OLIVER W J, COHEN E L, NEEL J V (1975), 'Blood pressure, sodium intake, and sodium related hormones in the Yanomamo Indians, a "no-salt" culture', *Circulation*, 52(1), 146–151.

PAGE L B, DAMON A, MOELLERING R C, JR. (1974), 'Antecedents of cardiovascular disease in six Solomon Islands societies', *Circulation*, 49(6), 1132–1146.

PAGE L B, VANDEVERT D E, NADER K, LUBIN N K, PAGE J R (1981), 'Blood pressure of Qash'qai pastoral nomads in Iran in relation to culture, diet, and body form', *Am J Clin Nutr*, 34(4), 527–538.

PERRY I J, BEEVERS D G (1992), 'Salt intake and stroke: a possible direct effect', *J Hum Hypertens*, 6(1), 23–25.

POULTER N R, KHAW K T, HOPWOOD B E, MUGAMBI M, PEART W S, ROSE G, SEVER P S (1990), 'The Kenyan Luo migration study: observations on the initiation of a rise in blood pressure', *BMJ*, 300(6730), 967–972.

PRIMATESTA P, BROOKES M, POULTER N R (2001), 'Improved hypertension management and control: results from the health survey for England 1998', *Hypertension*, 38(4), 827–832.

SACKS F M, SVETKEY L P, VOLLMER W M, APPEL L J, BRAY G A, HARSHA D, OBARZANEK E, CONLIN P R, MILLER E R, III, SIMONS-MORTON D G, KARANJA N, LIN P H (2001), 'Effects on blood pressure of reduced dietary sodium and the Dietary Approaches to Stop Hypertension (DASH) diet. DASH-Sodium Collaborative Research Group', *N Engl J Med*, 344(1), 3–10.

SACN (2003). Scientific Advisory Committee on Nutrition, Salt and Health. 2003. The Stationery Office. Available at http://www.sacn.gov.uk/pdfs/sacn_salt_final.pdf. Accessed 22 March 2005.

SAFAR M E, CHAU N P, WEISS Y A, LONDON G M, MILLIEZ P L (1976), 'Control of cardiac output in essential hypertension', *Am J Cardiol*, 38(3), 332–336.

SAFAR M E, THUILLIEZ C, RICHARD V, BENETOS A (2000), 'Pressure-independent contribution of sodium to large artery structure and function in hypertension', *Cardiovasc Res*, 46(2), 269–276.

SCHMIEDER R E, MESSERLI F H, GARAVAGLIA G E, NUNEZ B D (1988), 'Dietary salt intake. A determinant of cardiac involvement in essential hypertension', *Circulation*, 78(4), 951–956.

SCHMIEDER R E, MESSERLI F H (2000), 'Hypertension and the heart', *J Hum Hypertens*, 14(10–11), 597–604.

SIMON G, ILLYES G (2001), 'Structural vascular changes in hypertension: role of angiotensin II, dietary sodium supplementation, and sympathetic stimulation, alone and in combination in rats', *Hypertension*, 37(2), 255–260.

SMITH S J, MARKANDU N D, SAGNELLA G A, MacGREGOR G A (1985), 'Moderate potassium chloride supplementation in essential hypertension: is it additive to moderate sodium restriction?' *Br Med J (Clin Res Ed)*, 290(6462), 110–113.

STAESSEN J, BULPITT C J, FAGARD R, JOOSSENS J V, LIJNEN P, AMERY A (1988), 'Salt intake and blood pressure in the general population: a controlled intervention trial in two towns', *J Hypertens*, 6(12), 965–973.

STAESSEN J A, WANG J G, THIJS L (2001), 'Cardiovascular protection and blood pressure reduction: a meta-analysis', *Lancet*, 358(9290), 1305–1315.

STAMLER J, ELLIOTT P, APPEL L, CHAN Q, BUZZARD M, DENNIS B, DYER A R, ELMER P, GREENLAND P, JONES D, KESTELOOT H, KULLER L, LABARTHE D, LIU K, MOAG-STAHLBERG A, NICHAMAN M, OKAYAMA A, OKUDA N, ROBERTSON C, RODRIGUEZ B, STEVENS M, UESHIMA H, HORN L V, ZHOU B (2003), 'Higher blood pressure in middle-aged American adults with less education-role of multiple dietary factors: the INTERMAP study', *J Hum Hypertens*, 17(9), 655–775.

SWALES J (2000), 'Population advice on salt restriction: the social issues', *Am J Hypertens*, 13(1 Pt 1), 2–7.

SWIFT P A, MARKANDU N D, SAGNELLA G A, HE F J, MacGREGOR G A (2005), 'Modest salt reduction reduces blood pressure and urine protein excretion in black hypertensives. A randomised control trial.' *Hypertension*, 46, 308–312.

TANAKA H, TANAKA Y, HAYASHI M, UEDA Y, DATE C, BABA T, SHOJI H, HORIMOTO T, OWADA K (1982), 'Secular trends in mortality for cerebrovascular diseases in Japan, 1960 to 1979', *Stroke*, 13(5), 574–581.

TAUBES G (1998), 'The (political) science of salt', *Science*, 281(5379), 898–901, 903–907.

TIAN H G, GUO Z Y, HU G, YU S J, SUN W, PIETINEN P, NISSINEN A (1995), 'Changes in sodium intake and blood pressure in a community-based intervention project in China', *J Hum Hypertens*, 9(12), 959–968.

TOBIAN L, HANLON S (1990), 'High sodium chloride diets injure arteries and raise mortality without changing blood pressure', *Hypertension*, 15(6 Pt 2), 900–903.

TSUGANE S, SASAZUKI S, KOBAYASHI M, SASAKI S (2004), 'Salt and salted food intake and subsequent risk of gastric cancer among middle-aged Japanese men and women', *Br J Cancer*, 90(1), 128–134.

TUOMILEHTO J, PUSKA P, NISSINEN A, SALONEN J, TANSKANEN A, PIETINEN P, WOLF E (1984), 'Community-based prevention of hypertension in North Karelia, Finland', *Ann Clin Res*, 16 Suppl 43, 18–27.

TUOMILEHTO J, JOUSILAHTI P, RASTENYTE D, MOLTCHANOV V, TANSKANEN A, PIETINEN P, NISSINEN A (2001), 'Urinary sodium excretion and cardiovascular mortality in Finland: a prospective study', *Lancet*, 357(9259), 848–851.

VASAN R S, LARSON M G, LEIP E P, EVANS J C, O'DONNELL C J, KANNEL W B, LEVY D (2001), 'Impact of high-normal blood pressure on the risk of cardiovascular disease', *N Engl J Med*, 345(18), 1291–1297.

VERHAVE J C, HILLEGE H L, BURGERHOF J G, JANSSEN W M, GANSEVOORT R T, NAVIS G J, DE

ZEEUW D, DE JONG P E (2004), 'Sodium intake affects urinary albumin excretion especially in overweight subjects', *J Intern Med*, 256(4), 324–330.

WHELTON P K, HE J, CUTLER J A, BRANCATI F L, APPEL L J, FOLLMANN D, KLAG M J (1997), 'Effects of oral potassium on blood pressure. Meta-analysis of randomized controlled clinical trials', *JAMA*, 277(20), 1624–1632.

WHELTON P K, APPEL L J, ESPELAND M A, APPLEGATE W B, ETTINGER W H, JR., KOSTIS J B, KUMANYIKA S, LACY C R, JOHNSON K C, FOLMAR S, CUTLER J A (1998), 'Sodium reduction and weight loss in the treatment of hypertension in older persons: a randomized controlled trial of nonpharmacologic interventions in the elderly (TONE). TONE Collaborative Research Group', *JAMA*, 279(11), 839–846.

WHELTON P K, HE J, APPEL L J, CUTLER J A, HAVAS S, KOTCHEN T A, ROCCELLA E J, STOUT R, VALLBONA C, WINSTON M C, KARIMBAKAS J (2002), 'Primary prevention of hypertension: clinical and public health advisory from The National High Blood Pressure Education Program', *JAMA*, 288(15), 1882–1888.

WHO (2003). Joint WHO/FAO expert consultation on diet, nutrition and the prevention of chronic diseases. Geneva, Available at http://www.who.int/hpr/NPH/docs/who_fao_experts_report.pdf. Accessed 22 March 2005.

WONG B C, LAM S K, WONG W M, CHEN J S, ZHENG T T, FENG R E, LAI K C, HU W H, YUEN S T, LEUNG S Y, FONG D Y, HO J, CHING C K (2004), 'Helicobacter pylori eradication to prevent gastric cancer in a high-risk region of China: a randomized controlled trial', *JAMA*, 291(2), 187–194.

XIE J X, SASAKI S, JOOSSENS J V, KESTELOOT H (1992), 'The relationship between urinary cations obtained from the INTERSALT study and cerebrovascular mortality', *J Hum Hypertens*, 6(1), 17–21.

ZHOU B F, STAMLER J, DENNIS B, MOAG-STAHLBERG A, OKUDA N, ROBERTSON C, ZHAO L, CHAN Q, ELLIOTT P (2003), 'Nutrient intakes of middle-aged men and women in China, Japan, United Kingdom, and United States in the late 1990s: the INTERMAP study', *J Hum Hypertens*, 17(9), 623–630.

3

Health issues relating to monosodium glutamate use in the diet

J. D. Fernstrom, University of Pittsburgh School of Medicine, USA

3.1 Introduction

Why should a book devoted to reducing salt in food include a chapter on the safety of monosodium glutamate (MSG)? The discovery of a link between high salt intake and an increased risk of serious disease (e.g., hypertension[1–3]) has led to the recommendation that dietary sodium intake be reduced. But the palatability of many foods declines as the salt content is reduced, making it difficult for individuals to undertake and then maintain a low-salt diet. MSG improves the taste acceptability of foods with lowered salt content,[4,5] suggesting that its use in low-salt foods will make it easier for people to establish and maintain a low-salt diet. In contemplating the use of MSG in such foods, however, one wants to know beforehand that this application would be safe.

Before discussing safety, however, it is helpful to gain some understanding of the use of MSG in foods, and also of glutamate functions in the body. The flavor properties of MSG were first described almost 100 years ago in Japan, when the amino acid was extracted and purified from sea weed and shown to account for its dominant taste properties.[4,6,7] The taste of MSG is not identified as salty, per se, though this is a component of its taste profile, but rather as savory or meaty. In Japan, the taste of MSG is called 'umami'.[4] In the past 25 years, the umami taste has been demonstrated to be unique, separate from the four basic tastes of sweet, sour, salty and bitter, and thus has been classed as a fifth basic taste.[5] Sensitivity to the taste of MSG is enhanced about 100-fold by the presence of nucleotide monophosphates.[8] Very recently, taste receptors have been cloned for the umami taste, though it is currently debated whether such receptors are specific to umami (MSG) or shared by a number of amino acids.[6,9,10] The binding of MSG (and other amino acids) to such receptors is enhanced by the presence of mononucleotides.

In addition to its unique taste properties, MSG can improve the palatability of many foods when the salt (sodium chloride) content is reduced. When tested in soups and meals, the estimate has been made that the sodium content of such foods can be reduced by 30–40%, with no loss of palatability and acceptability, with the proper combination of sodium chloride and MSG.[4,11] Roininen et al.[12] note, in particular, that the addition of MSG to low-salt foods may improve their palatability and acceptability *during the period when individuals are becoming accustomed to low-salt diets.*

Since MSG improves the palatability and acceptability of low-salt foods,[8,12–15] and thus may find use in helping individuals reduce the salt content of their diet, it is reassuring to know that MSG is not toxic, and indeed produces no reproducible adverse effects, even when consumed in very high amounts. This is not a particularly surprising insight, once it is recognized that MSG is not simply a flavoring agent or flavor enhancer. Quite the contrary: MSG is the sodium salt of glutamic acid (GLU), a non-essential amino acid for humans and other mammals (i.e., the body can synthesize it). It is the dominant amino acid in dietary proteins, comprising about 10% of the total amino acid content.[16] Quite a lot of GLU is thus ingested every day as a normal constituent of the diet (discussed below). In the body, GLU serves a number of important functions, aside from its role in taste. For example, it is: (a) a major energy source for the gastrointestinal tract during digestion,[17] (b) a key metabolic intermediate at the interface between amino acid and carbohydrate metabolism,[18] and (c) the most abundant neurotransmitter in the central nervous system.[19] Such functions underscore the importance of this amino acid in physiology and metabolism.

However, despite the 'indispensability' of GLU to the normal functioning of the body, it's use in food as a flavor enhancer (primarily as MSG) has been reputed in the past to be toxic, leading to brain damage and a host of other adverse effects in humans and animals. If such is the case, how can its use be permitted as a food additive, and more to the point, how can the body 'survive' the flood of GLU that enters every day as a *natural* constituent of foods?

The basic task of this chapter is to answer these questions. The issue of GLU (and MSG) safety in food was raised over 35 years ago.[20,21] In the intervening period, a great deal of research has been conducted on this issue, particularly in humans, and has led to the broad conclusion that GLU and MSG are safe to consume. The chapter will begin with an overview of the GLU content of foods, and then discuss the two broad issues that have been the primary foci of concern about dietary MSG safety over the past 35 years: (a) effects of dietary GLU and MSG on nervous system development and function, and (b) allergic and non-specific effects of dietary MSG. It will not discuss more standard aspects of glutamate safety, such as general testing for general toxicity, mutagenicity, teratogenicity, all of which has been done for MSG, and discussed at length by regulatory bodies (e.g., ref. 22).

3.2   Glutamate and MSG in food

Glutamate occurs naturally in foods in both free and protein-bound forms. GLU constitutes about 8 to 10% of the amino acid content of the average dietary protein (not counting glutamine).[16,23] If the average American consumes about 100 g protein/day,[23] the GLU intake is 8–10 g, or 110–140 mg/kg (assuming a 70 kg body mass). The free GLU content of foods varies widely, being low in most. However, it is high in some foods, such as tomatoes (246 mg/100 g food, or about 300 mg in a whole medium tomato), corn (106 mg/100 g food, or about 150 mg in a large ear of corn), and some cheeses (parmesan cheese, 1520–1680 mg/100 g food, or 152–168 mg in 2 tablespoons of grated cheese; Roquefort cheese, 1620 mg/100 g food, or about 500 mg/1 ounce serving; gouda cheese, 580 mg/100 g food, or about 175 mg/1 ounce serving).[8,24,25] Notwithstanding, consuming a tomato and a portion or two of the above cheeses during a meal might add up to 10 mg/kg MSG, a small 'dose' in comparison to the GLU consumed as a component of dietary protein. Hence, total daily GLU ingestion from *natural sources* in the diet is probably in the range of 100–150 mg/kg/day.

GLU is added to some foods (usually as MSG) to enhance flavor. Estimates of daily MSG intake *for this purpose* are 350–510 mg/day in the US population, 1120–1600 mg/day in the Japanese population, 1570–2300 mg/day in the Korean population, and 3000 mg/day in the Taiwanese population.[16,24] In the United Kingdom, where an assessment of MSG intake has actually been made, intake estimates range from 586 mg/day for the whole population to 2329 mg/day for 'extreme consumers' (defined as '3 × mean of consumption of households purchasing the specified food', in this case MSG-containing foods, and intended to capture individuals at the 97.5th percentile of consumption).[26] Hence, for the 70 kg human, daily GLU intake *from MSG* ranges from 5 mg/kg/day in the US population, to 43 mg/kg/day in the Taiwan population.

From this information, the amount of MSG consumed daily in the US population (5 mg/kg/day) is a modest fraction of the total GLU consumed (100–150 mg/kg/day). Such would also be true in the British population, taken as a whole. For the extreme British consumer (70 kg) about 33 mg/kg/day MSG might be added to the diet, or between 20 and 33% of total GLU intake. Estimates of extreme consumption have also been made for the US population, giving a value of 12 mg/kg/day for adults falling into the 99th percentile of intake (see table 4 in ref. 27). Hence, even for extreme consumers, the ingestion of added GLU as MSG in foods is still a fraction of the total GLU consumed daily as a *natural* constituent of foods. As discussed below, it is useful to keep such dosages in mind, when considering studies designed to explore for adverse effects of dietary MSG. It is also useful to keep in mind that despite the generally low doses of MSG consumed on a chronic basis in society, individuals may on occasion ingest food items that contain amounts of added MSG that seem large by comparison (as a single dose). For example, an 8-ounce bowl of a Chinese wonton soup, if flavored at 0.6% MSG, would deliver to a 70 kg individual a single oral dose of about 25 mg/kg MSG, quite a bit higher than the average daily dose for Americans of around 5 mg/kg/day. However, as discussed below, such doses of MSG in a food matrix neither raise plasma GLU concentrations nor elicit unpleasant effects.

3.3   Glutamate, MSG and the nervous system

Glutamate is an excitatory neurotransmitter in the central nervous system.[19] In addition, it is a natural constituent of the diet. This relationship might lead one to speculate that dietary GLU intake might directly influence brain GLU concentrations, and thus brain function. However, such is not the case. The blood-brain barrier normally prevents blood GLU from entering the brain.[28] The GLU present in brain cells is thus all synthesized in the brain, and highly compartmentalized metabolically, to control its ability to excite neurons.[29] If this is the case, then why are GLU and MSG a topic for discussion in a chapter about GLU safety in the diet, particularly in relation to the nervous system? In 1969, an article appeared in *Science* indicating that an injection of MSG could damage a portion of the brain (the hypothalamus), when given at a very high dose to infant mice.[20] A similar effect was seen when MSG (or glutamate) was orally administered to mice in high doses, and led to the speculation that MSG ingestion in infant foods might be dangerous.[21,30] The considerable flurry of publicity and subsequent formal review that followed ultimately led to the precautionary, voluntary removal of MSG from baby food.[31] In the intervening 30 years, a great deal of data has been collected on the effects of oral MSG (and GLU) on brain chemistry and function, mostly during the period between 1973 and 1985. The essential conclusion has been that very large, oral doses of GLU or MSG, administered to rodents, when they cause substantial elevations in plasma GLU concentrations, can cause toxicity to the animal's hypothalamus.[32] However, the threshold plasma GLU concentration for such effects appears to be in the range of 2000 nmol/mL, 15–20-fold above normal, and are produced by either oral gavage of between 500 and 1000 mg/kg MSG or subcutaneous injection of at least 200 mg/kg MSG.[32,33] This is an *enormous* rise in plasma GLU, one that has never been found to occur under normal (or even abnormal[34]) circumstances of MSG ingestion by humans.[35] Indeed, the dietary consumption of glutamate has never been reported to cause neuropathology in humans.[36] Accordingly, scientific reviews of MSG in relation to brain function have found there to be no hazard to humans.[37]

There are many reasons why dietary GLU and/or MSG pose no threat to the human brain (*in utero*, infant, or adult). First, the gastrointestinal tract metabolizes almost all (95+%) of the GLU (or MSG) it receives; very little makes it into the systemic circulation.[17] Hence, under normal dietary circumstances, plasma GLU does not increase appreciably following the ingestion of food, even when it contains a large amount of protein (and thus GLU). For example, when a high-protein (1 g/kg) meal is ingested, which would contain about 10% GLU, or 100 mg/kg GLU, plasma GLU rises no more than about two-fold.[27] When a meal containing about half this amount of protein is ingested, the increase in plasma GLU is at most 50% above baseline values.[27] Adding a very large MSG dose to the meal (100 or 150 mg/kg) produces only small, additional increases in plasma GLU (over that observed when the meal alone is ingested). Furthermore, when subjects consume MSG added at a very high dose (100 mg/kg/day) to their normal daily diet, plasma GLU concentrations over the 24-hr period are no different from those observed when the same diet is consumed without added MSG (Fig. 3.1).[38] Plasma glutamate concentrations do not rise following the ingestion of MSG with breakfast, lunch and dinner. The figure shows plasma glutamate concentrations over the 24-hr period in adult male subjects who have consumed a normal diet either containing or lacking added MSG. Subjects consumed each diet on a separate occasion. They were studied in a clinical research unit; all meals were prepared and provided. Meals and snacks were consumed at the times indicated in the figure (arrows). On the MSG test day, subjects received 100 mg/kg MSG in divided doses with meals (15 mg/kg with breakfast, 40 mg/kg with lunch, 45 mg/kg with dinner). The meals on the placebo day were the same, but lacked added MSG. The diet provided 40 kcal/kg/day; the composition was 15% protein, 55% carbohydrate, and 30% fat (% total energy).

It *is* possible pharmacologically to give MSG in a manner that will cause a large rise in plasma GLU concentrations: this is done by administering MSG in water on an empty stomach (i.e., a large, single oral dose in high concentrations

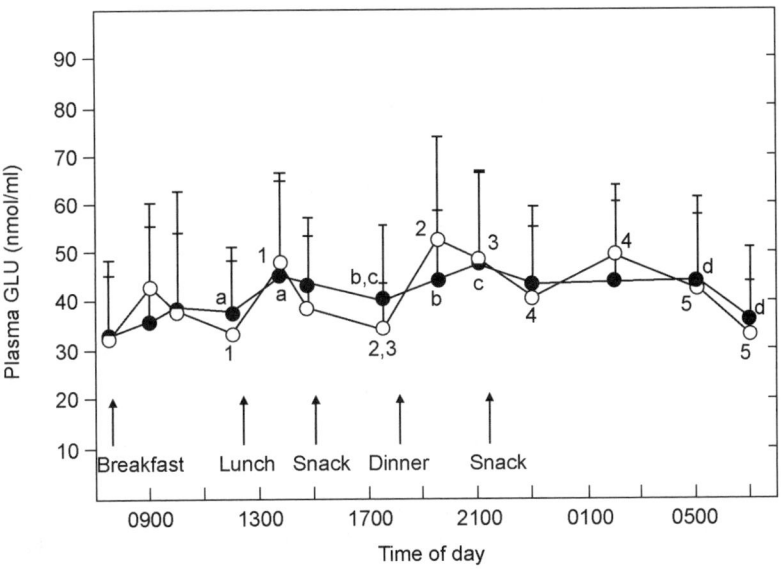

Fig. 3.1   Plasma glutamate concentrations over the 24-hr period in adult male subjects who have consumed a normal diet either containing or lacking added MSG. Data are means ±1 standard deviation ($n = 10$); the concentrations at different time points with different letters (no MSG added day, closed circles) or numbers (MSG added day, open circles) are significantly different ($P < 0.05$). Adapted from ref. 74.

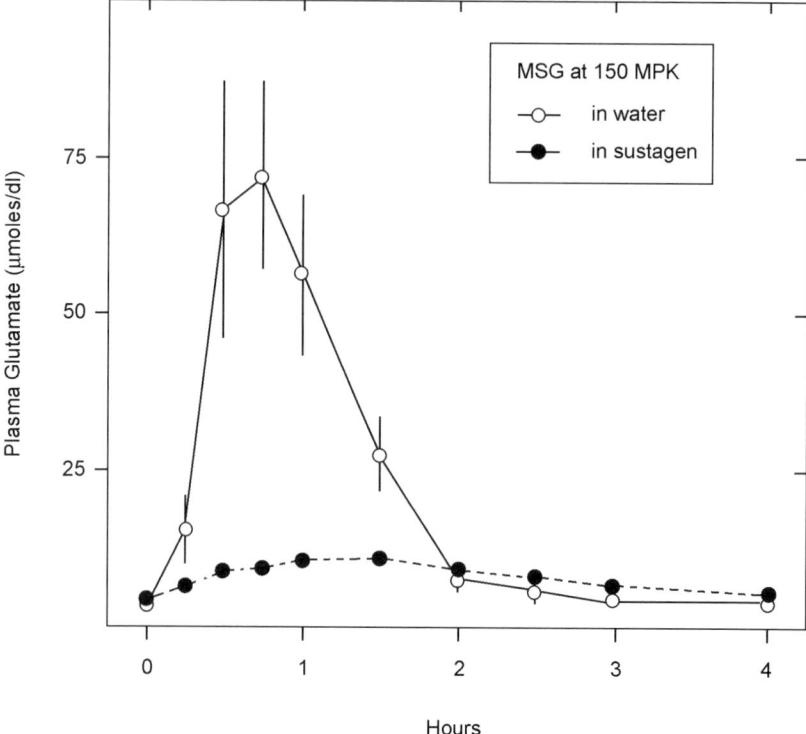

Fig. 3.2   Changes in plasma glutamate concentration in normal adult subjects ingesting MSG in either water or a liquid meal replacement. Adapted from ref. 75, with permission of the *American Journal of Clinical Nutrition.*

consumed all at once – not the manner in which MSG is used in foods). For example, a 150 mg/kg dose of MSG that causes almost no rise in plasma GLU when ingested as a part of a test meal, elicits a substantial rise in plasma GLU when ingested alone in water.[27] Figure 3.2 shows changes in plasma glutamate concentration in normal adult subjects ingesting MSG in either water or a liquid meal replacement. Overnight-fasted subjects (3 males, 3 females) consumed MSG (150 mg/kg) dissolved in 4.2 ml/kg of either water or a liquid meal replacement (Sustagen, Mead-Johnson, Evansville IN) in the morning. Blood samples were drawn at the indicated times. Data are means SEM ($n = 6$/group). Sustagen is 24.2% protein, 67.8% carbohydrate, and 8.0% fat (% energy); 4.2 ml/kg Sustagen delivers 6.61 kcal total energy/kg body weight. The area-under-the-curve for MSG in water differed significantly from that for MSG in Sustagen ($P < 0.05$, *t*-test). In practice, no human would knowingly ingest a 150 mg/kg dose of MSG in a typical volume of water, since it would be very unpalatable.[39] And even it they did, the peak plasma GLU concentration is not high enough to cause GLU penetration into brain. This fact is evidenced, for example, by the lack of an increase in plasma prolactin concentrations in humans following the ingestion of 150 mg/kg MSG, which produced an 11-fold rise in plasma GLU.[34] Stimulation of GLU receptors in the brain indirectly causes pituitary prolactin secretion;[40] the absence of a rise in plasma prolactin provides an indirect indication that GLU concentrations did not rise in the brain (hypothalamus) following this dose of MSG, despite the considerable rise in plasma GLU.

The reasons why brain GLU does not increase following a large rise in plasma GLU are most likely two-fold: first, the blood-brain barrier contains an amino acid transporter for acidic amino acids that functions to transport GLU out of, not into the brain.[28] Hence, there actually exists a 'barrier' to GLU penetration into brain from the circulation. Second, glial and neuronal cells in brain have transporters on their membranes that actively transport GLU out of the synapse and ECF and back into cells.[29] Figure 3.3 shows a glutamate synapse, showing glutamate trafficking between neurons and glial cells in the brain. Glutamate, synthesized via metabolic pathways in neurons, is concentrated into secretory vesicles (100,000 μmol/L; see Table 3.1) by a vesicular transporter. After release from the presynaptic terminal (element) following neuronal depolarization, glutamate concentrations in the synapse rise from very low to very high concentrations (2→1,000 μmol/L; see Table 3.1), and stimulate receptors. Glutamate is then rapidly cleared from the synapse by high-affinity membrane transporters located on the neurons and surrounding glial cells. In glial cells, glutamate is converted to glutamine, and diffuses and/or is transported out into the extracellular fluid, where it can be taken up into neurons and reconverted to glutamate and restored in vesicles. Glutamine shuttling of glutamate molecules between neurons and glia, along with the high-affinity glutamate transporters located on neurons and glia, keep extracellular fluid (not just synaptic) glutamate concentrations very low (0.5–2.0 μmol/L; see Table 3.1). In this manner, extracellular GLU concentrations are kept very low, except for the brief instants when GLU is purposefully released into a synapse

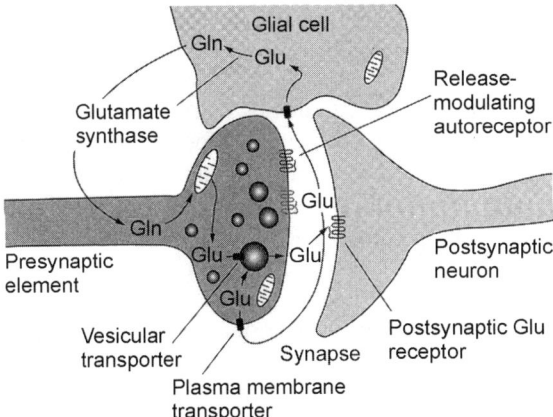

Fig. 3.3   Glutamate synapse, showing glutamate trafficking between neurons and glial cells in the brain. From ref. 76.

Table 3.1 Glutamate concentrations in body compartments and GLU transporter and receptor affinities.

Compartment	GLU concentration (μmol/L)
Plasma	30–100
Brain extracellular fluid	0.5–2.0
Synaptic vesicle (storage vesicle)	100,000
Synaptic cleft	2–1,000
GLT-1 transporter affinity	1–20
NMDA receptor affinity	2.5–3.0
AMPA receptor affinity	200–500
Metabotropic GLU receptor affinities	5–10

Abbreviations: GLT-1, rat glial glutamate transporter; NMDA, n-methyl-d-aspartate; AMPA, α-amino-3-hydroxy-5-methyl-4-isoxazoleproprionic acid; GLT-1, rat glial glutamate transporter; NMDA, n-methyl-d-aspartate; AMPA, α-amino-3-hydroxy-5-methyl-4-isoxazoleproprionic acid. Adapted from (73) with permission.

following neuronal depolarization, where it stimulates receptors on other neurons (its normal function). This GLU is then rapidly absorbed into cells and converted to a non-active amino acid, glutamine, which is then supplied back to neurons for reconversion to GLU.[29] This powerful mechanism for maintaining low brain ECF GLU concentrations should provide a mechanism by which any GLU that might penetrate into brain from the circulation can immediately be removed from the ECF. Indeed, in the small portions of the brain that lack a blood-brain barrier (e.g., the median eminence, located at the base of the hypothalamus), this cellular GLU transport mechanism, particularly in glial cells,[33] may in part explain why perfusion with GLU at plasma concentrations, which should be sufficient to stimulate some GLU receptors (Table 3.1), causes no adverse effects on neurons adjacent to these areas.[41] The potent glial and neuronal GLU transport mechanisms may conceivably also protect against non-catastrophic vagaries in blood-brain barrier function in some disease states (e.g., diabetes or hypertension[42,43]), that might be speculated to allow GLU penetration into the brain from the circulation. Aging by itself, however, does not seem to be associated with changes in blood-brain barrier GLU transport properties.[44] Of course, even if the blood-brain barrier did change under some circumstances, and allow GLU penetration from the circulation, because MSG consumption in the diet (even in high amounts) does not raise plasma GLU concentrations (Fig. 3.1), dietary MSG would *not* be a contributor to any changes in brain function that might result from the BBB alteration.

Table 3.1 shows glutamate concentrations in body compartments and GLU transporter and receptor affinities. Notice that the plasma GLU concentration is much higher than the GLU concentration in brain extracellular fluid, and is high in relation to the binding affinities for the listed GLU receptors (NMDA, AMPA, Metabotropic) in the brain. Hence, if a portion of the blood-brain barrier were to experience a catastrophic event, such as a stroke or hemorrhage, producing a tear or rupture and a loss of glial and neuronal GLU transporters (because they require energy to operate), then GLU concentrations in brain ECF might become high enough to stimulate GLU receptors excessively. This would not be expected in non-catastrophic situations, since the cellular GLU transporters would not be deprived of oxygen, and thus be able to maintain low extracellular GLU concentrations in the brain, even with increased GLU inflow from blood. A non-catastrophic situation would be a reduction in BBB effectiveness, such as is posited to occur in diabetes and hypertension.

Hence, in sum, the ingestion of even large amounts of MSG in the diet is unlikely to cause an increase in brain extracellular GLU concentrations, and is thus unlikely to influence brain function or produce adverse neural effects. First, the gastrointestinal tract blocks the penetration of almost all dietary GLU into the circulation. Second, the blood-brain barrier blocks GLU penetration into the brain. Third, membrane GLU transporters on glial and neuronal cell membranes in the brain keep extracellular and synaptic GLU concentrations very low. These cellular transporters would be expected to keep GLU concentrations low, even if GLU were to penetrate into the brain from the circulation (e.g., if the blood-brain barrier became leaky). Finally, because the ingestion of MSG with food does not elevate plasma GLU concentrations, dietary MSG would not contribute to any brain effects that might occur with its penetration into the brain under special medical circumstances (e.g., in association with possible blood-brain barrier disruption posited to occur in diabetes or hypertension).

Consideration should also be given to the fetal and infant brains. Might MSG ingestion by pregnant women expose the brains of their fetuses to elevated levels of GLU *in utero?* And, is the infant brain also at risk from dietary GLU (MSG)? First, the placenta forms a remarkable barrier to GLU penetration from the maternal circulation. Intravenous infusions of GLU into pregnant female monkeys, which cause enormous increases in maternal plasma GLU concentrations, produce no increases in fetal plasma GLU, except at an extremely high dose (400 mg/kg iv, a dose that raised maternal plasma GLU concentrations *70-fold*). At doses that raise maternal plasma GLU 10-20 fold, no increases in fetal plasma GLU are seen.[45] Figure 3.4 shows fetal plasma glutamate response to the intravenous infusion of glutamate into the maternal circulation. Catheters were placed into the maternal inferior vena cava and a vein in the arm, and into an interplacental vein (fetal circulation) of anesthetized female rhesus monkeys (*macaca mulatta*) during the last trimester of pregnancy ($n = 5$). Monosodium glutamate was infused into the maternal vena cava over a one-hour period, and blood was sampled periodically from both the maternal arm vein and the interplacental vein (times indicated in figure). The glutamate doses were: ▲, 150 mg/kg; X, 170–190 mg/kg (mean of two animals); ●, 220 mg/kg; ○, 400 mg/kg. The baseline level of plasma glutamate in the mothers was 4–5 μmol/100 mL. No rise in fetal plasma glutamate occurred up to a maternal plasma concentration of 100 μmol/100 mL (20 times normal); an increase in fetal plasma glutamate (to about 50 μmol/100 mL) occurred when the maternal plasma glutamate concentration reached 280 μmol/100 mL (70 times baseline).

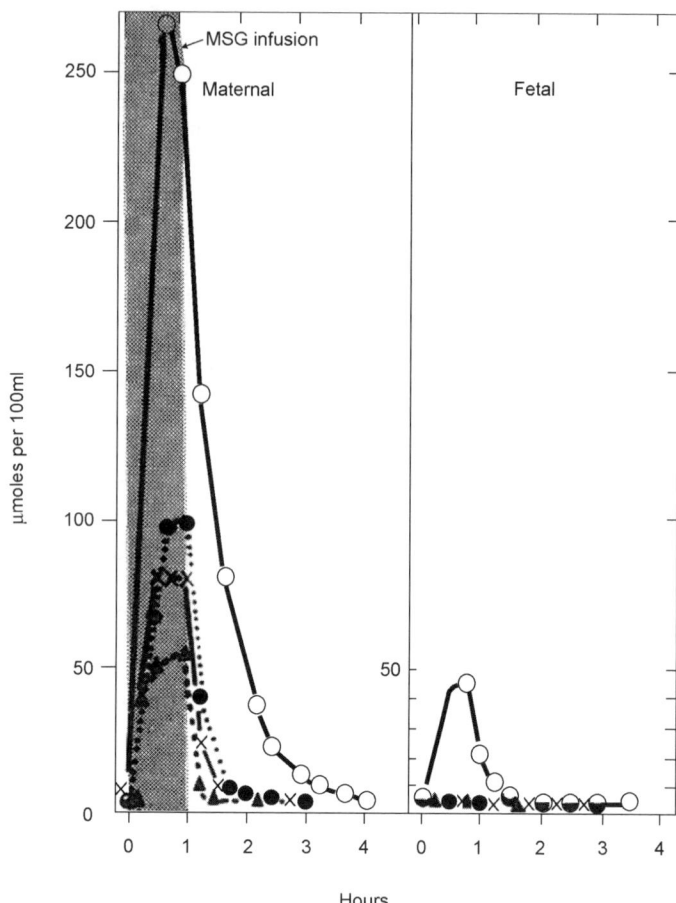

Fig. 3.4 Fetal plasma glutamate response to the intravenous infusion of glutamate into the maternal circulation. The glutamate doses were: ▲, 150 mg/kg; X, 170–190 mg/kg (mean of two animals); ●, 220 mg/kg; ○, 400 mg/kg. Figure reproduced from the *American Journal of Obstetrics and Gynecology*, 122 (1): 70–78, Stegink LD, 1975 with permission from Elsevier.

By way of comparison, Fig. 3.2 shows that 150 mg/kg MSG consumed in water by fasting subjects raises plasma glutamate 10-fold, while Fig. 3.1 shows that this dose produces no rise in plasma glutamate when ingested in the diet (lunch and dinner doses were 40 and 45 mg/kg, respectively). Hence, to attain a greater than 20-fold rise in plasma glutamate, at least 300 mg/kg MSG would be required, ingested in water on an empty stomach; in food, the dose would be much larger. Because of the unpleasant taste of MSG at high concentrations,[2,11] such extreme doses would never be consumed. What might explain this barrier to maternal-fetal GLU transfer? The results of recent studies show that GLU is a major energy source for the placenta, that the placenta extracts GLU from both fetal and maternal circulations to use in energy production, and indeed, that the fetal liver synthesizes GLU for export to placenta.[46] The barrier results from nothing more than the normal use of GLU by the placenta for energy production.

Concerning infants, the infant has been shown to metabolize GLU at the same rapid rate as adults.[47] The ingestion of a meal containing MSG (or GLU from any source, including dietary protein) should therefore produce the same plasma GLU responses as seen in adults for comparable doses (mg/kg). Hence, at doses likely to be palatable in food, MSG in the diet would not be expected to raise plasma GLU concentrations in infants, just as in adults. Hence, no effects would be expected in the brain. Commercially-prepared baby foods contain no added MSG;[31] hence infants fed such foods by their mothers would also experience no rise in plasma GLU because of MSG present in the foods. Finally, might nursing infants be exposed to an elevated intake of GLU, if breast milk GLU content rose with maternal GLU or MSG intake? No. The GLU content of breast milk is not affected by maternal diet, even when the mother ingests sufficient MSG (6 grams) to raise plasma GLU concentrations almost 3-fold.[48]

3.4   Adverse reactions to MSG

3.4.1   Chinese Restaurant Syndrome

Reports of adverse reactions to MSG have appeared in the scientific literature for almost 40 years, beginning in 1968 with the anecdotal letter-to-the-editor by Kwok in the *New England Journal of Medicine*.[49] The term 'Chinese Restaurant Syndrome' (CRS) was coined with this first 'case report', and was described as a group of three symptoms: (a) numbness at the back of the neck, arms and back, (b) general weakness, and (c) palpitations. The author, a physician, who was describing his own symptoms, noted that the effects occurred within 15–20 minutes of consuming Chinese food and persisted for about 2 hr, with no residual effects. Dr Kwok had no evidence that MSG was responsible, and indeed, noted that a number of ingredients in Chinese food might be responsible, only one of which was MSG. Within a few months, a report appeared in *Science* that described a number of trials of MSG in human subjects. The investigators offered a different set of symptoms said to be associated with MSG ingestion, which were: (a) burning, (b) facial pressure, and (c) chest pain. *All* subjects tested, who were described as 'normal', showed these responses. In addition, the symptoms were elicited only if MSG was ingested on an empty stomach.[50] Further, the effects were obtained, regardless of the form of glutamate (free glutamic acid, sodium glutamate, potassium glutamate), in oral doses of up to 12 g. Within a year of this report, another had appeared in the same journal (*Science*), finding *no* effects of MSG in normal subjects who had been given *extremely* large daily doses of glutamate (up to 147 g/day!) for 30 or more days.[51] Hence, all within one year: (a) a phenomenon (CRS) and its symptoms had been described;[49] (b) the phenomenon had been redefined to have a different set of symptoms than that originally assigned, with attribution specifically to MSG (none of the other ingredients in Chinese food was ever tested);[50] and (c) the attribution of symptoms to MSG had been questioned.[51] To add to this confusion, symptoms continued to multiply.

In 1972, Kenney and Tidball[52] conducted an acute study, in which MSG (5 g) was consumed in tomato juice (tomato juice with added salt was the placebo). One-third of the subjects (not all of them) reported one or more of a list of sensations: warmth/burning, stiffness/tightness, limb weakness, pressure, tingling, headache, light-headedness, or heartburn/gastric discomfort. Kenney described these symptoms as being consistent with local, upper esophageal irritation, and were also produced by such foods as black coffee or spiced tomato juice.[53,54] Other studies found some immediate responses to an MSG solution,[55,56] but the symptoms did not correspond to those originally described by Kwok[49] or Schaumburg.[50] Indeed, Kerr *et al.* noted some years after the initial CRS reports that 'such a wide variety of non-specific symptoms have been included in subsequent reports, that it is now difficult to find food-associated symptoms which have not been identified with the CRS at some time'.[57] These investigators were also aware that because of the high public visibility of the food additive safety issue, studies of food additives like MSG might inadvertently become biased. Indeed, when queried in questionnaires specifically about CRS and MSG, 25% of respondents (not 100%, as claimed by Schaumburg[50]) reported being susceptible to CRS.[58] However, when questionnaires did not bias subjects regarding CRS or MSG, asking them simply to associate any of a list of symptoms with particular ethnic cuisines, none associated all three symptoms said to define CRS (and MSG causality),[50] and only 2.6% associated one or two of these symptoms with Chinese food.[59]

When it sifted through these studies, the Joint WHO/FAO Expert Committee on Food Additives (JECFA) came to the conclusion that MSG is not the causal agent in CRS.[22] Yet in 1992, the US Food and Drug Administration commissioned further evaluation of adverse reactions to MSG by the Life Sciences Research Office (LSRO) of the Federation of American Societies for Experimental Biology (FASEB). LSRO convened a panel of experts to evaluate available information; a report was published in 1995.[16] The panel concluded that causality had been demonstrated between MSG and CRS, and that there was a subgroup of healthy individuals in the population that could manifest CRS (termed 'MSG symptom complex' in their report) in response to an oral dose of 3 g or more of MSG. The MSG symptom complex was defined to include a large number of symptoms, namely warmth/burning, stiffness/tightness of face and chest, limb weakness, pressure, tingling, headache, light-headedness and gastric discomfort. Further, the report stated that documented evidence indicated that MSG could provoke an asthma attack in some asthmatic individuals.[16] From these conclusions, the panel recommended that additional studies be conducted to identify those individuals in the population who are sensitive to MSG, and to demonstrate this sensitivity using a double-blind paradigm. Further, 'double-blind placebo-controlled challenges on separate occasions must reproduce symptoms with the ingestion of MSG and produce no response with the placebo'.[16] Three such replications were recommended when MSG was examined in a food matrix. MSG doses of up to 5–6 grams were suggested for single-challenge, and dose-response studies.

Soon after these recommendations were made, a study appeared that examined adverse responses in individuals *self-identified* as MSG-sensitive.[60] To qualify for the study, candidates had to have two or more symptoms within 3 hr of ingesting a meal. The symptoms were: general weakness, muscle tightness or twitching, flushing or sweating, burning, headache/migraine, chest pain, palpitation/heart pounding, numbness/tingling, or one of these symptoms plus one other of their own description. Such symptoms, identified before challenge, were designated index symptoms. Hence, on admission to the study, the subjects knew which symptoms were of interest, which may have entered an element of bias into the outcome. Of over 600 inquiries, 61 subjects entered the study, the first phase of which was a double-blind challenge with 5 g MSG or placebo (in a liquid that masked the distinct taste of MSG) on an empty stomach; the ordering of the challenges was randomized. A positive response was defined as a report of two or more of the above symptoms following challenge. Considering the responses of all 61 subjects, there was no significant difference between MSG and placebo in the total number of symptoms reported, though reported severity was significantly different. An ordering effect was present: placebo responses were greater in subjects receiving placebo first than in those who received it as the second challenge. Of the 61, 22 responded to MSG (reported two or more index symptoms) but not to placebo. The other 39 either responded to placebo alone, placebo *and* MSG, or neither. Hence, of the 61 who met the criteria of MSG responder (self-reported) at screening, only 22 (38%) actually gave a response to MSG (and not placebo) under double-blind conditions.

The blind was not broken at the end of the first trial, so all subjects who responded to one treatment (either MSG or placebo), but not the other ($n = 37$) were asked to participate in the second phase of the study; 36 agreed. In the second challenge, subjects received on separate occasions either placebo or MSG on an empty stomach at one of three doses (1.25, 2.5 or 5.0 g) in liquid. The investigators reported that even in the second phase, an ordering effect of placebo was present. Nevertheless, they observed a significant, dose-related increase in the total number of symptoms reported, and in their severity. Statistical significance was achieved (relative to placebo) at the 2.5 and 5 g dose levels. Several of the individual symptoms also showed this dose-related effect (headache, muscle tightness, numbness/tingling).

By the LSRO/FASEB standard,[16] the key limitation of this study was that it only reported group responses. That is, it did not assess if individual subjects showed the same reactions to MSG with two or more challenges, but no reactions to placebo. Hence, we do not know, for example, if a subject who experienced muscle tingling following the first MSG challenge experienced muscle tingling on MSG rechallenge, but not when challenged with placebo. The ordering effect with placebo (subjects receiving placebo first reported more index symptoms than those receiving it second) is also problematic, as it suggests that while subjects may not have been able to distinguish MSG from placebo in the first trial, they may have been able to do so by the second. Such issues are important in a situation where the subject may desire to demonstrate his/her responsiveness to a test agent.

A few years later, another study appeared,[61] which used the LSRO/FASEB recommendations regarding protocol design.[16] Geha et al.[61] screened subjects, based on their reporting to have experienced a reaction to an Asian meal thought to contain MSG, and to have experienced at least two of the following symptoms: general weakness, muscle tightness, muscle twitching, feeling of flushing, sweating sensation, burning sensation, headache-migraine, chest pain, palpitations, feeling of numbness-tingling. The study was conducted in three centers: Boston, Chicago and Los Angeles. Despite this large population base, and repeated advertisements, only 178 inquiries were made, and 132 finally entered the study.

The study had four phases. In phase A, overnight-fasted subjects ingested on separate occasions 5 g MSG or placebo (in the same liquid formulation used by Yang et al.[60] to mask the taste of MSG). Treatment order was randomized; the study was double-blind. After ingesting the test material, subjects were queried at 15 min intervals for 2 hr if any of the above symptoms had occurred. Of the 130 subjects who completed phase A, 50 (38%) reported two or more symptoms after MSG ingestion, and none or one symptom after placebo. The other subjects either reported responses to both placebo and MSG challenge ($n = 19$), to neither treatment ($n = 44$), or to placebo but not MSG ($n = 17$). Considering the responses of the entire group, the frequency of reported occurrences of five of the symptoms (weakness, muscle tightness, flushing, burning sensation, and headache-migraine) was significantly greater after MSG than after placebo.

In phase B, fasting subjects were challenged with each of several doses of MSG (0, 1.25, 2.5, 5.0 g, on separate occasions). Treatment order was randomized. All subjects were invited to participate, except for those who showed no response in phase A to either treatment ($n = 44$). The eligible group was thus 86 individuals; 69 entered and completed this trial. (The double-blind was maintained; all responders were included, even if responding to both treatments, to keep the group size as large as possible.) As before, following challenge, the subjects were queried for symptoms at 15 min intervals for 2 hr. Compared to the placebo response, no significant rise in the occurrence of any symptom was noted at the 1.25 g MSG dose; at the 2.5 g dose, a significant increase was noted for only one symptom, muscle tingling. At the 5 g dose, significant increases occurred for six symptoms (weakness, muscle tightness, flushing, sweating, headache-migraine, numbness-tingling). Of particular interest, however, when the blind was broken at the end of phase B, was the finding that while 37 of the subjects who had responded to MSG (5 g) but not placebo in phase A had participated in phase B, only 19 of these subjects reported two or more symptoms following MSG (5 g) but not placebo, in phase B. Moreover, at this point the LSRO/FASEB report had been released, and the investigators looked to see how many of the 19 had experienced the same symptoms in both trials.

Fourteen had reported the same symptoms. Hence, on repeated challenging of self-identified MSG-sensitive individuals with a very large dose of MSG (5 g), only about 10% reported the same response each time.

Using the LSRO/FASEB report as a guide, Geha *et al.* invited the 19 individuals who had responded to MSG (5 g) but not placebo on two occasions (to maximize group size, all 19 were invited, not just the 14 who had reported the same responses) to participate in phase C. Twelve agreed to participate. In phase C, fasting subjects received MSG (5 g) or placebo (sucrose, 5 g) in capsules with water, to minimize the chance that the MSG could be tasted. The order was randomized, and the trial (MSG vs placebo) was conducted twice. In this phase, subjects were asked simply to write down all symptoms experienced (no list was provided), at 15 min intervals for 2 hr after dosing. At the conclusion of the study, only two of 12 subjects reported two or more symptoms after 5 g MSG but not placebo in *both* of the challenges. Moreover, these two subjects did not report the same symptoms in trials A, B and C. Hence, at the conclusion of this phase of the study, designed to meet the LSRO/FASEB guidelines, none of the 130 self-identified MSG-sensitive individuals met the established criteria for MSG sensitivity.

Phase D was designed to evaluate the MSG response when consumed with food. However, only two subjects remained, and they did not meet the LSRO/FASEB criteria for MSG sensitivity. Nonetheless, they were asked to participate. Because the trial had no statistical power, the outcome is not meaningful. This phase consisted of testing placebo vs MSG (5 g) capsules in three separate trials (six total tests), when consumed with a breakfast meal. In the three tests with MSG, the subjects reported two or more symptoms on only one occasion, and the symptoms were not the same as those they had reported in the earlier trials.[61]

Hence, over the 30 years that MSG has been argued to be the agent in Chinese food that produces the 3–10 symptoms reported to define CRS, the predicted incidence of CRS in the US population has gone from 100%[50] to essentially 0%.[61] While individuals may persist in their belief that they have an unpleasant sensitivity to MSG, when tested under conditions where neither testor nor testee is aware of the treatment (as far as possible), such MSG-sensitive subjects fail to demonstrate this sensitivity, even subjectively.

3.4.2   Asthma

In 1981, in a letter-to-the-editor, Allen and Baker[62] described two patients who reported asthma attacks 12 hr after eating in a Chinese restaurant. When these individuals were challenged with 2.5 g MSG in capsules, they experienced asthma attacks 10–12 hr later, with a measured reduction in peak expiratory flow rates (PEFR). Some years later, Allen *et al.*[63] reported on 32 subjects, who submitted to a single-blind challenge with MSG. Fourteen of the subjects reported having wheezing attacks after Chinese meals; the other 18 had unstable asthma, with sudden, unexplained attacks and a sensitivity to chemicals. The investigators measured PEFR for 12 hr after an MSG challenge. Fourteen of the 32 patients experienced asthma attacks after ingesting MSG; one patient reacted to 1.5 g MSG, while the other 13 responded to a 2.5 g (but not 1.5 g) dose. The times of onset were distributed over the 12 hr study period, though no subject had an attack the first hour after MSG ingestion, the time when plasma glutamate (GLU) concentrations would have been elevated. The design of this study has been criticized by Stevenson.[64] He argued that because patients were asked to discontinue anti-asthmatic medications prior to and during the experiment, and the placebo trial always preceded the MSG trial, MSG testing may have yielded positive responses simply because anti-asthmatic medications were lower in the body when MSG was administered than when placebo was ingested. A similar concern was expressed about the study of Moneret-Vautrin.[65]

Four published studies have found no evidence of an asthmatic attack precipitated by MSG.[64] For example, in the study of Woods et al.,[66] 12 asthmatic subjects were recruited who reported that MSG caused them to have asthma attacks. They were maintained on their medications during the experiment, and their asthma was stable. On separate occasions, they were challenged with 0, 1.5 and 5 g MSG in capsules, ingested 30 minutes after a breakfast in a clinical setting. Airway resistance was measured at intervals over the succeeding 8 hours in the clinic, and then for an additional 4 hours at home with a portable device they had been trained to use. No effects were seen at either dose, compared to placebo.

Woessner et al.[67] recruited 30 asthmatic individuals who reported having asthma attacks in Chinese restaurants, and who believed MSG to be the cause. They recruited an additional 70 asthmatic patients, who reported no asthma attacks in Chinese restaurants, but who had a chemical sensitivity (to aspirin). All patients were maintained on their anti-asthma medications, and were asked to consume a low-MSG diet for several days prior to the study. The study was conducted in a clinical research center. In the morning of the first day, subjects were given a light, low-MSG breakfast, followed by several placebo capsules. Airway stability (forced expiratory volume, FEV1) was measured at frequent intervals. If FEV1 was stable through to the next day, the procedure was repeated the following morning, with the subject ingesting 2.5 g MSG in capsules, rather than placebo. If the subject showed no response to MSG, he/she was studied no further. If an airway response was observed, the patient was invited to undergo two additional, double-blind tests of MSG vs placebo. When the 70 patients reporting no sensitivity to MSG were tested, none had a response to MSG; they were studied no further. When the 30 patients were examined who reported MSG sensitivity, only one may have had a response to MSG. When this patient was then subjected to two sequential placebo-MSG trials, the effect was not reproduced. Therefore, the Woessner et al. study found no asthmatic subjects who showed an increase in airway resistance following MSG ingestion.[67]

3.5 Commentary on likely future trends

Monosodium glutamate is one of the most studied food additives in history. The findings, collected over more than 40 years, indicate that its use in the diet is safe. It is therefore unlikely that a great deal of energy will continue to be devoted to the study of safety issues. However, the debate over MSG led to increased scientific interest in studying GLU functions in the body, and the outcome, as noted in the introduction, has been remarkable. The most obvious example has been the role of GLU in brain function (it is a neurotransmitter), which was stimulated enormously and directly by Olney's observations in hypothalamus.[20,68] Interest in the taste properties of MSG was also stimulated by the safety controversy, and has led to the identification of umami (MSG taste) as a basic taste, and recently to the molecular identification of umami and amino acid taste receptors.[69,70] Work is thus just beginning in this area of investigation. Finally, the safety controversy required that much more be learned about GLU metabolism in the body (apart from protein synthesis and turnover), which produced remarkable new information about the metabolic handling of GLU in the gut, placenta, liver, and brain. More recently, the metabolic utilization of GLU in muscle has become a topic of increasing interest.[71,72] Interest in the intermediary metabolism of GLU is likely to continue unabated. Hence, the future trend in work on GLU is likely to be centered on learning more about the normal physiologic and metabolic functions of this remarkable amino acid.

3.6 Sources of further information and advice

There are no technical websites that offer information on MSG (all are consumer oriented). The following symposia and volumes provide the most comprehensive technical information, and provide citations in the primary literature:

Umami, a special issue of the journal *Food Reviews International* (**14** (2/3): 123–337, 1998), contains a number of technical articles offering a large body of information regarding the glutamate content of foods, the use of MSG in a variety of cuisines, and the sensory properties of umami.

International Symposium on Glutamate, proceedings of a symposium held October 1998 in Bergamo Italy, covering all contemporary issues of glutamate/ MSG safety, metabolism and physiology; published as a supplement to the *Journal of Nutrition*: *Journal of Nutrition* **130** (4S): 891S–1079S, 2000.

Analysis of Adverse Reactions to Monosodium Glutamate (MSG), prepared by the Life Sciences Research Office, Federation of American Societies for Experimental Biology, under contract (223-92-2185) to the US Food and Drug Administration, published by the American Institute of Nutrition in 1995, and available for purchase at *www.lsro.org*. This volume contains a fairly comprehensive review of issues relating to MSG safety at the time, and contains recommendations for further study. Most of the recommendations were subsequently taken up and examined, and published in the primary literature.

Glutamic Acid: Advances in Biochemistry and Physiology, Filer LJ, Garattini S, Kare MR, Reynolds WA, eds. New York, Raven Press, 1979, 400 pp (ISBN 0-89004-356-6). This book is out of print. It contains articles discussing MSG safety issues *circa* 1977, and the time of the US Select Committee on GRAS substances (SCOGS) activities. Based on this publication, the SCOGS committee reevaluated some of its initial conclusions regarding MSG safety *circa* 1975. The volume was the first to consider GLU safety, toxicity, metabolism and physiology, and the information it contains is still current.

Evaluation of the Health Aspects of Certain Glutamates as Food Ingredients, Supplemental Review and Evaluation. prepared by the Life Sciences Research Office, Federation of American Societies for Experimental Biology, under contract (FDA 223-75-2004) to the US Food and Drug Administration, published by the American Institute of Nutrition in 1996, and available for purchase at *www.lsro.org*. This is an addendum to the original SCOGS report on MSG, and updates the committee's recommendations in response to the information provided from a 1977 international glutamate symposium held in Milan Italy, and published as *Glutamic Acid: Advances in Biochemistry and Physiology* (see above).

Second International Symposium on Umami, proceedings of a symposium held October 1990 in Taormina Italy, covering all contemporary issues of umami taste physiology; published as a supplement to *Physiology and Behavior: Physiology and Behavior* **49**(5): 831–1030, 1991. Research in umami, at the physiological and molecular levels, has moved quickly in the past decade. Hence, current information is best obtained by seeking a recent journal review.

The 1987 JECFA report on glutamic acid, *L-Glutamic Acid and its Ammonium, Calcium, Monosodium and Potassium Salts*, which is a chapter in *Toxicological evaluation of certain food additives – Joint FAO/WHO expert panel on food additives* (WHO Food Additive Series, Volume 22, Cambridge University Press), is still in print. It may also be obtained from the following website: http://www.inchem.org/documents/jecfa/jecmono/v22je12.htm

3.7   References

1. HE FJ, MacGREGOR GA. Salt in food. *Lancet*, 2005; 365: 844–5.
2. HE FJ, MacGREGOR GA. How far should salt intake be reduced? *Hypertension*, 2003; 42: 1093–9.

3. HE FJ, MacGREGOR GA. Effect of modest salt reduction on blood pressure: a meta-analysis of randomized trials. Implications for public health. *J Human Hyperten*, 2002; 16: 761–70.

4. YAMAGUCHI S. Basic properties of umami and its effects on food flavor. *Food Rev Int*, 1998; 14: 139–76.

5. BELLISLE F. Glutamate and the UMAMI taste: sensory, metabolic, nutritional and behavioural considerations. A review of the literature published in the last 10 years. *Neurosci Biobehav Rev*, 1999; 23: 423–38.

6. LINDEMANN B, OGIWARA Y, NINOMIYA Y. The discovery of umami. *Chem Senses*, 2002; 27: 843–4.

7. IKEDA K. New seasonings. *Chem Senses*, 2002; 27: 847–9.

8. YAMAGUCHI S, NINOMIYA K. Umami and food palatability. *J Nutr*, 2000; 130(4S Suppl): 921S–6S.

9. NELSON G, CHANDRASHEKAR J, HOON MA, FENG L, ZHAO G, RYBA NJ, ZUKER CS. An amino-acid taste receptor. *Nature*, 2002; 416: 199–202.

10. CHAUDHARI N, LANDIN AM, ROPER SD. A metabotropic glutamate receptor variant functions as a taste receptor. *Nature Neuroscience*, 2000; 3: 113–19.

11. LOLIGER J. Function and importance of glutamate for savory foods. *J Nutr*, 2000; 130(4S Suppl): 915S–20S.

12. ROININEN K, LAHTEENMAKI L, TUORILA H. Effect of umami taste on pleasantness of low-salt soups during repeated testing. *Physiol Behav*, 1996; 60: 953–8.

13. YAMAGUCHI S, TAKAHASHI C. Interactions of monosodium glutamate and sodium chloride on saltiness and palatability of a clear soup. *J Food Sci*, 1984; 49: 82–5.

14. BALL P, WOODWARD D, BEARD T, SHOOBRIDGE A, FERRIER M. Calcium diglutamate improves taste characteristics of lower-salt soup. *Eur J Clin Nutr*, 2002; 56: 519–23.

15. CHI SP, CHEN TC. Predicting optimum monosodium glutamate and sodium chloride concentrations in chicken broth as affected by spice addition. *J Food Process Preserv*, 1992; 16: 313–26.

16. *Analysis of adverse reactions to monosodium glutamate (MSG)*. Bethesda, MD: FASEB Life Sciences Research Office, 1995.

17. REEDS PJ, BURRIN DG, JAHOOR F, WYKES L, HENRY J, FRAZER, EM. Enteral glutamate is almost completely metabolized in first pass by the gastrointestinal tract of infant pigs. *Am J Physiol*, 1996; 270: E413–18.

18. BROSNAN JT. Glutamate, at the interface between amino acid and carbohydrate metabolism. *J Nutr*, 2000; 130(4S Suppl): 988S–90S.

19. FONNUM F. Glutamate: A neurotransmitter in mammalian brain. *J Neurochem*, 1984; 42: 1–11.

20. OLNEY JW. Brain lesions, obesity, and other disturbances in mice treated with monosodium glutamate. *Science*, 1969; 164: 719–21.

21. OLNEY JW, HO OL. Brain damage in infant mice following oral intake of glutamate, aspartate or cysteine. *Nature*, 1970; 227: 609–11.

22. JECFA. *L-Glutamic acid and its ammonium, calcium, monosodium and potassium salts. Toxicological evaluation of certain food additives* – Joint FAO/WHO Expert Panel on Food Additives. Cambridge: Cambridge University Press, 1987, pp. 97–161.

23. MUNRO HN. Factors in the regulation of glutamate metabolism. In: Filer LJ, Jr., Garattini S, Kare MR, Reynolds WA, Wurtman RJ, editors. *Glutamic Acid: Advances in Biochemistry and Physiology*. New York: Raven Press, 1979, pp. 55–68.

24. GIACOMETTI T. Free and bound glutamate in natural products. In: Filer LJ, Garattini

S, Kare MR, Reynolds WA, Wurtman RJ, editors. *Glutamic Acid: Advances in Biochemistry and Physiology*. New York: Raven Press, 1979, pp. 25–34.

25. US DEPARTMENT OF AGRICULTURE, AGRICULTURAL RESEARCH SERVICE. USDA nutrient database for standard reference, Release 18. Nutrient Database Laboratory Home Page, http://www.ars.usda.gov/ba/bhnrc/ndl.

26. RHODES J, TITHERLEY AC, NORMAN JA, WOOD R, LORD DW. A survey of the monosodium glutamate content of foods and an estimation of the dietary intake of monosodium glutamate. *Food Add Contam*, 1991; 8: 663–72.

27. STEGINK LD, FILER LJ, BAKER GL, MUELLER SM, WU-RIDEOUT MY-C. Factors affecting plasma glutamate levels in normal human subjects. In: Filer LJ, Garattini S, Kare MR, Reynolds WA, Wurtman RJ, editors. *Glutamic Acid: Advances in Biochemistry and Physiology*. New York: Raven Press, 1979, pp. 333–51.

28. O'KANE RL, MARTINEZ-LOPEZ I, DEJOSEPH MR, VINA JR, HAWKINS RA. Na^+-dependent Glutamate Transporters (EAAT1, EAAT2, and EAAT3) of the Blood-Brain Barrier. A Mechanism for Glutamate Removal. *J Biol Chem*, 1999; 274: 31891–5.

29. DAIKHIN Y, YUDKOFF M. Compartmentation of brain glutamate metabolism in neurons and glia. *J Nutr*, 2000; 130: 1026S–31S.

30. OLNEY JW. Brain damage and oral intake of certain amino acids. *Adv Exp Med Biol*, 1976; 69: 497–506.

31. GILLETTE R. News and comment. Academy food committees: new criticism of industry ties. *Science*, 1972; 177: 1172–5.

32. AIROLDI L, BIZZI A, SALMONA M, GARATTINI S. Attempts to establish the safety margin for neurotoxicity of monosodium glutamate. In: Filer LJ, Garattini S, Kare MR, Reynolds WA, Wurtman RJ, editors. *Glutamic Acid: Advances in Biochemistry and Physiology*. New York: Raven Press, 1979, pp. 321–31.

33. HU L, FERNSTROM JD, GOLDSMITH PC. Exogenous glutamate enhances glutamate receptor subunit expression during selective neuronal injury in the ventral arcuate nucleus of postnatal mice. *Neuroendocrinology*, 1998; 68: 77–88.

34. FERNSTROM JD, CAMERON JL, FERNSTROM MH, MCCONAHA C, WELTZIN TE, KAYE, WH. Short-term neuroendocrine effects of a large oral dose of monosodium glutamate in fasting male subjects. *J Clin Endocrinol Metab*, 1996; 81: 184–91.

35. STEGINK LD, REYNOLDS WA, FILER LJ, BAKER GL, DAABEES TT, PITKIN RM. Comparative metabolism of glutamate in the mouse, monkey and man. In: Filer LJ, Garattini S, Kare MR, Reynolds WA, Wurtman RJ, editors. *Glutamic Acid: Advances in Biochemistry and Physiology*. New York: Raven Press, 1979, pp. 85–102.

36. MELDRUM B. Amino acids as dietary excitotoxins: A contribution to understanding neurodegenerative disorders. *Brain Res Rev*, 1993; 18: 293–314.

37. WALKER R, LUPIEN JR. The safety evaluation of monosodium glutamate. *J Nutr*, 2000; 130(4S Suppl): 1049S–52S.

38. TSAI PJ, HUANG PC. Circadian variations in plasma and erythrocyte concentrations of glutamate, glutamine, and alanine in men on a diet without and with added monosodium glutamate. *Metabolism*, 1999; 48: 1455–60.

39. YAMAGUCHI S. Fundamental properties of umami in human taste sensation. In: Kawamura Y, Kare MR, editors. *Umami: A Basic Taste*. New York: Marcel Dekker, 1987, pp. 41–73.

40. BRANN DW, MAHESH VB. Excitatory amino acids: evidence for a role in the control of reproduction and anterior pituitary hormone secretion. *Endocr Rev*, 1997; 18: 678–700.

41. PERUZZO B, PASTOR FE, BLAZQUEZ JL, SCHOBITZ K, PELAEZ B, AMAT P, RODRIGUEZ EM. A

second look at the barriers of the medial basal hypothalamus. *Exp Brain Res*, 2000; 132: 10–26.

42. STARR JM, WARDLAW J, FERGUSON K, MACLULLICH A, DEARY IJ, MARSHALL I. Increased blood-brain barrier permeability in type II diabetes demonstrated by gadolinium magnetic resonance imaging. *J Neurol Neurosurg Psychiatry*, 2003; 74: 70–6.

43. UENO M, SAKAMOTO H, TOMIMOTO H, AKIGUCHI I, ONODERA M, HUANG CL, KANENISHI K. Blood-brain barrier is impaired in the hippocampus of young adult spontaneously hypertensive rats. *Acta Neuropathologica*, 2004; 107: 532–8.

44. BUBNA-LITTITZ H. Age-related changes in the blood-brain barrier in the rat with reference to methionine, lysine, glutamic acid and N-methyl-N-nitrosourea. *Z Gerontol*, 1988; 21: 93–101.

45. STEGINK LD, PITKIN RM, REYNOLDS WA, FILER LJ, BOAZ DP, BRUMMEL MC. Placental transfer of glutamate and its metabolites in the primate. *Am J Obstet Gynecol*, 1975; 122: 70–8.

46. BATTAGLIA FC. Glutamine and Glutamate Exchange between the Fetal Liver and the Placenta. *J Nutr*, 2000; 130: 974S–7S.

47. STEGINK LD, FILER LJ, JR., BAKER GL, BELL EF. Plasma glutamate concentrations in 1-year-old infants and adults ingesting monosodium L-glutamate in consomme. *Pediatr Res*, 1986; 20: 53–8.

48. BAKER GL, FILER LJ, JR, STEGINK LD. Factors influencing dicarboxylic amino acid content of human milk. In: Filer LJ, Jr, Garattini S, Kare MR, Reynolds WA, Wurtman RJ, editors. *Glutamic Acid: Advances in Biochemistry and Physiology.* New York: Raven Press, 1979, pp. 111–23.

49. KWOK RHM. Chinese restaurant syndrome (letter). *New Engl J Med*, 1968; 278: 796.

50. SCHAUMBURG HH, BYCK R, GERSTL R, MASHMAN JH. Monosodium L-glutamate: its pharmacology and role in the Chinese restaurant syndrome. *Science*, 1969; 163: 826–8.

51. BAZZANO G, D'ELIA JA, OLSON RE. Monosodium glutamate: feeding of large amounts in man and gerbils. *Science*, 1970; 169: 1208–9.

52. KENNEY RA, TIDBALL CS. Human susceptibility to oral monosodium L-glutamate. *Am J Clin Nutr*, 1972; 25: 140–6.

53. KENNEY RA. Chinese Restaurant Syndrome. *Lancet*, 1980; i: 311–12.

54. KENNEY RA. The Chinese restaurant syndrome: an anecdote revisited. *Food Chem Toxicol*, 1986; 24: 351–4.

55. GORE ME, SALMON PR. Chinese restaurant syndrome: fact or fiction? *Lancet*, 1980; 1(8162): 251–2.

56. TARASOFF L, KELLY MF. Monosodium L-glutamate: A double-blind study and review. *Fd Chem Toxic*, 1993; 31: 1019–35.

57. KERR GR, WU LEE M, EL LOZY M, McGANDY R, STARE FJ. Objectivity of food-symptomatology surveys. Questionnaire on the 'Chinese restaurant syndrome'. *J Am Diet Assoc*, 1977; 71: 263–8.

58. REIF-LEHRER L. Possible significance of adverse reactions to glutamate in humans. *Fed Proc*, 1976; 35: 2205–11.

59. KERR GR, WU-LEE M, EL-LOZY M, MCGANDY R, STARE FJ. Prevalence of the 'Chinese restaurant syndrome'. *J Am Diet Assoc*, 1979; 75: 29–33.

60. YANG WH, DROUIN MA, HERBERT M, MAO Y, KARSH J. The monosodium glutamate symptom complex: assessment in a double-blind, placebo-controlled, randomized study. *J Allergy Clin Immunol*, 1997; 99: 757–62.

61. GEHA RS, BEISER A, REN C, PATTERSON R, GREENBERGER PA, GRAMMER LC, DITTO AM,

HARRIS KE, SHAUGHNESSY MA, *et al.* Multicenter, double-blind, placebo-controlled, multiple-challenge evaluation of reported reactions to monosodium glutamate. *J Allergy Clin Immunol*, 2000; 106: 973–80.

62. ALLEN DH, BAKER GJ. Chinese-restaurant asthma. *New Engl J Med*, 1981; 305: 1154–5.
63. ALLEN DH, DELOHERY J, BAKER G. Monosodium L-glutamate-induced asthma. *J Allergy Clin Immunol*, 1987; 80: 530–7.
64. STEVENSON DD. Monosodium Glutamate and Asthma. *J Nutr*, 2000; 130: 1067S–73S.
65. MONERET-VAUTRIN DA. Monosodium glutamate-induced asthma: study of the potential risk of 30 asthmatics and review of the literature. *Allergie et Immunologie*, 1987; 19: 29–35.
66. WOODS RK, WEINER JM, THIEN F, ABRAMSON M, WALTERS EH. The effects of monosodium glutamate in adults with asthma who perceive themselves to be monosodium glutamate-intolerant. *J Allergy Clin Immunol*, 1998; 101: 762–71.
67. WOESSNER KM, SIMON RA, STEVENSON DD. Monosodium glutamate sensitivity in asthma. *J Allergy Clin Immunol*, 1999; 104: 305–10.
68. OLNEY JW, SHARPE LG. Brain lesions in an infant rhesus monkey treated with monsodium glutamate. *Science*, 1969; 166: 386–8.
69. ZHAO GQ, ZHANG Y, HOON MA, CHANDRASHEKAR J, ERLENBACH I, RYBA NJ, ZUKER CS. The receptors for mammalian sweet and umami taste. *Cell*, 2003; 115: 255–66.
70. LI X, STASZEWSKI L, XU H, DURICK K, ZOLLER M, ADLER E. Human receptors for sweet and umami taste. *Proc Natl Acad Sci US*, 2002; 99: 4692–6.
71. MOURTZAKIS M, GRAHAM TE. Glutamate ingestion and its effects at rest and during exercise in humans. *J Appl Physiol*, 2002; 93: 1251–9.
72. GRAHAM TE, SGRO V, FRIARS D, GIBALA MJ. Glutamate ingestion: the plasma and muscle free amino acid pools of resting humans. *Am J Physiol*, 2000; 278: E83–E89.
73. MELDRUM BS. Glutamate as a neurotransmitter in the brain: Review of physiology and pathology. *J Nutr*, 2000; 130: 1007S–15S.
74. TSAI PJ, HUANG PC. Circadian variations in plasma and erythrocyte glutamate concentrations in adult men consuming a diet with and without added monosodium glutamate. *J Nutr*, 2000; 130(4S Suppl): 1002S–4S.
75. STEGINK LD, BAKER GL, FILER LJ. Modulating effect of Sustagen on plasma glutamate concentration in humans ingesting monosodium L-glutamate. *Am J Clin Nutr*, 1983; 37: 194–200.
76. FERNSTROM JD, FERNSTROM MH. Nutrition and the Brain. In: Gibney MJ, Macdonald IA, Roche HM, editors. *Nutrition and Metabolism*. Oxford, UK: Blackwell Science, 2003, pp. 145–67.

4

Dietary salt and flavor: mechanisms of taste perception and physiological controls

S. McCaughey, Monell Chemical Senses Center, USA

4.1 Introduction: overview of perception and intake of sodium chloride

Sodium is essential for the normal physiological function of human beings and other animals. It is the most prevalent cation in extracellular fluid, and decreases in sodium levels result in decreases in blood volume and pressure that can be fatal (Abraham *et al.*, 1975). Thus, there has been selective pressure for animals to be able to maintain sodium levels, and two processes have allowed this. The first is the conservation of sodium present in the body, which is accomplished through reabsorption in the kidney and other organs. The second is the ability to identify sodium when it is encountered in the environment and consume especially large amounts when necessary.

There are multiple ways for animals to know that they have sampled a source of sodium, but the sense of taste is the most immediate cue. In humans there is a unique sensory quality, saltiness, that is associated with sodium-containing compounds. A variety of animal species also treats substances with sodium as tasting qualitatively distinct from other substances. There are only a small number of other qualities that are recognized as serving as potential primary tastes. These include: bitterness, which serves to warn against toxins (Garcia and Hankins, 1975); sourness, which guards the body's pH levels (Beidler, 1975); sweetness, which helps to identify energy-rich carbohydrates (Pfaffmann, 1975); and umami, which may serve as a marker of protein (Mori *et al.*, 1991). Although there is debate over whether these terms can encompass all gustatory qualities (Erickson, 1982; Delwiche, 1996), they represent especially clear and salient sensations that are associated with critical functions of the taste system, and the existence of salty taste reinforces the importance of regulating sodium levels precisely. When sodium is lost rapidly due to events such as vomiting, blood loss, or diarrhea, there are not large stores in the body that can be accessed, as can be done for other minerals, such as calcium. It is therefore important for animals, including humans, to be able to deal with sodium deficits quickly, and salty taste provides a rapid means of recognizing external sodium.

Currently, many people live in a sodium-rich environment and consume table salt, primarily composed of NaCl, in excess of need (Loria *et al.*, 2001; Khaw *et al.*, 2004). Over the long term this may result in health problems such as hypertension (Beard *et al.*, 1997; Sacks *et al.*, 2001; Geleijnse *et al.*, 2004). Although care must be taken when generalizing between species, a full understanding of the mechanisms that have evolved to regulate, and often promote, sodium intake in animals will be useful in attempting to reduce sodium consumption in people. It will also be important to discover the full series of events that cause salty taste. Doing so will lead to opportunities to create perceptions of saltiness that are divorced from sodium content and that may therefore avoid the health problems associated with high salt intake.

4.2   Transduction of sodium by taste receptor cells

4.2.1   Sodium-specific mechanisms

The only compounds that taste primarily salty to humans are those that contain sodium or lithium, though other minerals, such as potassium and calcium, can have a salty component to their taste (van der Klaauw and Smith, 1995; Tordoff, 1996a). Among sodium-containing compounds, NaCl is the saltiest, and as the associated anion becomes larger, the perceived saltiness decreases (Schiffman *et al.*, 1980). Presumably, sodium is associated with a distinct taste quality because there is a unique sodium-related transduction mechanism; that is, there is a series of events that converts the chemical identity of sodium, but not of other ions except lithium, into an electrical signal that is propagated through the nervous system. Although specific transduction mechanisms for salty compounds have not yet been determined for humans, rats express epithelial sodium channels (ENaCs) that selectively allow the passage of sodium ions into taste tissue.

Gustatory transduction of NaCl and other compounds takes place throughout the oral cavity, including in taste papillae found on the tongue. There are several kinds of papillae, including fungiform and vallate, with each kind found in different locations. Within papillae are taste buds, each of which contains 50–150 taste receptor cells, and some of these cells are able to interact with taste solutions at their apical ends (Margolskee, 1995). Labeling techniques have confirmed that ENaCs are expressed on the apical ends, as well as other locations, in rat taste receptor cells (Kretz *et al.*, 1999; Lin *et al.*, 1999).

When a rat tastes NaCl, sodium is thought to flow passively down a concentration gradient, through ENaCs, and into taste receptor cells (see Fig. 4.1; Heck *et al.*, 1984). The sodium entry increases the membrane potential of the cell's interior relative to the outside, and this depolarization leads eventually to the release of neurotransmitter onto a peripheral nerve that transmits a signal to the brain. It is not known, though, how many intermediate steps are involved between depolarization and neurotransmitter release. In fact, multiple cells within a taste bud may need to be involved in the process, so that a cell expressing ENaCs may initiate a signal, which is then transmitted to adjacent cells through electrical or chemical coupling, and these other cells then contact a peripheral nerve (Lindemann, 1996).

The ENaCs expressed in rat tongue consist of three subunits. The alpha subunit is capable of acting as a functional channel by itself, but the functionality is increased by co-expression with the beta and gamma subunits (Kellenberger and Schild, 2002). In fungiform papillae, these ENaCs appear to be relatively selective for sodium and lithium ions over other monovalent cations, such as potassium. These channels can be blocked by amiloride (Heck *et al.*, 1984), and this compound and its more specific analog benzamil also partially block the responses elicited by NaCl in the chorda tympani, the nerve that innervates fungiform papillae on the anterior two thirds of the tongue (Brand *et al.*, 1985; DeSimone and Ferrel, 1985; Lundy and Contreras, 1997).

Is passage of sodium through ENaCs responsible for the unique taste of sodium-containing compounds in rats? Although these animals cannot verbalize their perceptions of taste quality, there are behavioral tests that can address this question. For example, rats are normally able to distinguish NaCl from KCl in an operant paradigm, in which they receive a reward for choosing correctly between the two solutions. However, when amiloride is added to NaCl, they are not able to distinguish it from KCl (Spector *et al.*, 1996). Work has also been done using generalization of a conditioned taste aversion (CTA) to NaCl. In this paradigm, rats are given access to a solution (the conditioned stimulus, or CS) and then made sick, so that in the future they avoid the CS and other solutions that taste similar to it. When the CS is NaCl mixed in amiloride, rats generalize their CTA to non-salty chloride-containing molecules, such as HCl or KCl (Hill *et al.*, 1990). Presumably, this is because amiloride has blocked the salty component of NaCl's taste, and what remains is a small sour or bitter component that is caused by ENaC-independent mechanisms.

Amiloride and related compounds do not fully eliminate the neural response to NaCl in rats, suggesting that other mechanisms are involved in sodium transduction. One proposal is that sodium also passes through the tight junctions between adjacent taste receptor cells and then through ENaCs found on the cells' basolateral membranes (Fig. 4.1; DeSimone *et al.*, 1993). These basolateral channels would be inaccessible to blockers such as amiloride and therefore could be responsible for the portion of the NaCl response that remains after amiloride application. This mechanism also explains why NaCl evokes a larger response than do other sodium compounds. In order for the sodium ion to cross the tight junctions, it needs to be accompanied by its anion to preserve electroneutrality; chloride's small size allows it to penetrate through the tight junctions easily, whereas larger anions cross with less facility. Alternatively, NaCl may differ from other sodium compounds because of chloride entry into taste receptor cells (Formaker and Hill, 1988).

Although there is some controversy, amiloride appears to be ineffective at blocking salty taste in humans. The first study to investigate this issue reported that amiloride blocked the overall intensity of NaCl (Schiffman et al., 1983). However, in later experiments in which subjects rated NaCl on a full range of taste qualities, amiloride did not affect the perceived saltiness of NaCl, but it did cause a reduction in the small amount of sourness generated by tasting NaCl (Ossebaard and Smith, 1995; Ossebaard et al., 1997). This does not eliminate the possibility that ENaCs are responsible for salty taste in humans. ENaCs found in human taste tissue may simply differ from ENaCs found in rat fungiform papillae. For example, human ENaCs may be similar to those found in rat vallate papillae in the posterior of the tongue; these channels are amiloride-insensitive, apparently due to high expression of the alpha subunit by itself, without co-expression of the beta and gamma subunits (Kretz et al., 1999; Lin et al., 1999).

A blocker of salty taste does exist for humans. The compound chlorhexidine reduces the perceived saltiness of NaCl and other compounds, but does not affect ratings of sweetness or sourness (Breslin and Tharpe, 2001; Frank et al., 2001). The mechanism of action of chlorhexidine is not known, though, and it may act through a non-specific process, rather than by blocking a channel. Thus, the primary transduction events for salty taste transduction in humans remain to be determined, and the answer may depend on understanding how chlorhexidine exerts its effects, or it may require the discovery of new compounds that influence salty taste and whose actions are well understood.

4.2.2   Non-specific mechanisms

Although the predominant taste quality of NaCl is salty, there is also a non-salty component to its taste in humans and animals. It is therefore likely that sodium activates multiple transduction pathways that vary in terms of their selectivity for sodium ions.

In humans the lowest concentrations of NaCl that can be detected taste mildly sweet, not salty (Bartoshuk et al., 1978). It is not known, though, whether this occurs because of the involvement of the proposed sweetener-binding receptor, a dimer of the proteins T1R2 and T1R3 (Nelson et al., 2001). Even at higher concentrations, NaCl elicits a small amount of perceived sourness. The mechanism for this is not known, but it appears to be amiloride-sensitive in some people (see Section 4.2.1). This is not necessarily due to the involvement of ENaCs, though, as amiloride may also affect other kinds of channels (Kellenberger and Schild, 2002).

In rats and at least one mouse strain, sodium passes through a non-selective cation channel found in taste receptor cells, which also allows passage of calcium and potassium ions (Fig. 4.1; Lyall et al., 2004). This channel, which is a variant of the VR-1 receptor that binds the irritant capsaicin, is responsible for the amiloride-insensitive portion of the NaCl response in the chorda tympani. It is not known what effect this mechanism has on the perception of NaCl in these animals, though it may be especially important at low concentrations of NaCl. Ten and 30 mM NaCl can be detected by rats, but they appear to have a taste quality that is different from that associated with NaCl concentrations greater than 100 mM (Yamamoto et al., 1994). In the chorda tympani nerve the fibers that respond most specifically to NaCl respond only at concentrations of 30 mM or greater, whereas fibers that are sensitive to other cations in addition to sodium respond to 1 mM NaCl (Frank et al., 1983; Ninomiya and Funakoshi, 1988). These results are consistent with the VR-1 channel being the primary transduction mechanism for NaCl at low concentrations, but this has not been tested directly yet.

4.3 Brain areas activated by salty taste

4.3.1 Taste quality

After NaCl causes the depolarization of taste receptor cells, the gustatory signal is sent to the brain via branches of the facial, glossopharyngeal, and vagus nerves (Smith and Frank, 1993). In rats, all of these nerves are responsive to NaCl, but they are not equally important for the perception of saltiness. The chorda tympani branch of the facial nerve appears to be more crucial than other nerves, such as the glossopharyngeal. The ability of rats to discriminate NaCl from KCl is impaired by transection of the former, but not of the latter (Spector and Grill, 1992). Most likely, this is due to the fact that fibers that respond selectively to salty stimuli are found only in the chorda tympani. Studies indicate that these cells, usually called N-fibers, receive signals from taste receptor cells that express amiloride-blockable ENaCs, whereas less selective fibers (H-fibers) in this nerve do not (see Fig. 4.1; Ninomiya and Funakoshi, 1988). The glosso-pharyngeal, although activated by NaCl application, does not contain individual fibers that respond selectively to salty stimuli or that are amiloride-sensitive (Frank, 1991; Formaker and Hill, 1991). Thus, when the chorda tympani nerve is transected, oral NaCl application is able to influence gustatory neurons in the brain, but it appears to do so in a way that is indistinguishable from the activation caused by other minerals such as KCl.

This distinction between amiloride-sensitive and -insensitive neurons is maintained in central gustatory areas, such as the nucleus of the solitary tract (NST). In rats and hamsters, NST neurons can be categorized based on their profiles of responding to representative sweet, salty, sour, and bitter stimuli. Neurons that are responsive primarily to salty stimuli (usually referred to as N-cells), or those that are both salt- and sugar-sensitive (S-cells), show reduced responses to NaCl following amiloride application. Cells that are highly respon-sive to acids or bitter stimuli, in addition to NaCl (H-cells), are not affected by amiloride application (Scott and Giza, 1990; Smith et al., 1996). Similar results were reported for salt- and acid-sensitive cells in the next gustatory relay in the rat, the parabrachial nucleus, though there were too few sugar-sensitive cells to make strong conclusions (Lundy and Norgren, 2001). One synapse away, in the gustatory thalamus, results are found that are comparable to those for responses in the NST. Responses to NaCl in thalamic N- or S-cells are reduced by amiloride, whereas responses in H-cells are not affected (Verhagen *et al.*, 2005). Cells in gustatory thalamus then project to the cortical gustatory region in agranular insular cortex, but the effects of amiloride on cortical taste responses have not been examined.

What these data suggest is that when a rat tastes NaCl, there is a gustatory signal that is carried through segregated pathways across at least four synapses. One of these pathways is relatively specific for salty compounds, though some of these cells are also affected by the taste of sweeteners, and its initial step involves passage of sodium through amiloride-sensitive ENaCs. The other pathway can also be activated by sour and bitter compounds, and the initial transduction event likely involves passage of sodium through a VR-1-related channel or through ENaCs expressing only the alpha subunit.

Figure 4.1 shows a simplified picture of some of the important events that take place when a rat tastes sodium. In receptor cells found in fungiform papillae at the front of the tongue, sodium passes through amiloride-sensitive epithelial sodium channels (ENaCs) that are found on the apical end of the cells and that exclude most other cations (top, solid arrow). Sodium may also pass through the tight junctions between adjacent taste receptor cells and then through ENaCs found on the basolateral surface of the cells (top, dashed arrow). These cells then send taste information to the brain via N-fibers of the chorda tympani nerve, which synapse on cells in the nucleus of the solitary tract (NST) that are selectively responsive to NaCl (N-cells) or that respond to NaCl and sweeteners (S-cells). Sodium can also enter fungiform taste receptor cells via a VR-1-related channel that allows passage of other cations, such as potassium (top, dotted arrow); these cells send a signal to the brain via H-fibers of the chorda tympani nerve, which synapse on cells in the NST that respond to acids and bitter stimuli, in addition to NaCl (H-cells). Sodium probably enters receptor cells in vallate papillae on the back of the tongue via non-specific channels that also allow passage of other cations. These channels are likely to be the alpha subunit of ENaC, expressed without the beta or gamma subunits. Sodium may also pass through VR-1 channels in vallate papillae, but this has not been tested directly. These receptor cells send information to H-cells in the NST via the glossopharyngeal nerve. From the NST cells project to the parabrachial nucleus and then to gustatory thalamus, and segregation is maintained between amiloride-sensitive (N-cells and S-cells) and insensitive (H-cells) pathways.

Activation of the sodium-specific pathway is necessary for the perception of salty taste, but it is not sufficient for it. That is, a rat may be unable to perceive salty taste if this pathway is blocked, but an increase in the firing rate of these cells cannot be the only event that generates perceptions of saltiness. This is due, in part, to the fact that even 'sodium-specific cells' often respond to non-salty

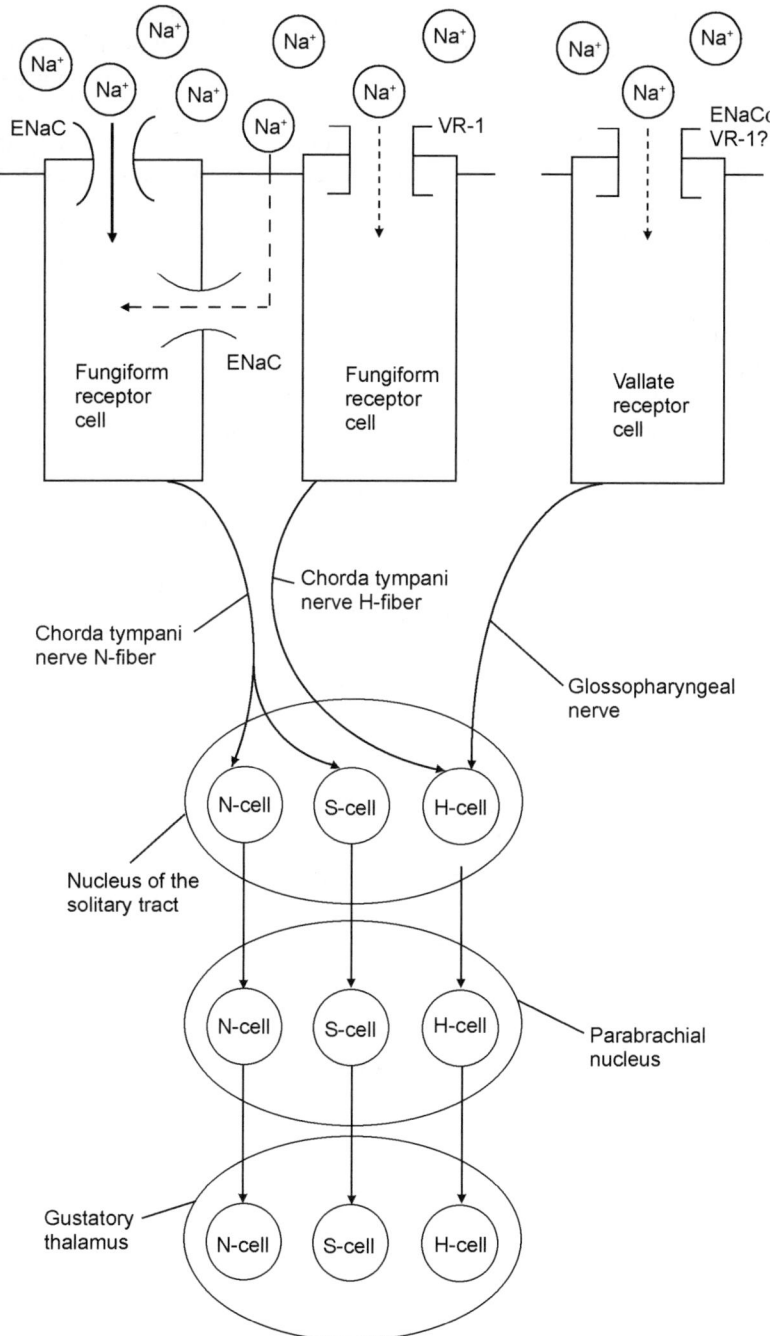

Fig. 4.1 An overview of some of the important events that take place when a rat tastes sodium. (See text for process description.)

compounds if they are presented at a high enough concentration (Ganchrow and Erickson, 1970; Scott and Perrotto, 1980). There is a larger issue here, which is the neural basis for taste quality in general. Some investigators have emphasized the role of particular subsets of cells in generating particular taste qualities (Zhang et al., 2003). However, the fact that gustatory neurons in the brain rarely respond exclusively to one class of compounds (e.g., sweeteners) suggests that the perception of saltiness and other qualities depends on patterns of responding across many cells that vary in their response sensitivities (Pfaffmann, 1941; Smith et al., 2000).

These patterns can be studied quantitatively by measuring the responses of many individual gustatory neurons to different chemicals, which allows for the creation of an 'across-neuron profile' for each chemical; correlations between the across-neuron profiles of pairs of chemicals are then calculated. When this is done for responses in the NST and other taste-responsive brain areas of the rat, compounds that contain sodium or lithium tend to have across-neuron profiles that are highly correlated with each other (e.g., $r > +0.90$ for 0.1 M NaCl vs 0.1 M LiCl). When the profiles of salty and non-salty compounds are compared, the correlations tend to be lower (e.g., $r \leq +0.60$ for 0.1 M NaCl vs 1 M sucrose; Doetsch and Erickson, 1970; Scott and Perrotto, 1980). The across-neuron profile of responding to NaCl is changed by amiloride application, so that it becomes similar to that of sour and bitter stimuli, including KCl (Scott and Giza, 1990; Smith et al., 1996). This provides further evidence that passage of sodium through ENaCs is responsible for the unique taste of sodium in rats, and the non-salty aspects of its taste occurs through separate transduction mechanisms that are also activated by other ions such as potassium.

In human subjects the taste of NaCl causes increased blood flow to primary gustatory cortex and other brain areas, according to work with functional brain imaging (Small et al., 1999). However, these methods lack the resolution to answer whether there are sodium-specific and non-specific cells that respond to the taste of NaCl within these areas. Recordings from the gustatory cortex of macaques suggest that the principles involved are similar to those for taste-responsive areas of the rat brain. That is, there is a subset of neurons that responds relatively specifically to salty stimuli, but NaCl also generates a unique across-neuron profile of responding that differs from the profiles generated by non-salty compounds. Moreover, the similarity of profiles for different salts in the macaque gustatory cortex closely matches the similarity of saltiness ratings in human subjects (Scott et al., 1994). It is not known whether there are segregated pathways in the brain for the salty and non-salty components of NaCl's taste in primates, as there are in rats.

Under normal conditions, of course, it is rare for pure NaCl to be consumed, and sodium would be encountered as part of a mixture. The interactions involved in taste mixtures can be complex in humans and rodents, and the resulting taste quality does not necessarily represent a perfect blending of the mixture's individual components (Erickson, 1982; Frank et al., 2003). One of the more striking examples of this is the ability of sodium to block bitter taste

(Breslin and Beauchamp, 1995). The exact mechanism that underlies this effect is not known, although gustatory nerve recordings in hamsters suggest that it must arise in the periphery of the taste system (Formaker and Frank, 1996).

4.3.2 Palatability

Salty compounds generate not only perceptions related to taste quality and intensity, but also hedonic perceptions that range from highly palatable to highly unpalatable, and different brain areas are involved in the two kinds of evalua-tions. For example, cells in the nucleus of the solitary tract and parabrachial nucleus of rats respond with similar across-neuron profiles of activity to NaCl concentrations ranging from 0.01 to 0.60 M (Ganchrow and Erickson, 1970; Scott and Perrotto, 1980), but this range includes both preferred and avoided concentrations in two-bottle tests (Weiner and Stellar, 1951; Flynn et al., 1993). Thus, while the activity of these neurons may help to determine sodium's unique taste, it does not correlate well with a rat's preference for or avoidance of NaCl. Other brain areas found in the ventral forebrain, such as the lateral hypothalamus and central nucleus of the amygdala, appear to be involved in the motivation to consume NaCl, but do not influence its taste quality (Kawamura et al., 1970; Galaverna et al., 1991, 1993; Seeley et al., 1993).

The palatability of NaCl, of course, can vary at different times. Under normal conditions, rats prefer NaCl over water in two-bottles tests and consume it in excess of need, but it is not as highly appetitive as sugars. For example, rats forgo mild brain stimulation reward delivered to the ventral forebrain in order to gain access to sucrose, but not to NaCl (Conover and Shizgal, 1994; Conover et al., 1994).

However, when rats are sodium-deprived they show a compensatory appetite for NaCl, and this behavior is expressed rapidly and does not depend on learning about the consequences of their intake (Epstein and Stellar, 1955; Nachman, 1962). Instead, the appetite seems to be driven by the taste of NaCl, which acquires an enhanced ability to activate neurons involved in hedonics and reward. Sodium-deprived rats will forgo brain stimulation reward delivered to the ventral forebrain in favor of drinking NaCl, suggesting that salt intake and reward activate a shared neural pathway in these animals (Conover et al., 1994). The ingestion of NaCl in sodium-deprived rats may also stimulate release of dopamine in the nucleus accumbens, an event associated with other rewarding behaviors (Roitman et al., 1997, 1999). There are also changes in measures of taste reactivity when rats are sodium-deprived. These measures include a number of stereotyped facial and body movements that are associated with the taste of either palatable or unpalatable substances and are labeled 'ingestive' or 'aversive', respectively. When sodium-deprived rats receive oral infusions of 0.5 M NaCl, they show more ingestive reactions and less aversive reactions than do non-deprived rats, suggesting that sodium status affects the perceived palatability of NaCl (Berridge et al., 1984).

There are also changes in responses to NaCl in gustatory areas when rats are sodium-depleted or when sodium appetite is created through other means, but it is unclear whether these changes stimulate increased NaCl intake. The most commonly observed effect of sodium appetite on neural responses in the chorda tympani, nucleus of the solitary tract, and parabrachial nucleus has been a decrease in sensitivity to NaCl (Contreras, 1977; Contreras et al., 1984; Jacobs et al., 1988; Bernstein and Taylor, 1992; Vogt and Hill, 1993; Nakamura and Norgren, 1995; Shimura et al., 1997; McCaughey and Scott, 2000). In most cases this effect has been restricted to the subset of cells that responded most selectively to NaCl but not to non-salty stimuli (Contreras, 1977; Contreras and Frank, 1979; Contreras et al., 1984; Jacobs et al., 1988; McCaughey and Scott, 2000). In addition, these results have been observed most consistently for hypertonic concentrations of NaCl (Bernstein and Taylor, 1992; Contreras et al., 1984; McCaughey and Scott, 2000). This decrease in the neural response to concentrated NaCl may lead to less avoidance of it due to a decrease in perceived intensity (Contreras, 1977), but this explanation alone is not adequate to explain the avidity with which deprived rats consume NaCl. An alternate explanation was suggested by one study, in which a decrease in the responses of salt-responsive NST neurons was accompanied by an *increase* in the response to NaCl by sucrose-responsive neurons (Jacobs et al., 1988). Thus, NaCl may have acquired an enhanced ability to activate brain areas that normally are activated strongly only by the taste of sugars, leading to an increase in its palatability.

The brain areas involved in determining the palatability of NaCl in humans remain to be determined. They will likely include areas such as orbitofrontal cortex that are thought to be involved in hedonic responses to other taste qualities (O'Doherty et al., 2001; Zald et al., 2002).

The pleasantness of NaCl depends not only on the sense of taste, but also on other sensory attributes such as texture. Unlike rats (Bare, 1949; Beauchamp and Bertino, 1985), most people prefer salted food over pure NaCl solution, although some liquid foods, such as soups, are made more palatable by the addition of substantial amounts of NaCl (Beauchamp et al., 1983). The exact mechanisms for this are not known, although the anatomical substrate is likely to involve areas such as the orbitofrontal cortex, which contains cells that are sensitive to both taste and touch in non-human primates (Kadohisa et al., 2005).

4.4   Physiological factors that influence salt intake

4.4.1   Need-based sodium appetite

Given the physiological importance of sodium, it is essential for animals to be able to be able to replenish this mineral when they are depleted of it. In fact, an extremely wide variety of species is known to respond to sodium deficiency by developing an appetite for NaCl (Denton, 1961, 1982; Denton et al., 1993; Rowland et al., 2004).

A number of factors influence sodium appetite in rats, but the most important ones are the central action of angiotensin II and aldosterone, molecules that also act to cause sodium conservation in the kidney and other organs (Epstein, 1991a). Angiotensin II is thought to act on brain areas near the third ventricle, such as the organum vasculosum of the lamina terminalis (Fitts and Masson, 1990), median preoptic area (Fitts *et al.*, 1990), and magnocellular areas of the paraventricular nucleus of the thalamus (Zardetto-Smith *et al.*, 1993). Aldosterone, on the other hand, exerts its actions on areas such as the medial amygdala (Nitabach *et al.*, 1989). Moreover, these two molecules act synergistically, so that if low concentrations of angiotensin II are infused into the third ventricle of rats pretreated with aldosterone or its precursor, deoxycorticosterone acetate, the animals drink large amounts of NaCl, even though they are not in need of it (Fluharty and Epstein, 1983; Sakai, 1986). This behavior is thought to arise because the treatment mimics the normal physiological response to sodium deprivation (Epstein, 1991a).

There are also other mechanisms that promote a need-based sodium appetite in rats. For example, it is thought that central oxytocin normally acts to inhibit sodium intake, and this inhibition is released by sodium deprivation (Stricker and Verbalis, 1996). In addition, recent evidence suggests that angiotensin II may exert some of its effects on NaCl intake by being converted to angiotensin III (Wilson *et al.*, 2005).

Do these same mechanisms operate in humans? Although sodium deficiency is not encountered normally, there have been attempts to induce it experimentally in human volunteers, and there is evidence that a need-based appetite can be stimulated. In one study, the amount of salt added to food increased after a sodium depletion regimen (DeWardener and Herxheimer, 1957). In another, subjects increased their preferred concentration of sodium added to soup or crackers during a period of sodium depletion, and they also rated highly salted foods as being more desirable (Beauchamp *et al.*, 1990). Increases in the amount of salt added to soup have also been observed following exercise, which stimulates sodium loss through perspiration (Leshem *et al.*, 1999), and as a result of hemodialysis, which lowers plasma sodium levels (Leshem and Rudoy, 1997).

There is also evidence that disorders of angiotensin II or aldosterone can influence the desire for salt. There is a widely cited report of a child developing a pronounced salt craving as a result of Addison's disease, which involves reduced secretion of aldosterone (Wilkins and Richter, 1940). Patients with congenital adrenal hyperplasia have also demonstrated an increased preference for salt that was related to negative sodium balance (Kochli *et al.*, 2005). Although these conditions may not have relevance for the general population, they help to confirm that similar mechanisms can control sodium intake in both humans and other animals.

4.4.2   Need-free salt intake

Some species, such as humans and rats, will consume more sodium than is necessary to maintain a positive sodium balance. It has been suggested that this overconsumption is useful, in that it helps animals to learn about potential sources of sodium that can be accessed during later periods of need (Epstein, 1991b). However, it is also possible that it is a side effect of other physiological consequences.

One potential factor is a prior history of severe sodium depletion. Adult rats that have been sodium-depleted two or more times will maintain an elevation in daily NaCl intake for months after they have restored their sodium balance to normal (Falk and Lipton, 1967; Sakai et al., 1989; Frankmann et al., 1991). Sakai and colleagues (1987) showed that the actions of angiotensin II and aldosterone are sufficient for this continued need-free appetite. That is, simply administering the hormones by themselves, with no accompanying sodium depletion, will have the same effect. Thus, it appears that exposure to high levels of sodium-regulating hormones can have long-term, possibly permanent, effects on salt intake by exerting organizational effects on the brain.

People rarely encounter extended periods of restricted access to NaCl, but there are a number of events, including vomiting and diarrhea, that are associated with acute sodium loss. Crystal and Bernstein investigated whether the severity of a mother's morning sickness affects her child's subsequent preference for salt. They found that infants whose mothers experienced moderate to severe vomiting during pregnancy had higher preferences for 0.2 M NaCl than did infants whose mothers experienced little or no vomiting (Crystal and Bernstein, 1998). In another study, they found higher preferences for salty snack foods in college students whose mothers had experienced more severe morning sickness (Crystal and Bernstein, 1995). Similar results were reported in a paper by Leshem (1998), in which higher preferences for salt in adolescence were related to greater frequency of morning sickness by the subjects' mothers, as well as to more frequent childhood vomiting or diarrhea by the subjects themselves. There is also a report of elevated salt preferences in adolescents who had been fed a formula deficient in chloride as infants (Stein et al., 1996). In this case, the subjects would not have experienced sodium-deficiency, but the lack of chloride should have stimulated a hormonal response similar to that associated with sodium depletion.

The amount of salt added to foods also depends on the recent history of salt use. In addition to studies in which severe sodium depletion was induced in human subjects, there have been experiments involving smaller reductions in dietary salt levels that were unlikely to induce a need for sodium. The evidence suggests that if the low-sodium diets are maintained for 8–12 weeks, there is a reduction in the most preferred concentration of NaCl, and formerly preferred concentrations are found to be unpleasant (Bertino et al., 1982; Beauchamp et al., 1983; Blais et al., 1986). Different effects may be found, though, with a shorter period of sodium restriction, including a reduction in the perceived intensity of salt in food and an increase in perceived pleasantness (Beauchamp et al., 1983).

Although most people overconsume sodium relative to recommended daily amounts, the opposite is true for some other minerals, including calcium and potassium. In both cases, many people consume them in smaller amounts than the recommended daily allowance (Miller and Weaver, 1994; Geleijnse et al., 2004). When rats are deprived of either of these other minerals, they show elevated intake of NaCl relative to non-deprived animals (Adam and Dawborn, 1972; Tordoff et al., 1990). The exact causes of these need-free sodium appetites are not known. In the case of calcium-deprivation, though, there is some evidence that an increase in the palatability of sodium is involved (McCaughey et al., 2005), and the appetite does not appear to be stimulated by the actions of the angiotensin II or aldosterone (Tordoff et al., 1993), as is the case for need-based sodium appetite. This raises the issue of whether underconsumption of calcium contributes to an increased liking for salt in humans, as Tordoff has suggested (1996b). More data will be needed to answer this question, but working to increase people's intake of calcium and potassium is a worthy goal in its own right due to the long-term health problems that are associated with chronic underconsumption of these minerals (McCarron, 1985; Barger-Lux and Heaney, 1994; Geleijnse et al., 2004).

4.5 Implications for food product development

Currently, no effective sodium-free salt substitutes are available. This contrasts markedly with the widespread use of artificial sweeteners. Most likely, this is due to the different natures of the transduction mechanisms for saltiness and sweetness. Sweeteners are thought to bind to a heterodimer of two proteins with long extracellular domains (Nelson et al., 2001), which provides for many potential interaction sites for artificial compounds. Saltiness, in contrast, is thought to involve passage of sodium ions through a narrowly-gated ion channel, and it will likely be impossible to find another substance, apart from toxic lithium ions, that mimics sodium's passage through this channel.

A more promising approach will be to find substances that enhance the perceived saltiness of a given amount of NaCl. Several compounds have been proposed to accomplish this, but none are widely used. The antifibrillary drug bretylium tosylate enhances the passage of sodium through ENaCs (Ilani et al., 1982) and increases the perceived intensity of NaCl in human subjects (Schiffman et al., 1986). However, its cardiac effects preclude widespread use, and it appears to act in ways that are complementary to amiloride and therefore may not increase perceived saltiness. The detergent cetylpyridiunium chloride has also been shown to enhance perceived saltiness in humans when used at the proper concentration, but may exert the opposite effect at other concentrations (DeSimone et al., 2001). There is also a report of choline chloride enhancing the saltiness of NaCl in rats (Locke and Fielding, 1994), but the effects on human subjects are unknown.

The amount of sodium in foods could also be reduced by limiting its use to that needed for generating saltiness and eliminating other uses, such as serving as a preservative or bitter-blocker. The need for sodium as a bitter-blocking agent will decrease as researchers gain a better understanding of bitter taste. This will include the development of other bitter-blockers to take the place of sodium, but may also involve a better exploitation of individual differences in bitter sensitivities. It is becoming increasingly clear that different individuals may vary widely in their sensitivities to a particular bitter compound, possibly due to genetic variation in the taste receptor proteins responsible for binding the compound (Delwiche *et al.*, 2001; Mennella *et al.*, 2005). Naturally, it may be difficult to sell a particular food if even a small percentage of the population finds it to be intensely bitter, but the possibility remains that foods could be sold that are mildly bitter to some people but not bitter to others.

4.6   Future trends

The causes of high salt intake may vary between individuals. Although genetic differences may be involved, environmental factors may also play a large role. For example, people who have experienced a single episode of severe sodium depletion in the past may have permanently elevated salt intake (see Section 4.4.2), and they may require methods that are different from those employed for reducing the salt intake of the general population.

In terms of modifying salty taste, the best opportunity will be to alter the initial sequence of events for sodium transduction, which still need to be determined for humans. If a method can be found to enhance the saltiness of NaCl, then people may achieve their preferred levels of saltiness without using harmful amounts of sodium. In theory, it could also be possible to affect salty taste perception by affecting taste-responsive neurons in the brain, though naturally this would be more technically challenging than affecting events at the periphery of the taste system. This approach will be easier if there are segregated pathways for salty taste in humans, as there are in rats (see Section 4.3.1), and if the neurons that respond most selectively to sodium are found to differ from other cell types in their expression of particular proteins.

The involvement of angiotensin II and aldosterone in controlling the motivation to consume salt (see Section 4.4.1) provides specific targets for manipulating the desire for salt in humans, even if those molecules are not the cause of the high salt intake. The risk of serious side effects with such an approach is considerable, though, because angiotensin II and aldosterone are involved in a wide array of systems in the brain and elsewhere, including regulation of blood pressure and heart rate (Phillips, 1987).

It will also be important for researchers to investigate not only need-based salt appetite in animal models, but also expand their research into need-free influences on salt intake that may be more relevant to human consumption. The mechanisms that are responsible for the sodium appetite of animals deprived of calcium or potassium still need to be determined. Even if those mechanisms are not responsible for the high salt intake of most people, this research may help to increase intakes of these other minerals to healthier levels.

Regardless of which methods are used to reduce overconsumption of salt, the process will be aided by the fact that an extended period on a low-salt diet changes perception of NaCl, so that formerly preferred concentrations taste more intense and unpleasant (see Section 4.4.2). This means that low-sodium diets may be difficult and require active effort for only a few months. If people are made more aware of this, it should help with compliance on these diets.

4.7   Sources of further information and advice

Derek Denton's book *The Hunger for Salt* remains the definitive comprehensive work on salt intake. A detailed review of the neuropeptides and brain areas involved in controlling salt appetite can be found in Jay Schulkin's book *Sodium Hunger: The Search for a Salty Taste*.

There has been extensive research into the role of the angiotensin II and aldosterone in controlling need-based sodium appetite. Only a brief description of it was presented here, since this mechanism is unlikely to be the cause of high salt intake in most people, and also because this work has been summarized thoroughly elsewhere (Epstein, 1991a; Johnson and Thunhorst, 1997; Fitzsimons, 1998; Daniels and Fluharty, 2004; Sakai, 2004).

There are two reviews that discuss the initial transduction events for sodium in more detail than is presented here (DeSimone *et al.*, 1993; Lindemann, 1997). For a broader summary of gustatory transduction mechanisms, see papers by Margolskee (1995) and Lindemann (1996). Issues related to neural responses to the taste of NaCl are covered more thoroughly in a previously published article (McCaughey and Scott, 1998). This topic is also part of a larger question of how neural activity generates perceptions of taste quality ('gustatory coding'). This is a complex topic that was covered only briefly here, but more complete considerations of it have been published (Woolston and Erickson, 1979; DiLorenzo, 2000; Frank, 2000; Scott and Giza, 2000; Smith *et al.*, 2000).

A summary of investigations into salt intake and preference in humans can be found in reviews by Mattes (1997) and Beauchamp and colleagues (1983), and Michael Tordoff (1996b) has reviewed the role of low calcium intake in causing sodium appetite.

4.8   References

ABRAHAM S F, BLAINE E H, DENTON D A, MCKINLEY M J, NELSON J F, SHULKES A, WEISINGER R S and WHIPP G T (1975), 'Phylogenetic emergence of salt taste and appetite', in Denton D A and Coghlan J P, *Olfaction and Taste V*, New York, Academic Press, 253–260.

ADAM W R and DAWBORN J K (1972), 'Effect of potassium depletion on mineral appetite in the rat', *J Comp Physiol Psychol*, 78(1), 51–58.

BARE J K (1949), 'The specific hunger for sodium chloride in normal and adrenalectomized white rats', *J Comp Physiol Psychol*, 42, 242–253.

BARGER-LUX M J and HEANEY R P (1994), 'The role of calcium intake in preventing bone fragility, hypertension, and certain cancers', *J Nutr*, 124, 1406S–1411S.

BARTOSHUK L M, MURPHY C and CLEVELAND C T (1978), 'Sweet taste of dilute NaCl: psychophysical evidence for a sweet stimulus', *Physiol Behav*, 21(4), 609–613.

BEARD T C, BLIZZARD L, O'BRIEN D J and DWYER T (1997), 'Association between blood pressure and dietary factors in the dietary and nutritional survey of British adults', *Arch Intern Med*, 157(2), 234–238.

BEAUCHAMP G K and BERTINO M (1985), 'Rats (*Rattus norvegicus*) do not prefer salted solid food', *J Comp Psychol*, 99, 240–247.

BEAUCHAMP G K, BERTINO M and ENGELMAN K (1983), 'Modification of salt taste', *Ann Intern Med*, 98, 763–769.

BEAUCHAMP G K, BERTINO M, BURKE D and ENGELMAN K (1990), 'Experimental sodium depletion and salt taste in normal human volunteers', *Am J Clin Nutr*, 51(5), 881–889.

BEIDLER L M (1975), 'Phylogenetic emergence of sour taste', In Denton D A and Coghlan J P, *Olfaction and Taste V*, New York, Academic Press, 71–76.

BERNSTEIN I L and TAYLOR E M (1992), 'Amiloride sensitivity of the chorda tympani response to sodium chloride in sodium-depleted Wistar rats', *Behav Neurosci*, 106, 722–725.

BERRIDGE K C, FLYNN F W, SCHULKIN J and GRILL H J (1984), 'Sodium depletion enhances salt palatability in rats', *Behav Neurosci*, 98, 652–660.

BERTINO M, BEAUCHAMP G K and ENGELMAN K (1982), 'Long-term reduction in dietary sodium alters the taste of salt', *Am J Clin Nutr*, 36(6), 1134–1144.

BLAIS C A, PANGBORN R M, BORHANI N O, FERRELL M F, PRINEAS R J and LAING B (1986), 'Effect of dietary sodium restriction on taste responses to sodium chloride: a longitudinal study', *Am J Clin Nutr*, 44(2), 232–243.

BRAND J G, TEETER J H and SILVER W L (1985), 'Inhibition by amiloride of chorda tympani responses evoked by monovalent salts', *Brain Res*, 334, 207–214.

BRESLIN P A and BEAUCHAMP G K (1995), 'Suppression of bitterness by sodium: variation among bitter taste stimuli', *Chem Senses*, 20(6), 609–623.

BRESLIN P A and THARP C D (2001), 'Reduction of saltiness and bitterness after a chlorhexidine rinse', *Chem Senses*, 26(2), 105–116.

CONOVER K L and SHIZGAL P (1994), 'Competition and summation between rewarding effects of sucrose and lateral hypothalamic stimulation in the rat', *Behav Neurosci*, 108, 537–548.

CONOVER K L, WOODSIDE B and SHIZGAL P (1994), 'Effects of sodium depletion on competition and summation between rewarding effects of salt and lateral hypothalamic stimulation in the rat', *Behav Neurosci*, 108, 549–558.

CONTRERAS R J (1977), 'Changes in gustatory nerve discharges with sodium deficiency: a single unit analysis', *Brain Res*, 121, 373–378.

CONTRERAS R J and FRANK M (1979), 'Sodium deprivation alters neural responses to gustatory stimuli', *J Gen Physiol*, 73, 569–594.

CONTRERAS R J, KOSTEN T and FRANK M E (1984), 'Activity in salt taste fibers: peripheral mechanism for mediating changes in salt intake', *Chem Senses*, 8, 275–288.

CRYSTAL S R and BERNSTEIN I L (1995), 'Morning sickness: impact on offspring salt preference', *Appetite*, 25(3), 231–240.

CRYSTAL S R and BERNSTEIN I L (1998), 'Infant salt preference and mothers' morning sickness' *Appetite*, 30(3), 297–307.

DANIELS D and FLUHARTY S J (2004), 'Salt appetite: a neurohormonal viewpoint', *Physiol Behav*, 81(2), 319–337.

DELWICHE, J (1996), 'Are there "basic" tastes?', *Trend Food Sci Tech*, 7, 411–415.

DELWICHE J F, BULETIC Z and BRESLIN P A (2001), 'Covariation in individuals' sensitivities to bitter compounds: evidence supporting multiple receptor/transduction mechanisms', *Percept Psychophys*, 63(5), 761–776.

DENTON, D A (1961), 'The selective appetite for Na$^+$ shown by Na$^+$ deficient sheep', *J Physiol*, 157, 97–116.

DENTON D A (1982), *The Hunger for Salt*, New York, Springer-Verlag.

DENTON D A, EICHBERG J W, SHADE R and WEISINGER R S (1993), 'Sodium appetite in response to sodium deficiency in baboons', *Am J Physiol*, 264, R539–R543.

DESIMONE, J A and FERRELL F (1985), 'Analysis of amiloride inhibition of chorda tympani taste response of rat to NaCl', *Am J Physiol*, 249, R52–R61.

DESIMONE J A, YE Q and HECK GL (1993), 'Ion pathways in the taste bud and their significance for transduction', *Ciba Found Symp*, 179, 218–229.

DESIMONE J A, LYALL V, HECK G L, PHAN T H, ALAM R I, FELDMAN G M and BUCH R M (2001), 'A novel pharmacological probe links the amiloride-insensitive NaCl, KCl, and NH$_4$Cl chorda tympani taste responses', *J Neurophysiol*, 86(5), 2638–2641.

DEWARDENER H E and HERXHEIMER A (1957), 'The effect of a high water intake on salt consumption, taste thresholds and salivary secretion in man', *J Physiol*, 139, 53–63.

DILORENZO P M (2000), 'The neural code for taste in the brain stem: response profiles', *Physiol Behav*, 69, 87–96.

DOETSCH G S and ERICKSON R P (1970), 'Synaptic processing of taste-quality information in the nucleus tractus solitarius of the rat', *J Neurophysiol*, 33(4), 490–507.

EPSTEIN A N (1991a), 'Neurohormonal control of salt intake in the rat', *Brain Res Bull*, 27, 315–320.

EPSTEIN A N (1991b), 'Thirst and salt intake: a personal view and some suggestions', In Ramsey D J and Booth D A, *Thirst – Physiological and Psychological Aspects*, Berlin, Springer-Verlag, 481–501.

EPSTEIN A N and STELLAR E (1955), 'The control of salt preference in the adrenalectomized rat', *J Comp Physiol Psychol*, 46, 167–172.

ERICKSON R P (1982), 'Studies on the perception of taste: do primaries exist?', *Physiol Behav*, 28(1), 57–62.

FALK J L and LIPTON J M (1967), 'Temporal factors in the genesis of NaCl appetite by intraperitoneal dialysis', *J Comp Physiol Psychol*, 63(2), 247–251.

FITTS D A and MASSON D B (1990), 'Preoptic angiotensin and salt appetite', *Behav Neurosci*, 104, 643–650.

FITTS D A, TJEPKES D S and BRIGHT R O (1990), 'Salt appetite and lesions of the ventral part of the ventral median preoptic nucleus', *Behav Neurosci*, 104, 818–827.

FITZSIMONS J T (1998), 'Angiotensin, thirst, and sodium appetite', *Physiol Rev*, 78(3), 583–686.

FLUHARTY S J and EPSTEIN A N (1983), 'Sodium appetite elicited by intracerebroventricular infusion of angiotensin II in the rat: II. Synergistic interaction with systemic mineralocorticoids', *Behav Neurosci*, 97, 746–758.

FLYNN F W, SCHULKIN J and HAVENS M (1993), 'Sex differences in salt preference and taste reactivity in rats', *Brain Res Bull*, 32, 91–95.

FORMAKER B K and FRANK M E (1996), 'Responses of the hamster chorda tympani nerve to binary component taste stimuli: evidence for peripheral gustatory mixture interactions', *Brain Res*, 727, 79–90.

FORMAKER B K and HILL D L (1988), 'An analysis of residual NaCl taste response after amiloride', *Am J Physiol*, 255, R1002–R1007.

FORMAKER B K and HILL D L (1991), 'Lack of amiloride sensitivity in SHR and WKY glossopharyngeal taste responses to NaCl', *Physiol Behav*, 50(4), 765–769.

FRANK M E (1991), 'Taste-responsive neurons of the glossopharyngeal nerve of the rat', *J Neurophysiol*, 65, 1452–1463.

FRANK M E (2000), 'Neuron types, receptors, behavior, and taste quality', *Physiol Behav*, 69, 53–62.

FRANK M E, CONTRERAS R J and HETTINGER T P (1983), 'Nerve fibers sensitive to ionic taste stimuli in chorda tympani of the rat', *J Neurophysiol*, 50, 941–960.

FRANK M E, GENT J F and HETTINGER T P (2001), 'Effects of chlorhexidine on human taste perception', *Physiol Behav*, 74, 85–99.

FRANK M E, FORMAKER B K and HETTINGER T P (2003), 'Taste responses to mixtures: analytic processing of quality', *Behav Neurosci*, 117(2), 228–235.

FRANKMANN S P, ULRICH P and EPSTEIN A N (1991), 'Transient and lasting effects of reproductive episodes on NaCl intake of the female rat', *Appetite*, 16(3), 193–204.

GALAVERNA O G, DELUCA L A JR, SCHULKIN J, YAO S-Z and EPSTEIN A N (1991), 'Deficits in NaCl ingestion after damage to the central nucleus of the amygdala in the rat', *Brain Res Bull*, 28, 89–98.

GALAVERNA O G, SEELEY R J, BERRIDGE K C, GRILL H J, EPSTEIN A N and SCHULKIN J (1993), 'Lesions of the central nucleus of the amygdala: I. Effects on taste reactivity, taste aversion learning, and sodium appetite', *Behav Brain Res*, 59, 11–17.

GANCHROW J R and ERICKSON R P (1970), 'Neural correlates of gustatory intensity and quality', *J Neurophysiol*, 33, 768–783.

GARCIA J and HANKINS W G (1975), 'The evolution of bitter and the acquisition of toxiphobia', In Denton D A and Coghlan J P, *Olfaction and Taste V*, New York, Academic Press, 39–46.

GELEIJNSE J M, KOK F J and GROBBEE D E (2004), 'Impact of dietary and lifestyle factors on the prevalence of hypertension in Western populations', *Eur J Public Health*, 14(3), 235–239.

HECK G I, MIERSON S and DESIMONE J A (1984), 'Salt taste transduction occurs through an amiloride-sensitive sodium transport pathway', *Science*, 223, 403–405.

HILL D L, FORMAKER B K and WHITE K S (1990), 'Perceptual characteristics of the amiloride-suppressed sodium chloride taste response in the rat', *Behav Neurosci*, 104, 734–741.

ILANI A, LICHTSTEIN D and BACANER M B (1982), 'Bretylium opens mucosal amiloride-sensitive sodium channels', *Biochim Biophys Acta*, 693, 503–506.

JACOBS K M, MARK G P and SCOTT T R (1988), 'Taste responses in the nucleus tractus solitarius of sodium-deprived rats', *J Physiol*, 406, 393–410.

JOHNSON A K and THUNHORST R L (1997), 'The neuroendocrinology of thirst and salt appetite: visceral sensory signals and mechanisms of central integration', *Front Neuroendocrinol*, 18(3), 292–353.

KADOHISA M, ROLLS E T and VERHAGEN J V (2005), 'Neuronal representations of stimuli in the mouth: the primate insular taste cortex, orbitofrontal cortex and amygdala', *Chem Senses*, 30(5), 401–419.

KAWAMURA Y, KASAHARA Y and FUNAKOSHI M (1970), 'A possible brain mechanism for rejection behavior to strong salt solution', *Physiol Behav*, 5, 67–74.

KELLENBERGER S and SCHILD L (2002), 'Epithelial sodium channel/degenerin family of ion channels: a variety of functions for a shared structure', *Physiol Rev*, 82(3), 735–767.

KHAW K T, BINGHAM S, WELCH A, LUBEN R, O'BRIEN E, WAREHAM N and DAY N (2004), 'Blood pressure and urinary sodium in men and women: the Norfolk Cohort of the

European Prospective Investigation into Cancer (EPIC-Norfolk)', *Am J Clin Nutr*, 80(5), 1397–1403.

KOCHLI A, TENENBAUM-RAKOVER Y and LESHEM M (2005), 'Increased salt appetite in patients with congenital adrenal hyperplasia 21-hydroxylase deficiency', *Am J Physiol*, 288(6), R1673–1681.

KRETZ O, BARBRY P, BOCK R and LINDEMANN B (1999), 'Differential expression of RNA and protein of the three pore-forming subunits of the amiloride-sensitive epithelial sodium channel in taste buds of the rat', *J Histochem Cytochem*, 47(1), 51–64.

LESHEM M (1998), 'Salt preference in adolescence is predicted by common prenatal and infantile mineralofluid loss', *Physiol Behav*, 63(4), 699–704.

LESHEM M and RUDOY J (1997), 'Hemodialysis increases the preference for salt in soup', *Physiol Behav*, 61(1), 65–69.

LESHEM M, ABUTBUL A and EILON R (1999), 'Exercise increases the preference for salt in humans', *Appetite*, 32(2), 251–260.

LIN W, FINGER T E, ROSSIER B C and KINNAMON S C (1999), 'Epithelial Na+ channel subunits in rat taste cells: localization and regulation by aldosterone', *J Comp Neurol*, 405(3), 406–420.

LINDEMANN B (1996), 'Taste reception', *Physiol Rev*, 76(3), 718–766.

LINDEMANN B (1997), 'Sodium taste', *Curr Opin Nephrol Hypertens*, 6(5), 425–429.

LOCKE K W and FIELDING S (1994), 'Enhancement of salt intake by choline chloride', *Physiol Behav*, 55(6), 1039–1046.

LORIA C M, OBARZANEK E and ERNST N D (2001), 'Choose and prepare foods with less salt: dietary advice for all Americans', *J Nutr*, 131, 536S–551S.

LUNDY R F JR and CONTRERAS R J (1997), 'Temperature and amiloride alter taste nerve responses to Na^+, K^+, and NH_4^+ salts in rats', *Brain Res*, 744(2), 309–317.

LUNDY R F JR and NORGREN R (2001), 'Pontine gustatory activity is altered by electrical stimulation in the central nucleus of the amygdala', *J Neurophysiol*, 85(2), 770–783.

LYALL V, HECK G L, VINNIKOVA A K, GHOSH S, PHAN T H, ALAM R I, RUSSELL O F, MALIK S A, BIGBEE J W and DESIMONE J A (2004), 'The mammalian amiloride-insensitive non-specific salt taste receptor is a vanilloid receptor-1 variant', *J Physiol*, 558, 147–159.

MARGOLSKEE R F (1995), 'Receptor mechanisms in gustation', in Doty R L, *Handbook of Olfaction and Gustation*, Marcel Dekker, New York, 575–595.

MATTES R D (1997), 'The taste for salt in humans', *Am J Clin Nutr*, 65, 692S–697S.

McCARRON D A (1985), 'Is calcium more important than sodium in the pathogenesis of essential hypertension?', *Hypertens*, 7, 607–627.

McCAUGHEY S A and SCOTT T R (1998), 'The taste of sodium', *Neurosci Biobehav Rev*, 22, 663–676.

McCAUGHEY S A and SCOTT T R (2000), 'Rapid induction of sodium appetite modifies taste-evoked activity in the rat nucleus of the solitary tract', *Am J Physiol*, 279, R1121–R1131.

McCAUGHEY S A, FORESTELL C A and TORDOFF M G (2005), 'Calcium deprivation increases the palatability of calcium solutions in rats', *Physiol Behav*, 84, 335–342.

MENNELLA J A, PEPINO M Y and REED D R (2005), 'Genetic and environmental determinants of bitter perception and sweet preferences', *Pediatrics*, 115(2), e216–222.

MILLER G D and WEAVER C M (1994), 'Required versus optimal intakes: a look at calcium', *J Nutr*, 124, 1404S–1405S.

MORI M, KAWADA T, ONO T and TORII K (1991), 'Taste preference and protein nutrition and

L-amino acid homeostasis in male Sprague-Dawley rats', *Physiol Behav*, 49(5), 987–995.

NACHMAN M (1962), 'Taste preferences for sodium salts by adrenalectomized rats', *J Comp Physiol Psychol*, 55, 1124–1129.

NAKAMURA K and NORGREN R (1995), 'Sodium-deficient diet reduces gustatory activity in the nucleus of the solitary tract of behaving rats', *Am J Physiol*, 269, R647–R661.

NELSON G, HOON M A, CHANDRASHEKAR J, ZHANG Y, RYBA N J and ZUKER C S (2001), 'Mammalian sweet taste receptors.' *Cell*, 106(3), 381–390.

NINOMIYA Y and FUNAKOSHI M (1988), 'Amiloride inhibition of responses of rat single chorda tympani fibers to chemical and electrical stimulations', *Brain Res*, 451, 319–325.

NITABACH M N, SCHULKIN J and EPSTEIN A N (1989), 'The medial amygdala is part of a mineralocorticoid-sensitive circuit controlling NaCl intake in the rat', *Behav Brain Res*, 35, 127–134.

O'DOHERTY J, ROLLS E T, FRANCIS S, BOWTELL R and McGLONE F (2001), 'Representation of pleasant and aversive taste in the human brain', *J Neurophysiol*, 85(3), 1315–1321.

OSSEBAARD C A and SMITH D V (1995), 'Effect of amiloride on the taste of NaCl, Na-gluconate and KCl in humans: implications for Na+ receptor mechanisms', *Chem Senses*, 20(1), 37–46.

OSSEBAARD C A, POLET I A and SMITH D V (1997), 'Amiloride effects on taste quality: comparison of single and multiple response category procedures', *Chem Senses*, 22(3), 267–275.

PFAFFMANN C (1941), 'Gustatory afferent impulses', *J Cell Comp Physiol*, 17, 243–358.

PFAFFMANN C (1975), 'Phylogenetic origins of sweet sensitivity', in Denton D A and Coghlan J P, *Olfaction and Taste V*, New York, Academic Press, 3–10.

PHILLIPS M I (1987), 'Functions of angiotensin in the central nervous system', *Ann Rev Physiol*, 49, 413–435.

ROITMAN M F, SCHAFE G E, THIELE T E and BERNSTEIN I L (1997), 'Dopamine and sodium appetite: antagonists suppress sham drinking of NaCl solutions in the rat', *Behav Neurosci*, 111(3), 606–611.

ROITMAN M F, PATTERSON T A, SAKAI R R, BERNSTEIN I L and FIGLEWICZ D P (1999), 'Sodium depletion and aldosterone decrease dopamine transporter activity in nucleus accumbens but not striatum', *Am J Physiol*, 276, R1339–1345.

ROWLAND N E, FARNBAUCH L J and CREWS E C (2004), 'Sodium deficiency and salt appetite in ICR: CD1 mice', *Physiol Behav*, 80(5), 629–635.

SACKS F M, SVETKEY L P, VOLLMER W M, APPEL L J, BRAY G A, HARSHA D, OBARZANEK E, CONLIN P R, MILLER E R 3RD, SIMONS-MORTON D G, KARANJA N, LIN P H, DASH-SODIUM COLLABORATIVE RESEARCH GROUP (2001), 'Effects on blood pressure of reduced dietary sodium and the Dietary Approaches to Stop Hypertension (DASH) diet', *N Engl J Med*, 344(1), 3–10.

SAKAI R R (1986), 'The hormones of renal sodium conservation act synergistically to arouse a sodium appetite in the rat', in de Caro G, Epstein A N and Massi M, *The Physiology of Thirst and Sodium Appetite*, New York, Plenum Press, 425–430.

SAKAI R R (2004), 'The future of research on thirst and salt appetite', *Appetite*, 42(1), 15–19.

SAKAI R R, FINE W B, EPSTEIN A N and FRANKMANN S P (1987), 'Salt appetite is enhanced by one prior episode of sodium depletion in the rat', *Behav Neurosci*, 101(5), 724–731.

SAKAI R R, FRANKMANN S P, FINE W B and EPSTEIN A N (1989), 'Prior episodes of sodium depletion increase the need-free sodium intake of the rat', *Behav Neurosci*, 103(1), 186–192.

SCHIFFMAN S S, MCELROY A E and ERICKSON R P (1980), 'The range of taste quality of sodium salts', *Physiol Behav*, 24(2), 217–224.

SCHIFFMAN S S, LOCKHEAD E and MAES F W (1983), 'Amiloride reduces the taste intensity of Na+ and Li+ salts and sweeteners', *Proc Nat Acad Sci USA*, 80, 6136–6140.

SCHIFFMAN S S, SIMON S A, GILL J M and BEEKER T G (1986), 'Bretylium tosylate enhances salt taste', *Physiol Behav*, 36, 1129–1137.

SCHULKIN J (1991), *Sodium Hunger: The Search for a Salty Taste*, Cambridge, Cambridge University Press.

SCOTT T R and GIZA B K (1990), 'Coding channels in the taste system of the rat', *Science*, 249, 1585–1587.

SCOTT T R and GIZA B K (2000), 'Issues of gustatory neural coding: where they stand today', *Physiol Behav*, 69, 65–76.

SCOTT T R and PERROTTO R S (1980), 'Intensity coding in pontine taste area: gustatory information is processed similarly throughout rat's brainstem', *J Neurophysiol*, 44, 739–750.

SCOTT T R, PLATA-SALAMAN C R and SMITH-SWINTOSKY V L (1994), 'Gustatory neural coding in the monkey cortex: the quality of saltiness', *J Neurophysiol*, 71(5), 1692–1701.

SEELEY R J, GALAVERNA O G, SCHULKIN J, EPSTEIN A N and GRILL H J (1993), 'Lesions of the central nucleus of the amygdala: II. Effects on intraoral NaCl intake', *Behav Brain Res*, 59, 19–25.

SHIMURA T, KOMORI M and YAMAMOTO T (1997), 'Acute sodium deficiency reduces gustatory responsiveness to NaCl in the parabrachial nucleus of rats', *Neurosci Lett*, 236, 33–36.

SMALL D M, ZALD D H, JONES-GOTMAN M, ZATORRE R J, PARDO J V, FREY S and PETRIDES M (1999), 'Human cortical gustatory areas: a review of functional neuroimaging data'. *Neuroreport*, 10(1), 7–14.

SMITH D V and FRANK M E (1993), 'Sensory coding by peripheral taste fibers', in Simon S A and Roper S D, *Mechanisms of Taste Transduction*, Boca Raton, CRC Press, 295–338.

SMITH D V, LIU H and VOGT M B (1996), 'Responses of gustatory cells in the nucleus of the solitary tract of hamster after NaCl or amiloride adaptation', *J Neurophysiol*, 76, 47–58.

SMITH D V, ST. JOHN S J and BOUGHTER J D (2000), 'Neuronal cell types and taste quality coding', *Physiol Behav*, 69, 77–85.

SPECTOR A C and GRILL H J (1992), 'Salt taste discrimination after bilateral section of the chorda tympani or glossopharyngeal nerves', *Am J Physiol*, 263, R169–R176.

SPECTOR A C, GUAGLIARDO N A and ST. JOHN S J (1996), 'Amiloride disrupts NaCl versus KCl discrimination performance: implications for salt taste coding in rats', *J Neurosci*, 16, 8115–8122.

STEIN L J, COWART B J, EPSTEIN A N, PILOT L J, LASKIN C R and BEAUCHAMP G K (1996), 'Increased liking for salty foods in adolescents exposed during infancy to a chloride-deficient feeding formula', *Appetite*, 27(1), 65–77.

STRICKER E M and VERBALIS J G (1996), 'Central inhibition of salt appetite by oxytocin in rats', *Regul Pept*, 66, 83–85.

TORDOFF M G (1996a), 'Some basic psychophysics of calcium salt solutions', *Chem Senses*, 21, 417–424.

TORDOFF M G (1996b), 'The importance of calcium in the control of salt intake', *Neurosci Biobehav Rev*, 20(1), 89–99.

TORDOFF M G, ULRICH P M and SCHULKIN J (1990), 'Calcium deprivation increases salt intake', *Am J Physiol*, 259, R411–419.

TORDOFF MG, HUGHES R L and PILCHAK D M (1993), 'Independence of salt intake from the hormones regulating calcium homeostasis', *Am J Physiol*, 264, R500–512.

VAN DER KLAAUW N J and SMITH D V (1995), 'Taste quality profiles for fifteen organic and inorganic salts', *Physiol Behav*, 58(2), 295–306.

VERHAGEN J V, GIZA B K and SCOTT T R (2005), 'Effect of amiloride on gustatory responses in the ventroposteromedial nucleus of the thalamus in rats', *J Neurophysiol*, 93(1), 157–166.

VOGT M B and HILL D L (1993), 'Enduring alterations in neurophysiological taste responses after early dietary sodium deprivation', *J Neurophysiol*, 69, 832–841.

WEINER I H and STELLAR E (1951), 'Salt preference of the rat determined by a single-stimulus method', *J Comp Physiol Psychol*, 44, 394–401.

WILKINS L and RICHTER C P (1940), 'A great craving for salt by a child with corticoadrenal insufficiency', *JAMA*, 114, 866–868.

WILSON W L, ROQUES B P, LLORENS-CORTES C, SPETH R C, HARDING J W and WRIGHT J W (2005), 'Roles of brain angiotensins II and III in thirst and sodium appetite', *Brain Res*, 1060, 108–117.

WOOLSTON D C and ERICKSON R P (1979), 'Concept of neuron types in gustation in the rat', *J Neurophysiol*, 42(5), 1390–1409.

YAMAMOTO T, SHIMURA T, SAKO N, YASOSHIMA Y and SAKAI N (1994), 'Some critical factors involved in formation of conditioned taste aversion to sodium chloride in rats', *Chem Senses*, 19, 209–217.

ZALD D H, HAGEN M C and PARDO J V (2002), 'Neural correlates of tasting concentrated quinine and sugar solutions', *J Neurophysiol*, 87(2), 1068–1075.

ZARDETTO-SMITH A M, THUNHORST R L, CICHA M Z and JOHNSON A K (1993), 'Afferent signaling and forebrain mechanisms in the behavioral control of extracellular fluid volume', *Ann NY Acad Sci*, 689, 161–176.

ZHANG Y, HOON M A, CHANDRASHEKAR J, MUELLER K L, COOK B, WU D, ZUKER C S and RYBA N J (2003), 'Coding of sweet, bitter, and umami tastes: different receptor cells sharing similar signaling pathways', *Cell*, 112(3), 293–301.

5

Dietary salt and the consumer: reported consumption and awareness of associated health risks

J. Purdy and G. Armstrong, University of Ulster, UK

5.1   Introduction

High intakes of dietary salt have long been associated with high blood pressure and an increased risk of heart disease and stroke. Since 1994, the evidence of an association between dietary salt intake and blood pressure has increased (SACN, 2003). 'High blood pressure is one of the major modifiable causal factors in the development of cardiovascular disease' (FSAI, 2005). The relationship between high salt intake and high blood pressure is supported by many observational and experimental studies (Dahl and Love, 1954, 1957; INTERSALT, 1988; Elliot *et al.*, 1996). Although the link between high salt intake and high blood pressure is well established, what is less well known is the amount of salt consumed and the differences in salt intake according to demographic and social factors.

It is important to determine the patterns of salt intake, both discretionary intake and consumption of high salt foods being consumed frequently by the various consumer groups. Acquiring this data will enable health educators to target advice at the relevant consumer groups. It can also be used to lobby government and food manufacturers in relation to understanding how to reduce the salt content of processed foods frequently consumed by particular groups of consumers.

As part of a review and investigation of current trends in salt consumption and related consumer attitudes it is important to consider the factors which have led to high salt intakes. These are discussed in the following section.

5.2   Changing consumer trends

Changing consumer lifestyle patterns have had a significant impact on dietary intake in recent years, and especially in relation to salt consumption. An increased consumption of processed convenience foods, with high salt contents, has had serious implications on the nutritional status of consumers' diets and ultimately a negative impact on health. 'High intakes of processed foods, such as ready meals, pizzas, savoury snacks and bread have led to an excessive consumption of dietary salt' (Wheelock, 1998).

An awareness of the influences which motivate consumer food choice is important in developing a better understanding of consumer attitudes and behaviour along with the barriers faced by consumers which may discourage any desired changes in behaviour (Woolfe, 2000). Such an understanding is particularly relevant in light of reducing current high salt intakes and implementing improved dietary practices.

Increased consumption of processed foods has come about as a result of ever changing consumer lifestyle patterns. With longer working hours, greater numbers of working mothers, more time and money being spent on leisure activities, the demand for convenience food has greatly increased (Hitchman *et al.*, 2002; Mintel 2002). However, the demand for convenience has come as a cost both financially and nutritionally. The cost financially has been met with higher levels of disposable income. The cost nutritionally has resulted in an increased risk of hypertension, heart disease and stroke. Despite the poor nutritional content of many processed convenience foods, their demand is predicted to continue (MacGregor and de Wardener, 1998) as convenience is clearly still more important to consumers than nutrition (Stitt, 1998; Katz, 1999).

Consequently, there is a need to combine the concepts of convenience and health/nutrition and this presents a significant challenge to food product developers. In order to merge health and convenience, food product developers must take account of the key factors influencing consumer food choice and consider how these can be integrated with the changing lifestyle patterns and dietary needs of today's consumer. Research and development must form a central part of the food product development process in order to achieve consumer acceptability in terms of demands for convenience and sensory quality. Furthermore, 'an understanding of eating patterns, meal formats and the changing social nature of eating is the key not only to new product development and product acceptability but also to the adoption of healthful dietary advice, good nutrition and health promotion' (Marshall, 2000).

5.2.1   Food choice – key influences

Lifestyle changes are among a comprehensive range of factors which influence food choice. This section highlights the key factors which influence food choice in relation to increasing salt consumption.

Physiological factors

Bertino *et al.* (1982) support the view that the preferred level of salt in food is dependent on the amount of salt consumed. Consumers with a habitual high salt intake become less sensitive to salt and therefore need a higher salt concentration to obtain the same taste. Consequently, reducing salt in certain food items may be unacceptable to some consumers. However, 'a gradual reduction in dietary salt intake can increase consumers' sensitivity to salt and ultimately increase their preference for reduced salt products' (MacGregor, 2002). Therefore, a gradual reduction of salt in processed foods may be the most effective means of reducing salt intake.

Social factors

The immediate social context within which a meal is eaten can influence food choice and consumption (Shepherd, 1999) and for most consumers, this is the home and family environment.

The home environment is perhaps one of the most influential for children and will determine the eating patterns they adopt and implement throughout their own lives. In relation to food consumption, the influence of the family is becoming less powerful than it once was. Changing social trends have eroded the tradition of family mealtimes, leading to greater variation in the times at which food is consumed and an increase in snacking patterns (Reuters Business Insight, 2003).

One of the over-riding influences on consumption patterns is the size and make-up of the population. A decline in birth rate, an increase in the average age of the population and a decrease in household size (ONS, 2005) have contributed to changes in eating patterns and the subsequent demand for convenience foods. The UK has an ageing population, with an increase of 28% in the number of people aged 65 and over between 1971 and 2003. Projections suggest that the number of people aged 65 and over will exceed those aged under 16 from 2013 (ONS, 2005). 'The gradual ageing of the population and associated decreasing physical ability and dexterity with age may contribute to an inability or lack of motivation to cook' (Furey *et al.*, 2000), resulting in greater reliance on convenience foods.

The average household size has fallen from 2.9 persons before 1971, to 2.4 persons in 2003. More lone-parent families, smaller family sizes, and the increase in one-person households have driven this decrease (ONS, 2005). The decline in household size has contributed to some extent to the restructuring of family meals. Increasingly, different members of the family eat different things (Henchion, 2000). Furthermore, fewer families eat together and children are leading ever more complicated lives with an increase in the number of out of school activities which encourages a pattern of eating wherever and whenever is convenient (Bardsley, 2000; Davies, 2001). Changes in the lifestyle of the whole family have had an impact on the development and growth of convenience and semi-prepared foods. Ultimately this has led to an increase in salt consumption, due to the already high salt content of processed foods and frequency with which they are eaten.

Perhaps the single most important influence on food provisioning has been the changing role of women in the workplace (Marshall, 1995). The majority of women now work, whilst maintaining the bulk of responsibility for home and family life (Marshall, 1995; Davies, 2001). Consequently, there is less time for food preparation and family meals. In addition, the increased number of women at work creates additional income for many households (McHugh, *et al.*, 1993), enabling women to spend more money on convenience foods. This has resulted in a marked increase in the consumption of convenience foods, primarily as part or whole prepared ready meals (Bardsley, 2000; Mintel, 2002).

With less time for food preparation and diverse dietary preferences, it is inevitable that less time will be spent preparing a meal from raw ingredients. In many homes the main meal of the day is prepared in less than 30 minutes, as opposed to 2½ hours during the 1930s. In fact, there has been a decrease of 40 minutes in the time spent in home cooking, over the last decade (Harrision, 2001). This is partly due to the greater availability of labour saving devices, but also a reassessment of the opportunity cost of time. For example, some mothers place more value in talking to their children than preparing meals from basic ingredients.

Much of the increased consumption of convenience foods can be attributed to the de-skilling and de-domestication of consumers today. 'The decline in cooking skills can be attributed to the exclusion of cooking skills from the National Curriculum and optionalisation of home economics in England and Wales' (Furey *et al.*, 2000). Coupled with a lack of cooking facilities (Caraher *et al.*, 1999), many consumers are often restricted in their ability to prepare a home-cooked meal. Furey *et al.* (2000) reported that without cooking skills, control over what one eats is diminished and somewhat manipulated by the growing power of the retailer, irrespective of the high cost of convenience foods. Consequently, home-cooked meals are being substituted with convenience foods, such as ready meals with high salt content, which may have serious long-term implications for health status. This suggestion has been confirmed by a study in which Caraher and Lang (1999) reported that 'health is unequivocally linked to food skills and cooking'.

Cultural and ethnic influences
Due to increased travel and emigration, ethnic influences are now widely reflected in consumers' diets. This is most apparent from the variety of ethnic style ready meals currently available (Marshall, 1995). However, the growth and popularity of ethnic style ready meals (Henchion, 2000) is of considerable concern, given that these products are often heavily loaded with salt. It has been reported that 'Indian and Chinese foods are particularly problematic in relation to reducing salt' (Sainsbury's representative, cited by Whitworth, 2001). Therefore, food product developers must make a concerted effort to reduce the salt content of ready meals, especially ethnic style products.

Sensory factors
Of the many factors influencing food choice, sensory quality arguably has the greatest impact in determining consumer consumption of food. Booth and

Shepherd (1988) have noted that sensory influences are among the strongest and most immediate influences on an individual's acceptance of a food or beverage. Sensory quality is important in food choice for two reasons. Firstly, a product will either be accepted or rejected based on its appearance, aroma, flavour or texture. Secondly, sensory attributes can be seen as a key area in which food manufacturers can differentiate their products (Clark, 1998).

If food manufacturers can optimise the perceived sensory attributes of a product, it will help increase the perceived value of the product (Clark, 1998) which is important in the development and optimisation of the sensory quality and consumer acceptance of reduced salt foods. However, 'unless food products possess the sensory properties and quality attributes that the consumer wants, he/she will resist change, and the new food is unlikely to be purchased on a continuing basis' (Sloan, 2001). Therefore, food manufacturers must achieve satisfactory sensory quality in reduced salt foods, to ensure any long-term success and yet offer health benefits to consumers.

Economic factors
Food consumption is very much an economic activity reflected in both the price of food and by socio-economic differences, which determine the amount of money available to spend on food. As consumers' monetary resources and commitments vary, only a certain amount of money is available to spend on food (Williams, 1983). Furst et al. (1996) revealed that 'monetary considerations, consisting of price and the perceived worth of food to be bought, comprised a very salient value for many people, and indeed often dominated food choices'. Price has a strong influence on food choice and interacts with factors such as sensory quality in determining ultimate food purchase and consumption. Higher disposable incomes and a demand for more leisure time have resulted in an upsurge in consumer purchase of convenience style products.

As far as food consumption is concerned, people of higher socio-economic status are reported to consume a greater variety and range of foodstuffs, which are more likely to accord with whatever is thought to be nutritionally good (Mennell et al., 1992). Even more striking are the social class differences in food intake and indeed eating patterns among the less socio-economically privileged, where food intake is markedly further from the nutritional recommendations (Wardle, 1993). James et al. (1997) have reported that the diet of the lower socio-economic groups provides cheap energy from food such as meat products, full cream milk, fats, sugars, preserves, potatoes and cereals, but has little intake of vegetables, fruit and whole-wheat bread. Consequently consumers in these groups experience a greater incidence of premature and low birth weights, heart disease, stroke and some cancers in adults.

There is clear evidence to support the belief that income puts restraints on food choice as different social groups are forced to develop their own prefer-ences and tastes (National Consumer Council, 1992). As Leather (1992) suggests, 'better health will remain the prerogative of those who have access to and can afford a good diet'. In relation to salt intake, it is paramount that socio-economic differences in salt consumption are identified in order to implement effective salt reduction strategies.

Technological factors

Technological advances in food preservation techniques have contributed to a vast increase in the variety of foods available to consumers. Whilst canning, dehydration and freezing have been have been used for many years, others such as chilling, provide a new dimension in food preservation. Although the shelf-life of chilled produce is still short, consumers associate chilled foods with freshness and high sensory quality (Henchion, 2000; Mintel, 2000). Further-more, advancements in food production technology have facilitated the develop-ment of many complex and intricate food products, previously unavailable to consumers.

'Rapid development of technology in the food industry has contributed to an increased variety of foods for consumers' (Bogue *et al.*, 1999). With the benefit of advanced food production technology, and the necessary skills and expertise, manufacturers can produce complex food products such as ready meals. Whilst, the changes in technology have fuelled several generations of convenience foods with time-saving characteristics (Katz, 1999), they have, however, contributed to the increase in the consumption of fat, sugar and salt. In particular, large quantities of salt have been used in processed convenience foods to maintain shelf-life and enhance product flavour. In particular, salt has been used in chilled products for its preservation properties at any early stage in processing. In response to the urgent calls for a reduction in the salt content of processed food products, 'the food industry should attach high priority to research aimed at addressing technological, shelf-life, preservation and taste issues in relation to the reduction of the salt content of processed foods' (FSAI, 2005).

As the literature suggests, food choice is complex and there are many factors which have led to an increase in the consumption of processed convenience foods. Whilst these foods are known to have a high salt content, little is known about the frequency with which they are consumed by various consumer groups.

To date there has been a deficit of research in the area of salt consumption and related consumer awareness. This has been redressed by the findings of the study discussed in this chapter. A questionnaire was conducted among a representative sample of consumers in Northern Ireland to determine salt intake from discretionary and processed food sources.

5.3   Understanding the consumer and the salt issue

5.3.1   Assessing salt intake

Reducing salt intake of either an individual or a community presupposes knowledge of both the principal sources and total intake of salt. However, such information is very limited in Britain and other affluent countries (James *et al.*, 1987). One of the most common techniques used in assessing salt intake and determining salt consumption patterns is a consumer questionnaire. A range of questionnaires concerning dietary salt intake have been reported to provide a reliable estimation of salt intake (Hill *et al.*, 1980; Pietinen *et al.*, 1982; Shepherd *et al.*, 1985; Nakatsuka *et al.*, 1996). A questionnaire designed by Shepherd *et al.* (1985) appears to have best achieved this requirement. Despite the fact that urinary excretion remains the most precise method of measuring salt intake (Shepherd *et al.*, 1985), questionnaires have the distinct advantage of identifying the main sources of salt in the diet.

5.3.2   Use of questionnaires to determine salt intake

Shepherd *et al.* (1985) developed a questionnaire to determine discretionary salt usage and the frequency with which consumers were eating foods high in salt. On comparison of the questionnaire results with urinary excretion measurements, a highly positive correlation for total salt intake was reported (test-retest correlation of $r = 0.75$).

Salt intake measured from a food frequency questionnaire has been shown to correlate well with sodium excretion in urine (Ikeda *et al.*, 1988). Ikeda *et al.* (1988) concluded that food frequency scores and subsequent food intake patterns were useful in the estimation of salt intake. To a much lesser extent, studies by Hill *et al.* (1980) and Nagayama *et al.* (1986) reported accurate assessment of salt intake through questionnaire usage. A questionnaire (Hill *et al.*, 1980) on salty food and discretionary salt usage was found to be reliable, when analysed in a test-retest, but not shown to be predictive of sodium intake. A similar questionnaire (Nagayama *et al.*, 1986) revealed a significant correlation ($r = 0.389$, $p < 0.005$) only between sodium intake scores (as determined from the questionnaire) and urinary sodium values (24-hr urine sample). Differences between sodium intake and urinary sodium scores were observed for some foods, but not all. This would suggest that a urinary excretion measurement, used in conjunction with certain questionnaires, is a good predictor of sodium consumption in selected foods.

5.3.3   Categorisation of consumers based on salt intake

In addition to assessing salt intake, questionnaire data have also been used to categorise consumers as low, medium or high consumers of salt. Nakatsuka *et al.* (1996) categorised consumers based on their consumption of 'salty' foods. Pietinen *et al.* (1982), however, categorised consumers on the basis of the saltiness of the diet, achieved through salting habits, self-rating of salt use and frequency of use of seven salty foods. Categorisation of consumers in this way may provide an indication of salt consumption among specific types of consumers and can help determine the level of salt reduction required for these consumer groups. This data may be useful to health professionals and educators in the provision of dietary advice and by the food industry in the reformulation of products of low salt status for specific consumer segments.

5.3.4   Consumer awareness of salt and the effects on health status

Findings from a questionnaire by Tilston *et al.* (1993) suggested concern among consumers regarding the impact of salt on their health. Despite awareness of the potential harmful effects of salt on health, findings show a low use of salt alternatives and limited awareness of reducing salt consumption. Consumer knowledge of the salt content of specific processed foods was assessed revealing a poor level of knowledge. The same study reported high levels of awareness of low salt foods, but low levels of purchase.

The studies reviewed to date provide useful information in relation to salt consumption patterns and related issues between the early 1980s and 1990s. Despite the valuable contribution of this research in determining trends in salt consumption, the findings do not represent the dietary patterns of today's consumer and, ultimately, current salt intakes. With a lack of recent research, it was the objective of this research study to establish current trends in salt consumption among various consumer groups.

5.4   A consumer perspective

A consumer questionnaire examined trends in dietary salt intake from discretionary salt and processed food sources among consumers in Northern Ireland. The study investigated consumer knowledge of the salt content of processed foods, consumer awareness and purchase of low salt products and awareness of the health risks associated with habitual high salt intakes.

5.4.1   Determining consumer consumption of salt and awareness of associated health risks

The questionnaire was adapted from Shepherd *et al.* (1985) and Tilston *et al.* (1993) and included questions on the addition of salt at the table and during cooking, frequency of consumption of processed foods, awareness of the salt content of processed food items, awareness and consumption of low salt foods and consumer perception of the effect of salt intake on health. The questionnaire aimed to investigate the influence of demographic, economic, social and educational factors on salt consumption.

Consumer questionnaires ($n = 360$) were administered using the quota sampling technique in selected rural and urban/city areas of Northern Ireland. Sixty questionnaires were completed across six locations. Questionnaires were administered at the entrance to large retail food outlets, with the exception of Belfast, where questionnaires were administered in the main shopping areas within the city centre. All questionnaires were completed from Monday to Saturday, between 9am and 9pm.

Data were analysed using using SPSS for Windows, version 9.0 and statistical tests included the Kruskall-Wallis *H*-test and Mann Whitney *U*-test.

5.4.2   Consumer consumption of salt

Profile of consumer sample
The sample surveyed was closely representative of the population of Northern Ireland, based on socio-economic status as illustrated in Table 5.1.

The consumer sample comprised of 78.3% female and 21.7% male respondents. The greatest percentage of respondents (23.1%) were aged 40–49 years followed by those aged 50–59 years (21.7%) and 30–39 years (19.4%). Two-thirds of respondents (66.7%) were married or cohabiting, 19.2% single, 9% widowed and the remaining 4% separated or divorced. A total of 30.6% of respondents were educated to GCSE level, 22.5% of respondents had a further education/technical college qualification and 29.2% had a higher level qualification. These results are shown in more detail in Table 5.2. Despite the fact that time and resources limited the sampling as regards gender and age profile, educational and martial status, a satisfactory representation was obtained.

Discretionary salt consumption
Self-reported ratings of salt usage revealed that over half of respondents (50.6%) claimed to be light users of salt, 26.7% moderate users, 8% heavy users and 14.7% claimed never to use salt. Table 5.3 illustrates salt added at the table and during cooking. The greatest number of consumers claimed to add salt 'occasionally', when cooking (35.8%) and at the table (43.1%).

Data analysis revealed that male respondents reported a significantly higher usage of discretionary salt compared to females ($z = -2.7, p < 0.01$). Analysis of results showed that a significantly greater number of male respondents (51%) added salt at the table ($z = -2.5, p < 0.05$) 'frequently' or 'always' compared to 33% of female respondents. A similar trend was observed in relation to the addition of salt before or after tasting food, with 63% of male respondents adding salt before tasting their food ($z = -3.1, p < 0.01$), compared to 41% of female respondents. Male respondents were also more inclined to add a greater quantity of salt ('generous shake' and 'heavy covering') to their food than

Table 5.1   Comparison of sample population with actual population of Northern Ireland

Socio-economic groups	Percentage of Northern Ireland population*	Percentage of sample respondents	Total number of respondents
A	6.4	8.6	31
B	12.9	17.8	64
C1	12.1	13.3	48
C2	13.5	12.5	45
D	21.7	18.6	67
E	33.4	29.2	105
Total	100.0	100.0	360

* Northern Ireland Census Report (1991)

Table 5.2 Demographic profile of questionnaire respondents

Category		Number of respondents	Percentage of respondents
Gender	Male	78	21.7
	Female	282	78.3
Age	15–19	11	3.1
	20–29	44	12.2
	30–39	70	19.4
	40–49	83	23.1
	50–59	78	21.7
	60–69	50	13.9
	70+	24	6.7
Socio-economic	A	31	8.6
groupings	B	64	17.8
	C1	48	13.3
	C2	45	12.5
	D	67	18.6
	E	105	29.2
Marital status	Single	69	19.2
	Married/cohabiting	240	66.7
	Separated/divorced	16	4.4
	Widowed	33	9.2
Educational status	Primary	32	8.9
	GCSE	110	30.6
	A-Level	27	7.5
	Further	81	22.5
	Higher	105	29.2

female respondents ($z = -2.3$, $p < 0.05$) at the table. However, no significant differences were observed between males and females in relation to addition or quantity of salt used during cooking.

The trends in discretionary salt usage among men could be attributed to a lack of awareness of good dietary practice and little understanding of the effect of high salt consumption on health status. Furthermore, gender differences in sensitivity to salt may have contributed to an increased use of discretionary salt amongst male consumers. This suggestion is supported by the finding that male respondents were more inclined to add salt to their food before tasting, which would ultimately lead to a reliance on the addition of salt to flavour food.

Table 5.3 Discretionary salt usage (cooking and table salt usage)

Salt usage	Number (%) of consumers			
	Never	Occasionally	Frequently	Always
In cooking	110 (30.6)	129 (35.8)	28 (7.8)	89 (24.7)
At the table	73 (20.3)	155 (43.1)	55 (15.3)	77 (21.4)

Shepherd and Farleigh (1989) have commented that 'for many people adding salt to food may be more a habit than a conscious decision'. Consequently, habitual use of salt may increase tolerance to salt and in turn increase consumption.

Significant differences in the use of discretionary salt were observed across the various age groups, but no definitive patterns were evident, nor could be explained. Results revealed that consumers in the age groups 20–29 ($p < 0.05$) and 40–49 ($p < 0.01$) had a significantly higher intake of discretionary salt. This could be attributed to a high intake of processed foods, further increasing consumer tolerance to salt. It was expected that older consumers would have reported a higher intake of discretionary salt, having been accustomed to a lifetime of high salt consumption. However, given that elevated blood pressure is associated with increasing age, older consumers may have already reduced their salt intake in an effort to control blood pressure in conjunction with appropriate medication.

Statistically significant differences were noted in relation to self-reported salt usage among the various socio-economic groups ($\chi^2_5 = 13.1$, $p < 0.05$). A Mann-Whitney U-test revealed significant differences among socio-economic groups B and D ($z = 2.9$, $p < 0.05$), groups C1 and D ($z = 2.1$, $p < 0.05$) and groups D and E ($z = 3.5$, $p < 0.001$). The analyses revealed that consumers of lower socio-economic status (groups D and E) reported a heavier use of salt. However, no significant differences were observed between the socio-economic groups for addition of salt at the table and during cooking. This could be attributed to a greater reliance on processed foods and less preparation of meals using raw ingredients which might otherwise require the addition of salt either during cooking or at the table.

In view of the high level of self-reported discretionary salt usage, consider-able concern is raised about the need to reduce dietary salt intake. A previous study (Law, 1995) has already highlighted that individual consumers may consciously reduce discretionary salt intake, when prompted to do so by health educators. However, it is unrealistic to expect a whole population to reduce their salt intake by the recommended 30%. It therefore places greater onus on the government and the food industry to enforce and implement reductions in the salt content of processed foods, in an attempt to support the reduction of dietary salt intake across the population.

These findings give an indication of consumer trends in discretionary salt usage but provide no indication of actual salt intake. This is a potential limita-tion of the data, but the information still provides a useful indication of salt intake from discretionary sources according to demographic and socio-economic influences.

Addition of salt to single food commodities
Consumers were asked to identify if they added salt to a range of basic food commodities, either at the table or during cooking. Results presented in Table 5.4 show most consumers added salt to food at the table, compared to during

Table 5.4   Addition of discretionary salt to a range of basic food commodities

Food commodity	Addition of salt (percentage consumers)			
	During cooking	At the table	Do not add salt	Not eaten
Chips	0.6	63.6	25.3	10.3
Tomatoes	1.7	49.4	44.2	4.7
Potatoes	23.3	31.4	27.5	0.8
Vegetables	25.8	25.8	33.1	0.6
Pasta	21.1	7.8	50.6	18.6
Rice	25.3	8.3	53.9	11.4
Eggs	6.7	47.8	36.4	6.1
Soup	21.9	24.4	2.2	4.7
Red meat	12.2	37.8	40.0	6.9
Poultry	5.3	40.0	48.3	5.8
Fish	9.2	46.7	40.3	1.4

cooking. Salt was added at the table primarily to chips, tomatoes, eggs and fish. Addition of salt during cooking was most common amongst vegetables, rice and potatoes.

This part of the questionnaire revealed a strong tendency for consumers to add salt at the table and during cooking for the majority of food items listed and supports the already apparent high intake of salt among the consumer sample surveyed.

Frequency of consumption of processed foods
Consumers were asked to identify how frequently they consumed a range of popular processed foods, some of which were known to have a high salt content. Frequency scores showed that the majority of consumers (91.6%) consumed bread and breakfast cereals at least once a day. Ham (37.2%) and cheese (40.3%) were eaten by consumers 2–3 times per week whereas sausages, pizza, crisps, ready meals, soup, tinned vegetables, bacon and chips were consumed once a week by an average of 24% of consumers.

Data analysis revealed males had a significantly more frequent consumption of sausages ($z = -3.1$, $p < 0.01$), tinned vegetables ($z = -2.2$, $p < 0.05$), meat pies ($z = -3.1$, $p < 0.01$) and chips ($z = -3.4$, $p < 0.01$) compared with females. Gender differences could be attributed to the minimal preparation and cooking time required for these types of products. In a society where women still take primary responsibility for food preparation and cooking (Murcott, 2000) these findings may reflect an increased use and consumption of convenience type products amongst men. Furthermore, the products consumed most frequently, generally contain high quantities of fat and salt. This supports the earlier suggestion that men are less aware of good dietary practice and the importance of healthy eating messages to be targeted at this consumer group.

Consumers aged 15 to 29 had a significantly higher consumption of pizza and crisps compared to those aged 30 and over ($p < 0.001$), which are particularly popular among this age group as snack foods (Warwick, 1998). Consumption of chips was significantly more frequent among the youngest consumer group (15–19 years) ($\chi^2{}_6 = 15.7, p < 0.05$). It is worthy of note that frequent consumption of all of these products is likely to make a significant contribution to dietary salt intake (Tilston *et al.*, 1993).

Consumption of cheese and soup was more frequent among consumers over 40, whilst consumption of tinned vegetables and bacon was greatest amongst consumers aged over 60. These products would be considered to be more staple/traditional style products and therefore, as well as potential 'snack foods' for this particular age group, as compared to pizza and crisps for younger consumers. Again, frequent consumption could also contribute to an increased consumption of dietary salt intake.

Some interesting trends were observed in relation to educational status and frequency of consumption of processed foods. Data analysis revealed that consumers with educational qualifications above GCSE level consumed pizza and ready meals more frequently. It could be suggested that these consumers may have less time to prepare meals and therefore rely more heavily on convenience style products. Furthermore, consumers with professional occupations and higher disposable incomes can more easily afford convenience products such as ready meals and similar pre-prepared products (Bardsley, 2000).

In relation to consumers with a primary school education only, it was evident that these consumers had a higher consumption of sausages, bacon and tinned vegetables. In contrast to the more highly educated consumers, it could be assumed that these consumers have lower paid jobs and are less able to purchase more expensive convenience products such as ready meals. Furthermore, consumers with fewer educational qualifications may lack the cooking skills required to prepare fresh products and therefore select processed meat and vegetables, which require minimal preparation and cooking. Irrespective of their educational status however, it was apparent from this study that many consumers are frequently consuming highly salted processed foods such as pizza, ready meals, bacon and tinned vegetables.

Significant differences were noted amongst the socio-economic groups for frequency of consumption of sausages ($\chi^2{}_5 = 13.1, p < 0.05$), pizza ($\chi^2{}_5 = 24.6$, $p < 0.001$), crisps ($\chi^2{}_5 = 23.7, p < 0.001$), ready meals ($\chi^2{}_5 = 11.7, p < 0.05$) and chips ($\chi^2{}_5 = 23.0, p < 0.001$).

Sausages
Significance testing revealed that socio-economic group A had a significantly lower consumption of sausages, when compared to consumers in socio-economic groups C2, D and E. Socio-economic group B also had significantly less frequent consumption of sausages in comparison to groups C2 and D, whilst significant differences were also observed between groups C1 and D.

Sausages, in particular, are a low cost, convenience food, requiring minimal preparation in terms of cooking equipment and skill. Furey (2001) has suggested that 'lower income consumers generally have less education and therefore less knowledge and motivation to make healthy food choices'. In addition to this, dietary advice and health education may not be targeting or reaching all consumer segments. Consumers of lower socio-economic status in particular are less aware of the relationship between diet and health, as demonstrated within the sample surveyed. However, in selecting foods such as sausages, which are inexpensive and easily prepared, these consumers are increasing their salt intake due to the frequency with which they are consuming sausages (up to three times a week).

Crisps
In relation to consumption of crisps, significant differences were revealed between socio-economic groups A and C1 and C2. Socio-economic group E had a significantly lower consumption of crisps compared to groups B, C1, C2 and D. Given the frequency with which crisps and other highly salted processed snack foods are being consumed, consumers within the lower socio-economic groups are seriously increasing their salt intake and ultimately increasing the risk of elevated blood pressure (Stamler, 1997). Interestingly, consumers in socio-economic group E, which comprises mainly retired persons, had a significantly lower consumption of crisps. Processed snack foods, such as crisps, appear to be less popular among this age/socio-economic group. This could be because these products were not readily available when these consumers were children and young people and therefore they never acquired a preference for processed snack foods.

Pizza
The main differences observed in relation to consumption of pizza were noted amongst socio-economic group E. This group had significantly less frequent consumption of pizza when compared to groups B, C1, C2 and D. This could be explained by the fact that most respondents within group E were aged 60 or over and were unfamiliar with this product. Given the convenience, variety and popularity of pizza (Howitt, 1998), it is hardly surprising that such a broad range of consumers are eating this particular food as often as once per week.

Ready meals
The main differences in the consumption of ready meals were observed among socio-economic groups B and C1 compared with groups D and E. Socio-economic group B had a significantly more frequent consumption of ready meals when compared with D and E. Socio-economic group C1 also consumed ready meals significantly more frequently than group D.

Consumers (particularly consumers aged 25–34) within these socio-economic groups B and C1 have a higher disposable income and therefore can afford to purchase convenience foods such as ready meals (Reed et al., 2001). However, with the recognised high salt content of ready meals (Gibson et al., 2000) and the frequency with which they are consumed (Reed *et al.*, 2001), consumers in socio-economic groups B and C1 are further increasing their salt intake and consequently the long-term risk of hypertension.

Chips
Socio-economic group A had a significantly less frequent consumption of chips, compared to groups B, C1, C2, D and E. Significant differences were also observed between socio-economic group B and C1 and C2; C2 and D; and groups D and E. In each of these cases, groups C1, C2 and D had a more frequent consumption of chips.

The frequency with which chips are eaten is of considerable concern, given that almost two thirds of consumers reported to add salt to chips at the table, thus further increasing the salt intake of consumers of lower socio-economic status in particular.

In light of the frequency with which consumers from almost all socio-economic groups are consuming highly salted processed foods, there is sufficient evidence to support calls for a reduction in the salt content of existing processed foods. The evidence further warrants the need for ongoing product development to enhance the sensory quality and consumer acceptance of convenience food products, whilst significantly reducing the salt content.

5.4.3   Consumer awareness of salt intake and associated health risks
Awareness of salt content of processed foods
When consumers were asked to rate the salt content of popular processed food items (see Table 5.5) responses showed a considerable degree of accuracy. Respondents correctly rated ready meals, soup and sausages as having a high salt content. Bread and pizza, however, were incorrectly rated low and medium, instead of medium and very high respectively.

Female respondents were more aware of the salt content of processed foods than men, with significant differences in relation to the awareness of the salt content of ready meals, soup, pizza and sausages. This could be attributed to the fact that most food shopping is completed by women (Murcott, 2000) who may be more likely to be aware of product labelling, nutritional information and good dietary practices.

Table 5.5   Salt content of commonly eaten processed food products

Food item	Salt content (g/100g)	Salt content (g/serving)	Rating
Bread	1.3	0.5 (slice)	Medium
Ready meals	0.8	2.4 (300 g)	High
Sausages	3.0	3.0 (100 g)	High
Soup (canned)	1.0	3.0 (300 g)	High
Pizza	3.2	5.8 (180 g)	Very high

Consumers aged 40–59 were more aware of the salt content of prepared soups, than younger consumers. This could be largely due to the frequent consumption of these products among consumers in this age group.

Consumers with A-Level and third level educational qualifications were more likely to correctly rate ready meals as having a high to very high salt content ($\chi^2{}_6 = 20.6$, $p < 0.001$), compared to those with GCSE or fewer educational qualifications. This could be due to the fact that consumers with better educational qualifications were more inclined to consume ready meals and were therefore more aware of the salt content of these products from the sensory experience or by reading product labels.

Socio-economic status was demonstrated as having a significant effect on the awareness of the salt content of ready meals ($\chi^2{}_5 = 17.8$, $p < 0.01$). Significant differences were evident between socio-economic groups A and B compared with groups C2, D and E, and between groups C1 and C2, in their awareness of the salt content of ready meals. This finding could be considered characteristic of consumers with higher incomes and more hectic lifestyles, requiring the convenience of products such as ready meals and through usage and educational status, have an improved awareness of their nutritional content.

Results from this study demonstrated a low level of awareness of the salt content of popular processed foods, particularly bread and pizza. Thirty-eight per cent of consumers from this study scored the products correctly, which is comparable with a study by Tilston et al. (1993), who reported an overall poor level of knowledge as to the quantity of salt contained in processed foods. This relative lack of understanding may partially explain why the general UK salt consumption is so high. Consumers do not appear to realise which foods are high in salt and which foods are not (Tilston et al., 1993).

It is therefore essential that consumers are able to recognise highly salted processed foods and low salt alternatives where available. One way in which this can be achieved is by clearer salt/sodium labelling. There have been repeated calls for mandatory labelling of salt in food, with recent calls from the Chief Executive of FSAI (O'Brien, 2005). Whilst there has been some ambiguity surrounding the labelling of salt/sodium, however, there is evidence to suggest that retailers across Europe are increasingly aware of the need for clearer labelling and are beginning to become more pro-active in addressing this issue (Narhinen et al., 1998).

There is an urgent need for food manufacturers and retailers to make a concerted effort in relation to clearer labelling of salt to enable consumers to make a more informed choice, which may contribute towards an improved diet. If consumer knowledge and awareness of the salt content of processed foods was improved, consumers may be more reluctant to consume such highly salted foods frequently.

Awareness and purchase of low salt foods
Results of the study showed that over half (53.6%) of respondents had noticed low or reduced salt foods in the supermarket, but only 19.4% had actually

purchased these products. Of those consumers who bought low salt foods, the most frequently purchased were bacon, canned vegetables, margarine/spreads, bread and butter. Other foods included crisps, ready meals, soup and sauces.

Of the 70 consumers who purchased low salt foods, 49% indicated that they only bought them occasionally. However, a number of consumers did indicate that they bought low salt foods 'frequently' (25.5%) or 'always' (25.5%). A greater number of female consumers and those with A-level and higher level qualifications reported to purchase low salt products compared to consumers with primary school and GCSE level qualifications. However, no significant differences were observed in terms of purchasing patterns.

The main reason given for the purchase of low salt foods was a combination of 'better for you' (25.5%) and 'health conscious' (21.6%). However, there was evidence to suggest that consumers were not satisfied with the current range of low salt produce available, calling for more reduced salt products and better labelling of existing products.

Limited consumer awareness among less well educated consumers and those of lower socio-economic status could be attributed to a general lack of awareness and interest in diet and health-related issues. It was evident that consumers from lower socio-economic groups had a higher intake of processed foods and were less aware of the salt content of these foods.

Lack of consumer awareness of low salt products may have resulted in subsequent low levels of purchase. This could also be due to the limited range of low salt products available to certain consumer groups. Furthermore, poor product labelling may inhibit consumers from distinguishing low salt products from standard range products, resulting in low levels of purchase.

These findings support the need for food manufacturers and retailers to promote existing low or reduced salt foods, in an attempt to increase consumer awareness of low salt foods and subsequent consumption.

Awareness of health risks associated with high salt consumption
In this study over half the respondents (53.1%) thought that salt was harmful to health, and an additional 40% considered salt to be harmful when consumed in excessive quantities. Five per cent of respondents believed that salt was not at all harmful to health and one per cent was unsure. These results are comparable with previous studies by Tilston *et al.* (1993) and the British Market Research Bureau (1985). Findings from a study by Walsh and McGill (1983) reported that the majority of respondents believed excess salt consumption to be detrimental to health. This research study also revealed that the majority of respondents indicated that they would be prepared to reduce salt in their diet in order to improve their health.

In this study, educational status was the only factor which significantly impacted on consumer awareness of the health risks associated with habitual high salt intakes. Data analysis revealed that consumers with A-level and higher level qualifications were more aware of the effect of high salt intakes on health than those with lower level qualifications.

In light of these findings, it is paramount that all efforts to increase consumer awareness of the risks to health are enforced, especially among consumers of lower educational and socio-economic status. It was reported in this study that the majority of consumers (85.3%) would be prepared to reduce salt in their diet.

A recent study by the Food Standards Agency (Northern Ireland) (2004a) found an increase in the number of consumers (49% to 58%) reporting to be aware of the need to reduce their salt intake. Furthermore, there was an increase of 15% to 27% in the number of consumers claiming to eat less salt. Findings at UK level revealed the number of consumer correctly stating that one should eat less salt increased significantly from 2003 (increase of 3 percentage points to 54%) (FSA, 2004b). Almost two-thirds of consumers in socio-economic groups A and B, and half of consumers in groups C1, C2, D and E claimed to have reduced their salt intake. Central age bands (26–35; 36–49; 50–57) were more likely to suggest that salt intake should be reduced (FSA, 2004b).

In light of the findings of this study and that of the Food Standards Agency, innovative health promotion strategies coupled with a pro-active response from the food industry must be implemented, to effect a reduction in salt intake. With these measures in place, consumers could potentially reduce the associated risks of hypertension, heart disease and stroke.

5.5   Taking responsibility – implications for policy and food product development

Results from this study revealed that social and demographic factors impact significantly on dietary salt intake. The evidence presented would support a multi-faceted approach to salt reduction with the government and the food industry at the fore.

If any notable improvements in health are to be achieved, the proposed salt reduction strategy should include:

- mandatory labelling of the salt and sodium content of foods;
- a user-friendly food labelling scheme; and
- mandatory reduction of the salt content of processed foods.

Efforts to address salt reduction through health education have already commenced. The recent launch of 'Sid the Slug', the FSA's salt awareness campaign, has been used to raise consumer awareness of salt consumption and the effects of high salt intake on health status. Given the short time period since the launch of this campaign, it will be some time before health professionals will be able to determine the impact of this campaign in changing dietary patterns and ultimately reducing salt intake.

Bussell (2002) has reported that 'just adopting a healthy eating message does not always lead to a reduction in salt intake'. Therefore, a strategy to reduce dietary salt intake, must endeavour to raise consumer awareness of salt consumption and reduce the salt content of processed food products, especially convenience style products.

5.5.1   Mandatory labelling of the salt content of foods

As this study and other research (Tilston *et al.*, 1993; FSA, 2004a, 2005) has revealed, consumer awareness of the salt content of processed foods is limited. Findings also demonstrate low levels of awareness and purchase of low salt products. The evidence would suggest that urgent action is needed to raise consumer awareness of salt in food. This could be achieved in a number of ways. First, it should be the prerogative of government to ensure all processed food products are accurately labelled with salt and sodium content per serving and per 100g. Some manufacturers and retailers have taken the lead in salt labelling, but since this is not the case for all, mandatory labelling of salt in food becomes even more pertinent.

5.5.2   A user-friendly product labelling scheme

In relation to product labels, not all consumers take the time to read product labels and some may even have difficulty understanding them. For this reason, it may not be enough for the food industry to simply ensure the salt and sodium content is listed. A more user-friendly labelling scheme will be required. Efforts to create such a scheme have been evident through the design of the 'traffic light' system of labelling. This would appear to be the means by which consumers could easily identify the level of fat, salt and sugar in a food product based on low, moderate and high levels. Despite considerable support for the scheme, progress has been hindered by Tesco's decision not to use the FSA's proposed 'traffic light' labelling scheme (Landon, 2005; Institute of Consumer Sciences, 2005). The withdrawal of support from one of the major retailers could prove detrimental in terms of clearer labelling of foods which are high in salt, unless agreement can be reached by all stakeholders within the food industry.

In addition to clear, user-friendly product labelling, improved in-store promotion of low salt products would make a significant contribution in terms of raising consumer awareness of the salt content in food. Improved product labelling may help consumers identify low salt products more easily and increase subsequent purchase, but appropriate advertising, in-store promotions and purchase incentives, may further increase sales of low salt products.

Whilst raising consumer awareness is one way of endeavouring to reduce dietary salt intake, this alone will not guarantee a reduction in salt consumption at population level. It is therefore essential that food industry takes responsibility for reducing the salt content of processed food products, especially convenience style products.

5.5.3   Mandatory reduction of the salt content of processed foods

Evidence to date points towards the need for the food manufacturers to be more proactive reducing salt in processed food products. Given that up to 85% of salt intake can be attributed to processed foods (IFST, 1999), the onus is clearly on the food manufacturer to be more pro-active in terms of salt reduction. It must be acknowledged that there have been some positive moves forward by a select number of manufacturers in reducing the salt content of processed foods. The Food Standards Agency has been working closely with food manufacturers, retailers, caterers, suppliers, trade associations and public procurement bodies in order to reduce the salt content of foods produced and sold for public consumption (FSA, 2004b).

Whilst these developments are commendable and represent a significant move forward, it is essential that the salt content of all processed food products is reduced even further. In order for this to be implemented effectively, it is necessary that government legislate, putting in place an upper limit on the salt content of processed food products, as determined by the food category. Recognising that developing and implementing legislation would take some time, it would be the recommendation of these authors that the food industry establish a professional voluntary code of practice whereby food manufacturers make inroads into reducing the salt content of processed foods. It is suggested that action should be taken to reduce the salt content of foods currently on the market, whilst developing reduced salt formulations for new and improved product ranges.

The authors of this report recognise the practical and economic implications associated with salt reduction. Certain food categories, i.e. the bakery and chilled foods sector, where considerable product reformulation and alternative preservation techniques may be required, will need a longer time period and significant financial investment in order to achieve a reduced salt product which is safe, stable and acceptable to consumers. Given the extensive use of salt as a flavour-enhancing ingredient, food manufacturers should consider using higher quality raw ingredients, with better natural flavour, which could reduce the need for processing and the addition of large quantities of salt. Although this will have financial implications in terms of production, consumers may be prepared to pay a premium price for a product of high sensory quality, which has no negative effects on health.

Research by Purdy (2002) revealed that the salt content of a variety of processed ready meals could be reduced by up to 30% without affecting consumer acceptability of these products. Based on this study and other research (Wyatt, 1983; Norton, et al., 1991; Rodgers and Neale, 1999) food manufacturers can move forward in terms of addressing the technical and processing issues surrounding production of reduced salt products.

Taking account of the technological and sensory issues surrounding salt reduction, it remains pertinent that food manufacturers make concerted efforts to reduce the salt content of existing products. A reduction in the salt content of processed food will in turn lead to a reduction in dietary salt intake at population level.

The Scientific Advisory Committee on Nutrition (2003) reported that the greatest benefits are likely to be achieved by taking a population approach to reducing salt intakes, rather than through individual targeted advice. This has been endorsed by FSAI (2005) which reported that even a modest reduction in average dietary salt intake at the population level is likely to produce substantial falls in stroke and coronary heart disease mortality.

An urgent response is required from government and the food industry if any health benefits are to be achieved. Given the prevalence of hypertension, heart disease and stroke, the government and food industry need to respond to the recommendations outlined with immediate effect.

5.6   Sources of further information and advice

The following organisations and agencies are currently actively involved in work related to reducing dietary salt intake. The list below highlights some of the key agencies which will provide further information relevant to the contents of this chapter.

- Food Standards Agency – Salt Awareness Campaign (www.food.gov.uk and www.salt.gov.uk).
- Food Safety Authority of Ireland – www.fsai.ie. A recent publication by FSAI (Salt and Health: Review of the Scientific Evidence and Recommendations for Public Policy in Ireland – see reference section) is a particularly useful publication.
- Food Safety Promotion Board launched a salt awareness campaign in May 2005 to encourage consumers to be more conscious of what they eat and the levels of salt in their food. Further information can be obtained at www.safefoodonline.com
- Consensus Action on Salt and Health (CSAH) www.hyp.ac.uk/cash
- Northern Ireland Chest Heart and Stroke Association (NICHSA) www.nichsa.com. Also see CHS News, the journal of NICHSA.
- British Nutrition Foundation (www.nutrition.org.uk) is a useful source of information for both consumers and health professionals. This website is particularly useful for educational materials and resources regarding healthy eating.
- Blood Pressure Association (www.bpassoc.org.uk) provides practical information for individuals with high blood pressure, including advice on how to reduce dietary salt intake.
- Faculty of Public Health. Publication of 'Easing the Pressure: tackling hypertension – a toolkit for developing a local strategy to tackle high blood pressure' marks a significant move forward in terms of addressing dietary salt intake, at a national level, as a modifiable factor in the onset of hypertension.
- *Public Health News*. This fortnightly publication frequently publishes short articles updating readers on developments related to salt and efforts to reduce dietary salt intake. This publication can be accessed on line at www.publichealthnews.com

5.7   References

BARDSLEY N (2000), *Ready Meals*, 5th Edition, Key Note Market Report Plus.

BERTINO M, BEAUCHAMP G K and ENGELMAN K (1982), 'Long-term reduction in dietary sodium alters the taste of salt', *American Journal of Clinical Nutrition*, 36, 1134–1144.

BOGUE J C, DELAHUNTY C M, HENRY M K and MURRAY J M (1999), 'Market-oriented methodologies to optimise consumer acceptability of Cheddar-type cheeses', *British Food Journal*, 101 (4), 301–316.

BOOTH D A and SHEPHERD R (1988), 'Sensory influences on food acceptance – the neglected approach to nutrition promotion', *British Nutrition Foundation Bulletin*, 13, 39–54.

BRITISH MARKETING RESEARCH BUREAU (1985), *Consumers' attitudes to and understanding of nutritional labelling – summary report*, A research study on behalf of the Consumers' Association, MAFF and the National Consumers Council, Ealing, London.

BUSSELL G (2002), 'Children's diets: should they be taken with a pinch of salt?', *Nutrition and Food Science*, 32 (6), 231–236.

CARAHER M and LANG T (1999), 'Can't cook, won't cook: a review of cooking skills and their relevance to health promotion', *International Journal of Health Promotion and Education*, 37 (3), 89–100.

CARAHER M, DIXON P, LANG T and CARR-HILL R (1999), 'The state of cooking in England: the relationship of cooking skills to food choice', *British Food Journal*, 101 (8), 590–609.

CLARK J E (1998), 'Taste and flavour: their importance in food choice and acceptance', *Proceedings of the Nutrition Society*, 57 (4), 639–643.

DAHL L K and LOVE R A (1954), 'Evidence for relationship between sodium (chloride) intake and human essential hypertension', *Archives of Internal Medicine*, 94, 525–531.

DAHL L K and LOVE R A (1957), 'Etiological role of sodium chloride intake in essential hypertension in humans', *Journal of the American Medical Association*, 167, 397–400.

DAVIES D L (2001), 'Review of the UK marketplace for convenience foods', *Nutrition and Food Science*, 31 (6), 322–323.

ELLIOT P, STAMLER J, NICHOLS R, DYER A R, STAMLER R, KESTELOOT H and MARMOT M (1996), 'Intersalt revisited: further analyses of 24 hour sodium excretion and blood pressure within and across populations', *British Medical Journal*, 312 (7041), 1249–1253.

FOOD SAFETY AUTHORITY OF IRELAND (2005), *Salt and Health: Review of the Scientific Evidence and Recommendations for Public Policy in Ireland*, Dublin, FSAI.

FOOD STANDARDS AGENCY (2004a), *Consumer Attitudes to Food Standards*, London, TSN.

FOOD STANDARDS AGENCY (2004b), *Update of industry salt reduction activity (September, 2004)*, Available at www.food.gov.uk.

FOOD STANDARDS AGENCY (2005), *Consumer Attitudes to Food Standards*, London, MORI.

FUREY M S (2001), *An investigation to identify the characteristics, extent and location of 'food deserts' in rural and urban areas of Northern Ireland*, Doctor of Philosophy Thesis, University of Ulster, Northern Ireland.

FUREY S, STRUGNELL C and McILVEEN H (2000), 'Cooking skills: a diminishing art?', *Nutrition and Food Science*, 30 (5), 263–272.

FURST T, CONNORS M, BISOGNI C A, SOBAL J and WINTER FALK L (1996), 'Food choice: a conceptual model of the process', *Appetite*, 26 (3), 247–266.

GIBSON J M A, ARMSTRONG G A and McILVEEN H (2000), 'A case for reducing salt in processed foods', *Nutrition and Food Science*, 30 (4), 167–173.

HARRISON M (2001), 'Working wives lead to big chill', *Food Processing*, 70 (8), 17.

HENCHION M (2000), 'Meals for cash rich, time poor consumers', Teagasc Ready Meals Conference, available at http://www.teagasc.ie/publicatios/readymeals2000/paper01.htm (accessed 25 January 2002).

HILL L L, MONTANDON C, SCOTT L, HAMMOND G S, TRISTAN M P and BAER P E (1980), 'Validation of a salt intake questionnaire by urinary electrolyte excretion', *Preventive Medicine*, 9, 436.

HITCHMAN C, CHRISTIE I, HARRISON M and LANG T (2002), *Inconvenience Food*, London: DEMOS.

HOWITT S (1998), *UK Food Market*, Key Note Market Review. Key Note Ltd, Hampton, Middlesex.

IKEDA J, NAGATA H, HIGASHI A, WATANABE Y and KAWAI K (1988), 'A method for estimation of salt intake from the results of a food frequency questionnaire', *Japanese Journal of Hygiene*, 43, 907–916.

INSTITUTE OF CONSUMER SCIENCES (2005), 'Tesco dismisses "traffic light labelling"', *Consumer Sciences Today*, 6 (2), 18.

INSTITUTE OF FOOD SCIENCE AND TECHNOLOGY (1999), *Salt: Position Statement.*

INTERSALT RESEARCH COOPERATIVE (1988), 'Intersalt: an international study of electrolyte excretion and blood pressure. Results for 24 hour urinary sodium and potassium excretion', *British Medical Journal*, 297, 319–328.

JAMES W P, RALPH A and SANCHEZ-CASTILLO C P (1987), 'The dominance of salt in manufactured food in the sodium intake of affluent societies', *Lancet*, 21 Feb, 426–429.

JAMES W P, NELSON M, RALPH A and LEATHER S (1997), 'Socio-economic determinants of health: the contribution of nutrition to inequalities in health', *British Medical Journal*, 314 (7093), 1545–1550.

KATZ F (1999), 'How nutritious? Meets How convenient?', *Food Technology*, 53 (10), 44, 47, 48, 50.

LANDON, J (NATIONAL HEART FORUM) (2005), 'Senseless decision', *Public Health News*, 9 May 2005.

LAW M R (1995), 'Salt and blood pressure', *British Food Journal*, 97 (9), 33–34.

LEATHER S (1992), 'Less money: less choice. Poverty and the diet in the UK today', in National Consumer Council, *Your Food: Whose Choice*, London, HMSO.

MacGREGOR G A (2002), available at http://www.bbc.co.uk/hi/english/health.newsid_189000_189490.stm (accessed 19 February 2002).

MacGREGOR G A and DE WARDENER H E (1998), *Salt, Diet and Health*, Cambridge, Cambridge University Press.

MARSHALL D W (1995), 'Introduction: food choice, the food consumer and food provisioning', in Marshall D W, *Food Choice and the Consumer*, Glasgow, Blackie Academic and Professional.

MARSHALL DW (2000), 'British meals and food choice', in Meiselman H L, *Dimensions of the Meal. The science, culture, business and art of eating*, Gaithersburg, MD, Aspen Publishers, Inc.

McHUGH M, GREENAN K, KERRIGAN C and WIGHTMAN S (1993), 'Food shopping and cooking cycles: "time" – a critical dimension', *British Food Journal*, 5, 12–16.

MENNELL S, MURCOTT A and VAN OTTERLOO A H (1992), *The Sociology of Food, Eating, Diet and Culture*, London, Sage Publications.

MINTEL (2000), *Chilled Ready Meals*, UK, Mintel International Group Ltd.

MINTEL (2002), *Chilled Ready Meals*, UK, Mintel International Group Ltd.

MURCOTT A (2000), 'Is it still a pleasure to cook for him? Social changes in the household and the family', *Journal of Consumer Studies and Home Economics*, 24 (2), 78–84.

NAGAYAMA I, OSATO S, OZEKI S, KORA H, MATSUDA T, YASUTAKE R and WAKAHARA N (1986), 'A check on the accuracy of methods in estimating sodium intake', *Journal of the Japanese Society of Nutrition and Food Science*, 39, 89–93.

NAKATSUKA H, SATOH H, WATANABE T, IMAI Y, ABE K and IKEDA M (1996), 'Estimation of salt intake by a simple questionnaire', *Ecology of Food and Nutrition*, 35 (1), 15–23.

NARHINEN M, NISSINEN A and PUSKA P (1998), 'Salt labelling of food in supermarkets in Finland', *Agriculture and Food Science in Finland*, 7, 447–453.

NATIONAL CONSUMER COUNCIL (1992), 'Food Choice and the Consumer', in National Consumer Council, *Your Food: Whose Choice*, London, HMSO.

NORTHERN IRELAND CENSUS REPORT (1991), Department of Health and Social Services and Registrar General Northern Ireland, Belfast, HMSO.

NORTON V P, NOBLE J M and ROSTAN S (1991), 'Reduced sodium bakery products: consumer acceptance', *Journal of Foodservice Systems*, 6, 61–68.

O'BRIEN J (2005), Personal communication, 28 April 2005, Dublin.

OFFICE FOR NATIONAL STATISTICS (2005), *Social Trends 35*, Hampshire, Palgrave Macmillan.

PIETINEN P, TANSKANEN A and TUOMILEHTO J (1982), 'Assessment of sodium intake by a short dietary questionnaire', *Scandinavian Journal of Social Medicine*, 10, 105–112.

PURDY J M A (2002), *Sensory quality and consumer acceptance of reduced salt ready meals*, PhD Thesis, University of Ulster, Northern Ireland.

REED Z, McILVEEN H and STRUGNELL C (2001), 'The chilled ready meal market in Northern Ireland', *Nutrition and Food Science*, 31 (3), 103–109.

REUTERS BUSINESS INSIGHT (2003), *Identifying the convenience consumer*, UK, Datamonitor.

RODGERS A and NEAL B (1999), 'Less salt does not necessarily mean less taste', *The Lancet*, 353 (9161), 1332.

SCIENTIFIC ADVISORY COMMITTEE ON NUTRITION (2003), *Salt and Health*, Norwich, The Stationery Office.

SHEPHERD R (1999), 'Social determinants of food choice', *Proceedings of the Nutrition Society*, 58 (4), 807–812.

SHEPHERD R and FARLEIGH C A (1989), 'Sensory assessment of foods and the role of sensory attributes in determining food choice', in Shepherd R, *Handbook of Psychophysiology of Human Eating*, Chichester, John Wiley & Sons Ltd.

SHEPHERD R, FARLEIGH C A and LAND D G (1985), 'Estimation of salt intake by questionnaire', *Appetite*, 6, 219–233.

SLOAN A E (2001), 'Top 10 trends to watch and work on. 3rd Biannual Report', *Food Technology*, 55 (4), 38–58.

STAMLER J (1997), 'The Intersalt Study: background, methods, findings and implications', *American Journal of Clinical Nutrition*, 65 (suppl), 626s–642s.

STITT S (1998), 'Food for health or wealth in the 21st century', *Nutrition and Health*, 12, 203–213.

TILSTON C, NEALE F, GREGSON K and BOURNE S (1993), *Salt – a challenge to food manufacturers*. Food Marketing Research Group, University of Nottingham, Bradford, Horton Publishing.

WALSH G A and McGILL A E J (1983), 'Use, taste and awareness of salt in the diet', in McLaughlin, J V and McKenna, B M, *Proceedings of the 6th International Congress of Food Science and Technology*, Dublin, 19–23 September.

WARDLE J (1993), 'Psychology, food, health and nutrition', *British Food Journal*, 95 (9), 3–6.

WARWICK J (1998), '*Food choices of young people in Northern Ireland – the influences and health implications*', PhD Thesis, University of Ulster, Northern Ireland.

WHEELOCK V (1998), 'Salt – the next major food issue', *Food Industry News* April, 15–16.

WHITWORTH M (2001), 'Season of discontent', *Food Manufacture*, 76 (5), 32–33.

WILLIAMS A A (1983), 'Defining sensory quality in foods and beverages', *Chemistry and Industry*, 19 (3 October), 740–745.

WOOLFE J (2000), 'MAFF's food acceptability and choice research programme: an introduction', *Nutrition and Food Science*, 30 (1), 5–7.

WYATT C J (1983) 'Acceptability of reduced sodium in breads, cottage cheese and pickles', *Journal of Food Science*, 48, 1300–1302.

6

Consumer responses to low-salt food products

C Walsh, University of the Free State, South Africa

6.1 Introduction: importance of determining consumer responses to low-salt food products

Reduction of salt and sodium in the diet can have significant beneficial effects on the health of the broad public (discussed in Chapter 2 on health risks of excessive salt intake). Moderate restriction of salt intake has been proposed as a dietary goal for the general public, both healthy and ill (Krauss *et al.*, 1996; Adams *et al.*, 1995; Walker, 1995; WHO, 1999).

The degree to which sodium is restricted depends on the needs of the individual. The level of sodium prescribed should involve the least amount of restriction necessary to achieve the desired clinical response. Beauchamp *et al.* (1987) indicate that if sodium levels are severely restricted (more than 50 percent), 20 percent of the sodium is likely to be added again as table salt. This may be the reason why a moderate restriction of salt (6 gram or 100 mEq) intake has been set as a dietary goal to prevent hypertension (WHO, 1999).

According to Shepherd (2005) changing dietary behaviour is a complex issue. Possible reasons include poor knowledge of what foods contain and lack of motivation to change. Food choice is affected by physiological, social and cultural factors, all of which are interrelated.

Many individuals following sodium-restricted diets, have reported difficulties with compliance. Possible reasons include, first, that sodium-restricted diets are considered bland and tasteless. Many consumers consider changing habitual salt intake difficult and think that reduction of salt will make food unpalatable. Sensory studies have, however, shown that salt restriction can be achieved by simple dietary manipulation (discussed in Part III on reducing salt in particular foods).

A few studies have shown that sodium levels can be reduced by 30 to 40 percent, without affecting taste and consumer acceptability (Adams *et al.*, 1995; Malherbe *et al.*, 2003; Witschi *et al.*, 1985).

Another reason for poor compliance of a sodium-restricted diet may be cost. Low-sodium diets are more expensive than a regular diet if special low-sodium products are purchased. Implementation of recommendations to reduce sodium intake without additional cost require knowledge of the general dietary sources of sodium. A general knowledge of the rules for recipe modification, together with guidelines for the appropriate interpretation of food labels, can contribute to improved compliance.

Some researchers have reported that losses in salt-taste perception can make it difficult for elderly patients to comply with a low-sodium diet (Schiffman, 1993). In contrast, Drewnowski *et al.* (1996) have shown that perception of saltiness may be unrelated to sodium consumption in healthy elderly adults.

For recommendations to be made to consumers regarding reduction of sodium in the diet, it is necessary to determine both perception of saltiness and acceptability of low-salt food products. If the role played by sensory responses to salt can be determined, food preparation and consumption can be further modified and adherence to a sodium-restricted diet may be improved.

6.2   Methods to determine consumer responses to low-salt food products

6.2.1   Sensory evaluation

Sensory evaluation involves the development and use of principles and methods for measuring human responses to food (Sidel *et al.*, 1981). Sensory evaluation has been defined as a 'scientific method used to evoke, measure, analyze and interpret those responses to products as perceived through the senses of sight, smell, touch, taste and hearing' (Lawless and Heymann, 1998: 2; Stone and Sidel, 1993: 12; Charley and Weaver, 1998: 5).

The science of sensory evaluation relies on guidelines for the preparation and serving of samples under controlled conditions so that biasing factors are minimized. Sensory evaluation is a quantitative science in which numerical data are collected to determine specific relationships between product characteristics and human perception (Lawless and Heymann, 1998: 3).

Data generated from human observers are often highly variable, and panels of humans are by their nature a heterogeneous instrument for the generation of data. To assess whether the relationship observed between product characteristics and sensory responses are likely to be real, and not merely the result of uncontrolled variation in responses, statistics are used to analyze evaluation data (Lawless and Heymann, 1998: 3). Like other analytical test procedures, sensory evaluation is concerned with precision, accuracy, sensitivity and the avoidance of false positive results.

6.2.2   Classes of test methods

Sensory evaluation takes both analytical specifications of perceived product differences as well as predicting consumer acceptance in the real world into account (Lawless, 1991). Thus the test method should be appropriate to answer the questions being asked about the product in the test. These objectives have led to two 'styles' of performing sensory evaluation, namely Analytical (Type 1) and Affective (Type 2) sensory evaluation.

In Type 1 sensory evaluation, laboratory control is essential, in order to maximize the sensitivity of the measuring instrument. In Type 2, the ability to generalize to the consuming public is of greater concern. Different types of panelists, serving procedures, sensory methods and statistical models for analysis may be chosen, depending upon whether the objective of the test falls in the Type 1 or Type 2 class (Lawless, 1991).

Analytical methods evaluate differences or similarity as well as quality and/or quantity of sensory characteristics of a product. Affective methods evaluate preference and/or opinions of the product. Analytical methods can be either discriminative (measure whether samples are different or measure the ability of individuals to detect sensory characteristics) or descriptive (measure qualitative and/or quantitative characteristics). Panelists included in analytical sensory evaluation are usually trained, while panelists included in affective evaluation are usually untrained and representative of the target population (Charley and Weaver, 1998: 11). Due to the fact that this main objective of this discussion is related to determining consumer responses to low-salt food products, preference will be given to Type 2 or affective methods.

6.2.3   Affective (preference and acceptability) tests

Consumer acceptance and/or consumer preference testing is a common application of affective sensory analysis (Charley and Weaver, 1998: 11). The measure of acceptability based on sensory properties of a product is logical and necessary before a product is marketed and substantial capital invested.

During affective evaluation, subjective responses, such as product acceptance and preference, are measured (Sidel, 1988). An attempt is thus made to quantify the degree of liking or disliking a product (Charley and Weaver, 1998: 11; Lawless and Heymann, 1998: 628).

Generally, a large number of respondents are required for such evaluations (Charley and Weaver, 1998: 11). Respondents or panel members are usually consumers who are selected in accordance with a number of criteria (Lawless and Heymann, 1998: 470), which frequently include previous use or potential users of the product; size of family or age of specific family members; occupation of head of household; economic or social level; and geographic area.

There are two main approaches to affective consumer sensory testing: the measurement of acceptability and the measurement of preference (Lawless and Heymann, 1998: 430). When using affective tests it should be remembered that acceptability and preference are not the same thing. For example, one sample may be preferred above another sample, but both may be found unacceptable (Lyon *et al.*, 1992: 31). The terms 'acceptance' and 'acceptability' are used to refer to degree of liking and disliking (Lawless and Heymann, 1998: 628).

The most common approach is to offer people a choice of alternative products, then determine whether there is a clear preference (Lawless and Heymann, 1998: 43). Preference tests refer to all affective tests based on a measurement of preference, or a measure from which relative preference may be determined, for example, pleasure–displeasure, like–dislike. Preference measurement may include choice of one sample over another (paired comparison), a ranked order of liking (using more than two samples), or an expression of opinion on a hedonic (like/dislike) scale (rating) (Charley and Weaver, 1998: 11). Preference from ranking tests is direct, while preference from hedonic ratings is implied.

A hedonic rating scale is used to measure the level of liking for food products by a population. The samples are rated as to the degree to which they are liked or disliked. This scale may be applied in testing for preference or acceptance. Consumers are asked to record their reactions on a seven-point or a nine-point scale that ranges from like extremely to dislike extremely. The responses can be given verbally or facial hedonic scales or an unstructured line scale may be used.

A food action rating scale may be used to measure the level of acceptance of food products by a population. The scale is not applicable for rating specific characteristics; rather it is a measure of general attitude toward a food product (Lawless and Heymann, 1998: 466).

Methods of scaling involve the application of numbers to quantify sensory experience. It is through this process that sensory evaluation becomes a quantitative science subject to statistical analysis (Charley and Weaver, 1998: 12; Lawless and Heymann, 1998: 208; Stone and Sidel, 1993: 66).

6.3   Acceptability of low-salt food products and implications for food product development

Although little is really known about the reason for the preference for salty food, the first step in reducing salt intake is to understand sensory responses to salt (Beauchamp *et al.*, 1990; Beauchamp and Engelman, 1991). Sensory responses include taste adaptation to a certain sodium concentration (Bourne *et al.*, 1993; Lawless and Heymann, 1998: 44), sodium interaction with other food components (Adams *et al.*, 1995; Kroeze, 1990: 48,49; Malherbe *et al.*, 2003) and the inhibitory or masking interaction in mixtures of different tastes (Lawless and Heymann 1998: 44; Mattes, 1987: 133).

6.3.1   Taste adaptation
Adaptation to a certain sodium concentration, because of the habitual intake of the same concentration level, is referred to as taste adaptation (Kroeze, 1990: 48,

49). It is believed that high salt ingestion is a consequence of habits learned during development (Beauchamp and Engelman, 1991; Bourne *et al.*, 1993). Thus, persons that habitually follow such a diet will eventually require a higher sodium concentration for a specific product to be acceptable.

A number of researchers (Bertino *et al.*, 1982a; Beauchamp *et al.*, 1982) have hypothesized that if a low-sodium diet is followed, higher concentrations of salt in food will become more intense and less pleasant. This hypothesis is based on reports that have indicated that, after individuals have been on a low-sodium diet, foods that used to taste just salty enough become too salty. That indicates that adaptation to salt has taken place. This hypothesis is strengthened by the fact that adults, placed on very low-sodium diets, are able to detect lower concentrations of salt in water than adults eating higher-sodium diets (Beauchamp *et al.*, 1982). The mechanism by which increased sensitivity could increase preference remains unclear.

Studies reported by Beauchamp *et al.*, (1982; 1990) and Beauchamp and Engelman (1991) found that sodium-depleted subjects subjectively reported an increased desire for food higher in salt and exhibited a tendency to judge higher amounts of salts in food as most preferred. In our society, however, the degree of depletion induced in the above studies (10 mmol/d) is unlikely to occur, except in extremely rare cases (Beauchamp *et al.*, 1990). In long-term studies in which individuals were required to maintain low-sodium diets for two to five months, results clearly demonstrate that, with longer periods of lowered dietary salt intake than the three weeks of restriction usually applied in studies, food with higher concentrations of salt become less pleasant (Beauchamp *et al.*, 1982; Bertino *et al.*, 1982a; 1982b: 150).

Adaptation to salt has two important effects on the taste of salt. Initially, salt is salty only at concentrations above the adapting concentration. Thereafter, concentrations below the adapting concentration evoke a bitter-sour or bland taste that increases in intensity as the concentration decreases (Beauchamp *et al.*, 1990; Lawless and Heymann, 1998: 44). With a two-thirds reduction in the sodium content of foods, Malherbe *et al.* (2003) reported a significant decrease in acceptability of mashed potatoes, vegetable soup and beef stew.

There is also a psychological mechanism that functions in direct opposition to the effect of deficiency on the taste of salt. Humans tend to judge sensory stimuli, including tastes, in terms of the contextual situation. Ratings of the intensity and pleasantness of a particular taste can be changed by manipulating the context by varying the frequency of presentation of various taste stimuli. If high concentrations of salt in soup are presented more frequently than low-concentrations, a particular high concentration will be rated as tasting less intense and more pleasant (Bertino *et al.*, 1982b: 150). This psychological mechanism can possibly be associated with changes in dietary and nutrient intakes that occur during urbanization, acculturation or westernization (Vorster *et al.*, 1995). It also explains the 'salt-seeking' behaviour that occurs, after adapting to the westernized lifestyle and diet (Bourne *et al.*, 1993).

6.3.2 Sodium interaction with other food components

Perceived saltiness does not only depend on the sodium concentration in a food, but also on the medium in which sodium is presented. In addition, the perception of saltiness is not the only factor that influences the acceptability of a food or dish (Adams *et al.*, 1995; Drewnowski *et al.*, 1996; Malherbe *et al.*, 2003). Although consumers may report a significant difference in the perception of saltiness of two dishes, neither may be acceptable. When eating, the acceptability of a dish is generally of greater importance than the perception of saltiness.

Food components that may have an effect on consumer acceptability of low-sodium foods include complexity and texture of the food, both of which are not mutually exclusive.

Complexity

Taste adaptation differs from partial adaptation, which occurs because of the complexity of the media in which the sodium concentration is presented and which gradually changes as chewing proceeds. Another form of adaptation which also occurs because of the complexity of the media in which the sodium concentration is presented, is considerable cross-adaptation, usually found between compounds with similar taste qualities (Kroeze, 1990: 48, 49).

McBurney *et al.* (as cited by Kroeze, 1990: 49) showed that this partial adaptation leads to the loss of absolute sensitivity (the sensitivity to a specific taste or one of the basic tastes, often associated with a specific dish), but in turn increases differential sensitivity (the sensitivity to all the tastes – the most prevalent/main taste in combination with the secondary tastes – present in a dish).

Although differences in perceived saltiness of foods can also be attributed to differences in solid or liquid composition of foods (Bertino *et al.*, 1982a; Beauchamp *et al.*, 1982; Mattes, 1987: 133), it is possible that the food used as a carrier also determines perceived saltiness (Adams *et al.*, 1995).

Malherbe *et al.* (2003) undertook a study evaluating both consumer acceptability and salt perception of food with a reduced sodium content. Acceptability was evaluated according to a nine-point Hedonic scale, ranging from 9 (dislike extremely) to 1 (like extremely). Salt perception was rated using a numerical rating five-point line scale, ranging from 1 (not salty at all) to 5 (too salty, no longer tasty). This study found that in dishes with normal and one-third sodium reduction content, where a main sodium carrier was present (such as beef in beef stew, potato in mashed potatoes, and porridge in crumbed porridge), the salt perception ratings were significantly lower than for dishes where a number of ingredients acted as carriers of sodium (vegetable soup). In contrast to the other dish combinations, the moderate reduction in sodium concentration in the vegetable soup did not affect perception of saltiness significantly.

In addition to the sodium carrier, it is possible that the complexity of the dish, as determined by the number of ingredients included, can also affect salt perception (Adams *et al.*, 1995). A significant decrease in salt perception can be observed with moderate sodium reduction in simple dishes such as crumbed porridge or mashed potatoes (Malherbe *et al.*, 2003). Simple dishes (which display only one or similar taste qualities in one dish) tend to display a higher level of cross-adaptation than complex dishes (such as beef stew and vegetable soup) (Kroeze, 1990: 50).

Although a one-third reduction in the sodium content of beef stew affected perception of saltiness, the acceptability ratings were not significantly affected. With a one-third reduction in sodium content, mean acceptability ratings of the beef stew (complex and coarse) and the vegetable soup (complex and smooth) remained unchanged (Malherbe *et al.*, 2003). When mixtures are developed from substances with different taste qualities, such as in the complex dishes where a number of ingredients are included, salt evokes more than one taste quality (Mattes, 1987: 133, Bertino *et al.*, 1982b: 1140). These other tastes evoked by salt may mask the one-third sodium reduction. Thus, dish complexity plays an important role in acceptability. In addition, Kroeze (1990: 48) suggests that partial adaptation (which occurs because of the complexity of the medium in which the sodium concentration is presented) contributes to the masking effect brought about by other tastes evoked by salt in mixtures with different taste qualities. Decrease in acceptability is prevented when there is a continuous perception of the different tastes in the dish, which masks the moderate sodium reduction.

Thus, in contrast to acceptability, which seems to be influenced largely by dish complexity and to a lesser degree by texture, perceived saltiness seems to depend on the presence of a carrier food, the composition of the food, and cross-adaptation.

It can be concluded that it is possible to reduce the sodium content in dishes with a complex composition by about 30 percent, without significantly changing acceptability. Possible reasons include the masking effect that takes place within complex dishes with reduced sodium content, partial adaptation, as well as the decreased absolute sensitivity and increased differential sensitivity.

Texture
Bertino *et al.* (1982a) has reported that in subjects maintaining a low-sodium diet, higher concentrations of salt in a solid food were more acceptable while the concentration that produced maximum pleasantness of salt in liquid food, such as soup, decreased.

Kroeze (1990: 41) indicates that the way food feels in the mouth is associated with texture. Thus, dish texture also plays an important role in the acceptability of different sodium concentrations in dishes, although to a lesser degree than complexity. This statement is supported by results of the study undertaken by Malherbe *et al.* (2003) that indicate that in contrast to perception of saltiness, the acceptability ratings of crumbed porridge (a simple dish with a coarse texture) remained the same for the full recipe as well as for the one-third sodium reduction. A significant decrease in acceptability occurred, however, when the sodium content of mashed potatoes (a simple dish with a smooth texture) was decreased by a third. In simple dishes where there is a lack of substances with different taste qualities, a course texture is more acceptable than a smooth texture when sodium concentration is decreased.

6.3.3   Inhibitory or masking interactions in mixtures of different tastes
Lawless and Heymann (1998: 44) suggest that one of the functional properties of taste function includes the tendency for mixtures of different tastes to show partially inhibitory or masking interactions. Some of these mixture inhibition effects, like the inhibition of bitterness by sweetness, appear to reside in the central nervous system, while others, such as the inhibition of bitterness by salt, are more likely due to peripheral mechanisms at the receptors themselves.

6.4   Recommendations and future trends

The association between hypertension and sodium intake stresses the requirement for food processing technologies that use less salt, and nutrition education strategies to promote decreased salt intake in home prepared meals.

Recommendations to reduce salt intake for the general public form an essential component of general dietary guidelines. Many proposed dietary changes are difficult to achieve in communities adopting a Westernized lifestyle where urbanization, economic improvement and social mobility are common (Bourne et al., 1993; Drewnowski and Popkin, 1997; Popkin and Doak, 1998).

Within this situation, the need exists for practical recommendations that can assist in decreasing sodium intake. In addition to reducing the sodium content by omitting added salt or by decreasing the amount added during food preparation, it is, however, necessary to consider salt interaction with sensory components and the media in which sodium is presented.

The impact of factors such as dish complexity and food texture on consumer acceptability of foods with a reduced sodium content are seldomly taken into account when making recommendations to reduce sodium in the diet or in the development of low-salt food products. Adjusting the composition of the diet or meal (e.g., food texture and dish complexity) to ensure maximum acceptability while at the same time reducing salt/sodium content may improve compliance necessary to achieve health benefits.

As far as acceptability of dishes with moderately reduced sodium content is concerned, complex dishes with a smooth texture such as vegetable soup, or with a coarse texture such as beef stew, are recommended above simple dishes. When simple dishes are ingested, those with a coarse texture (such as porridge) are more acceptable than those with a smooth texture (such as mashed potatoes), in cases where sodium content has been moderately reduced.

When salt perception is taken into account, however, dishes that comprise a number of ingredients with no specific carrier food seem to be rated more salty than dishes with a main carrier, in cases where the sodium content is decreased.

6.5 References

ADAMS SO, MALLER O and CARDELLO AV (1995), 'Consumer acceptance of food lower in sodium', *Journal of the American Dietetic Association,* 95(4), 447–453.

BEAUCHAMP GK and ENGELMAN K (1991), 'High salt intake. Sensory and bahavioural factors', *Hypertension,* 17, I176–I181.

BEAUCHAMP GK, BERTINO M and ENGELMAN K (1987), 'Failure to compensate decreased dietary sodium with increased table salt', *JAMA,* 258, 3275–3278.

BEAUCHAMP GK, BERTINO M, BURKE D and ENGELMAN K (1990), 'Experimental sodium depletion and salt taste in normal human volunteers', *American Journal of Clinical Nutrition,* 51, 881–889.

BEAUCHAMP GK, BERTINO M and MORAN M (1982), 'Sodium regulation: sensory aspects', *Journal of The American Dietetic Association,* 80(1), 40–45.

BERTINO M, BEAUCAMP GK and ENGELMAN MD (1982a), 'Long-term reduction in dietary sodium alters the taste of salt', *American Journal of Clinical Nutrition,* 36, 1134–1144.

BERTINO M, BEAUCHAMP KE and KARE MR (1982b), 'Dietary sodium and salt taste', in Fregly MJ and Kare RM, *The Role of Salt in Cardiovascular Hypertension,* New York, Academic Press.

BOURNE LT, LANGENHOVEN ML, STEYN K, JOOSTE PL, LAUBSCHER JA and VAN DER VYVER E (1993), 'Nutrient intake in the urban African population of the Cape Peninsula, South Africa. The Brisk study', *Central African Journal of Medicine,* 3(4), 238–246.

CHARLEY H and WEAVER C (1998), *Foods: a scientific approach,* New Jersey, Prentice-Hall, Inc.

DREWNOWSKI A and POPKIN BM (1997), 'The Nutrition Transition: New trends in the global diet', *Nutrition Reviews,* 55(2), 31–43.

DREWNOWSKI A, HENDERSON, DRISCOLL A and ROLLS BJ (1996), 'Salt taste perceptions and preferences are unrelated to sodium consumption in healthy older adults', *Journal of the American Dietetic Association,* 96, 471–474.

KRAUSS RM, DECKELBAUM RJ, ERNST N, FISHER E, HOWARD BV, KNOPP RH, KOTCHEN T, LICHTENSTEIN AH, MCGILL HC, PEARSON TA, PREWITT TE and STONE NJ (1996), 'Dietary guidelines for healthy American adults. A statement for health professionals from the Nutrition committee, Americal Heart Association', *Circulation,* 94, 1795–1800.

KROEZE JHA (1990), 'The perception of complex taste stimuli', in McBride RL and MacFie, *Psychological Basis of Sensory Evaluation,* New York, Elsevier Science Publishers.

LAWLESS HT (1991), 'Bridge the gap between sensory science and product evaluation', in Lawless HT and Klein BP, *Sensory Science Theory and Applications in Foods,* Amsterdam, Elsevier Publishers.

LAWLESS HT and HEYMANN H (1998), *Sensory Evaluation of Food: Principles and Practices,* New York, International Thomson Publishing.

LYON DH, FRANCOMBE MA, HASDELL TA and LAWSON K (1992), *Guidelines for Sensory Analysis in Food Product Development and Quality Control,* London, Chapman and Hall.

MALHERBE M, WALSH CM and VAN DER MERWE C (2003), 'Consumer acceptability and salt perception of food with a reduced sodium content', *Journal of Family Ecology and Consumer Sciences,* 31, 12–20.

MATTES RD (1987), 'Assessing salt taste preference and its relationship with dietary sodium intake in humans', in Solms J, Booth DA, Pangborn RM and Raunhardt O, *Food Acceptance and Nutrition*, London, Academic Press.

POPKIN BM and DOAK CM (1998), 'The obesity epidemic is a worldwide phenomenon', *Nutrition Reviews*, 56(4), 106–114.

SCHIFFMAN SS (1993), 'Perceptions of taste and smell in elderly persons', *Crit Rev Food Sci Nutr*, 33, 17–26.

SHEPHERD R (2005), 'Influences on food choice and dietary behaviour', *Forum Nutr*, 57, 36–43.

SIDEL JL (1988), 'Establishing a sensory specification', in Thomson DMH, *Food Acceptability*, London, Elsevier Science Publishers.

SIDEL JL, STONE H and BLOOMQUIST J (1981), 'Use and misuse of sensory evaluation in research and control', *Journal of Dairy Science*, 64, 2296–2302.

STONE H and SIDEL JL (1993), *Sensory Evaluation Practices*, San Diego, California, Academic Press.

VORSTER HH, OOSTHUIZEN W, STEYN HS, VAN DER MERWE AM and KOTZE JP (1995), 'Nutrition intakes of white South Africans – a cause for concern: the VIGTOR study', *South African Journal of Food Science and Nutrition*, 7(3), 119–126.

WALKER ARP (1995), 'Nutrition-related diseases in Southern Africa: with special reference to urban African populations in transition', *Nutrition Research*, 15(7), 1053–1094.

WITSCHI JC, ELLISON CR, DONALD DD, GAYE LV, SLACK WV and STARE FJ (1985), 'Dietary sodium reduction among students: feasibility and acceptance', *Journal of the American Dietetic Association*, 85(7), 816–821.

WORLD HEALTH ORGANISATION (WHO) (1999), 'International Society of Hypertension Guidelines for the Mangement of Hypertension. Guidelines subcommittee', *Journal of Hypertension*, 17, 151–183.

7

Improving the labelling of the salt content of foods

G. Bussell and M. Hunt, Food and Drink Federation (FDF), London, UK

7.1 What purpose does nutrition labelling serve and what are its limits?

Nutrition labels describe the nutrient content of a food and are intended to guide the consumer in food selection. Under current EU legislation (Directive 90/496/EC) the provision of nutrition information is voluntary unless a nutrition claim is made. If given however, it must cover certain prescribed nutrients in given formats (see page 136).

The Codex Guidelines on Nutrition Labelling (see page 137) interestingly set out their purpose as follows:

- To ensure that nutrition labelling is effective:
    1. in providing the consumer with information about a food so that a wise choice of food can be made;
    2. in providing a means for conveying information of the nutrient content of a food on the label;
    3. in encouraging the use of sound nutrition principles in the formulation of foods which would benefit public health;
    4. in providing the opportunity to include supplementary nutrition information on the label.
- To ensure that nutrition labelling does not describe a product or present information about it that is in any way false, misleading, deceptive or insignificant in any manner.
- To ensure that no nutritional claims are made without nutrition labelling.

The ability to give certain product information in nutrition labelling can provide incentives to manufacturers to develop products that promote public health and assist consumers in following dietary recommendations. An example of this would be the use of Guideline Daily Amounts (GDAs) (see page 141). Conversely, undue restrictions on the giving of information can inhibit innovation.

It is not the intention of nutrition labelling by itself to solve nutrition problems (see Section 7.2). Instead it provides essential information in a clear and consistent manner. It should be seen as one of the elements of nutrition policy within the wider context of information and education to enable appropriate dietary choices to be made. The space constraints on food labels mean that other means of communicating nutrition information must also be considered.

In the Environment, Food and Rural Affairs Committee's Report on Food Information (7th Report of 2004–2005 session), the following comment is recorded from Dr Susan Jebb (MRC Human Nutrition Research Unit):

> At the moment labelling is becoming bigger than it ought to be. We spend so much time and energy worrying about labelling when it is only one small part of the overall issue of how we are going to help consumers help themselves to a better diet.

7.2   The relationship between nutrition labelling and consumer health

There is a highly complex relationship between nutrition information and the endpoint of consumer health. It is not possible on the basis of existing research to quantify, or draw strong conclusions, as to health benefits associated with nutrition labelling alone. A range of factors, including socio-economic background, level of education, age, gender, interests in health, media awareness and mood also determine the extent to which nutrition information impacts positively on consumer health.

A fundamental element in the nutrition labelling/consumer health relationship is the first step in the process, namely ensuring the consumer reads the label. A number of factors such as time availability, reading and understanding ability, eyesight and motivational aspects will affect the extent to which the consumer will read the label. Nevertheless, several studies into food and nutrition labelling have indicated that the percentage of consumers sometimes reading nutrition labelling may range from 70–80% (EAS, 2004). But in reading the label, consumers commonly focus only on specific information that reflects their own concerns and interests. Overall it appears from experience in the United States, that introducing new (and believed to be improved) labelling can significantly benefit those consumers with a greater knowledge and interest in food labelling, but is of little benefit to sceptical or ill-informed consumers.

Isolating and evaluating the impact of nutrition labelling on an individual's broader nutrition knowledge is inherently problematic and there is currently inadequate research available to draw firm conclusions (European Heart

Network, 2003). It is essential that consumers take responsibility themselves. The label should give clear and objective information, but it cannot be the definitive guide as to how a food will fit into an individual's diet. (Although as can be seen on page 141, the use of Guideline Daily Amounts (GDAs) may be an aid to this.)

7.3   The current EU/UK nutrition labelling format

UK food labelling is based upon the requirements of the 1990 EU Nutrition Labelling Directive (90/496/EEC). This is currently under review to take account of numerous technical developments in the intervening years, both in ingredients and in the range of foods on offer.

The Nutrition Labelling Directive currently states that nutrition labelling is not required unless a claim is being made about the particular food. The Directive lays down a standardised, tabular format in which nutrition labelling must be presented (unless space on the label does not permit this) and two types of declaration are provided for:

- **Group 1** (basic information to be given when nutrition labelling is provided consisting of energy, protein, carbohydrate and fat)
- **Group 2** (extended information, required if a declaration is made for certain nutrients, consisting of energy, protein, carbohydrate, sugars, fat, saturates, sodium and fibre).

In addition information may also include the amounts of a number of other nutrients such as vitamins and minerals which are listed in the Annex of the Directive.

Whilst 'salt' is not officially identified in the Directive as a 'nutrient', sodium information on many labels is being supplemented by a 'salt equivalent' figure, based usually on the formula salt = sodium × 2.5. The current Directive requires the nutrition information to be given per 100 g/ml. It may also be given per portion (provided the number of portions in the package is stated) or per quantified serving, but this is in addition to, not instead of, per 100 g/ml.

7.4   How are the nutrient values arrived at?

The Directive allows for the values declared on the label to be arrived at in three ways. They are to be 'average values' based upon:

1. the manufacturer's analysis of the food;
2. calculations from the known or actual average values of the ingredients used;
3. a calculation from generally established and accepted data.

Manufacturers are free to use whichever method, or combination of methods, best fit the circumstances. However, for sodium values, manufacturers are increasingly choosing to use analysis. Some manufacturers who have an ongoing salt reduction programme may not always change the labels with each incremental decrease in salt content, preferring for cost reasons to wait until a significant reduction has been achieved (see page 140).

It has been estimated that the cost of laboratory analysis for the Group 1 nutrients (energy, protein, carbohydrate and fat) is an average 57 Euro per nutrient. If the requirement is increased from four to seven items by the addition of sugars, saturated fatty acids and sodium/salt, the cost may average 256 Euro per product (EAS, 2004).

Even for the method of calculating nutrient values on the basis of published data, the cost is estimated at more than 70 Euro per calculation (EAS, 2004). There is also the problem that published nutrient values of ingredients do not always cover all the main nutrients, or the ingredient definition may be too general.

Deriving the labelling value from generally established and accepted data is mostly used for single component food products, e.g. rice, flour, oils, milk, and commodities, fruits and vegetables. The cost of using this data is relatively small (EAS, 2004).

7.5   Current Codex guidelines on nutrition labelling

The Codex Alimentarius, or the 'food code' is a series of food standards that act as an international reference point for consumers, the food industry, national authorities and international bodies. It aims to provide consumer protection, while maintaining fair practice in international trade and presents a unique opportunity for countries to formulate and harmonise food standards.

The Codex standards are developed by the Codex Alimentarius Commission (CAC), which currently has representation from over 160 countries. Membership is open to member nations and associate members of FAO and WHO, although non-governmental organisations may attend Codex meetings as observers. Food Standards Agency (FSA) officials represent the UK on the CAC.

The code has had an enormous impact on the thinking of the food industry, as well as on the awareness of consumers. The standards set by CAC are also important because they may be used as international reference documents to assist in the settlement of trade disputes.

Agreed standards are not legally binding, but they carry a great deal of weight. Codex members (national governments) that agree to Codex standards are required, where possible, to align their domestic food legislation/standards with them. Therefore, there is a strong tendency for national and regional laws to take existing, relevant Codex standards as their starting point. This is now of little relevance to EU Member States, whose food law is already extensive. The EU became a member of Codex in 2004 and, as a general rule, the European Commission has exclusive competence in Codex discussions in areas of EU harmonised legislation, including nutrition labelling.

7.6   What changes may occur to the EU legislative framework on salt labelling?

There is recognition by the EU Commission that consumers find sodium difficult to understand and are often not aware that it is linked with salt. In one study it was found that 28% of those questioned thought that salt and sodium were the same (Co-op, 2002). The same study showed that when asked to interpret nutrition labels, only 14% could correctly interpret the sodium information. Despite believing that more education about labelling should take place, the food manufacturing industry recognises this difficulty in distinguishing between salt and sodium, which is why many manufacturers are moving to salt equivalent labelling (see page 139).

FSA hopes that the revised EU Nutrition Labelling Directive will allow just 'salt' to be labelled without sodium needing to be labelled at all. In the meantime, it encourages industry to provide levels of salt as well as sodium on packs, which is now being undertaken by the majority of the industry (see page 139). The food industry believes that, for the sake of accuracy and transparency, if salt equivalent labelling is to be used, it should be in addition to sodium labelling not instead of it, for the following main reasons:

• It would be misleading and inaccurate as a significant amount of sodium in food (around 30%) is not present as sodium chloride (see page 140).
• Increasing rates of diabetes will mean that more people may experience kidney problems. When kidney problems progress, patients have to watch the amount of total sodium they consume, and find it helpful to have sodium quantified in a precise way.

Another legislative change that may take place is for a new core of nutrient information to be mandatory on all food product labelling (not just on those making nutrient claims). This new core has been proposed by the EU Commission as: energy, fat, saturated fat, salt and sugars. It is not yet clear, however, what allowances might be made to accommodate some measure of nutrition labelling on small packs. An impact assessment carried out on behalf of DG Sanco on mandatory nutrition labelling showed that 17% of food products surveyed have potential problems with space availability for legible tabular labelling, including high salt products such as olives and seasonings (EAS, 2004).

On the basis of the current Group 1/Group 2 nutrition labelling requirements, some manufacturers want flexibility to allow shorter nutrient declarations where a claim would trigger a full Group 2 declaration, but on a product for which some of the nutrients are of little significance. This would allow declaration of Group 1 plus the nutrient for which the claim is made and thus reduce the total data burden on the label, including a number of zero or trace declarations. Given consumer interest in sodium/salt data, these would be less likely to be omitted under such arrangements. Their declaration would be secured if sodium/salt were part of a new core requirement.

There is some support for expressing the nutrient content additionally as a percentage of an agreed reference value (e.g., RDAs, Guideline Daily Amounts, Daily Reference Values etc.). In the UK many retailers and manufacturers are declaring the GDA for salt (set at 6 g based on the Scientific Advisory Committee on Nutrition (SACN) 2003 recommendation) and many more manufacturers, retailers and even some members of the hospitality industry, will soon follow. However, recommended levels of salt intake vary from country to country and so to be able to take a harmonised EU approach there would be a requirement first to set EU nutrient recommended intakes. This is being considered.

Industry hopes that future labelling regulations will allow sufficient flexibility to provide only that nutrition information most relevant to each type of product. This would also take into consideration the recommendations made by The Better Regulation Task Force Report of 2004 ('Make It Simple Make It Better') which recognised that more information on a label is not necessarily better. It stated:

> The current DG Sanco review (on labelling) should propose that no further mandatory label requirements should be issued, unless it is clear how extra space on the label is to be created.

7.7   Current voluntary nutrition labelling

7.7.1   Nutrition labelling

Although currently only required by law if a claim is made, around 75% of food manufacturers in the UK now provide nutrition labelling. This compares to 54% in Spain, 50% in Germany and 41% in Poland (EAS, 2004). The extent of the labelling may be constrained by available label space but many manufacturers label their products with the full eight Group 2 nutrients (see page 136). In the UK, tabular nutrition labelling is given on more than 75% of products in supermarkets due to the nutrition labelling of almost all of retailers' own-label products (EAS, 2004).

7.2.2   Salt equivalent labelling

The current EU legislation stipulates that only sodium needs to be labelled. However many retailers and manufacturers have moved to salt equivalent labelling. In its Delivering on our Commitments document of 2005, the Food and Drink Federation (FDF) set out its continued commitment to the labelling of salt (Food and Drink Federation, 2005):

More informative labelling
FDF members are committed to:

- working constructively with Government and other stakeholders to ensure the availability of clearer nutrition information under revised EU provisions.

Meanwhile FDF will encourage its members to provide on pack where practicable:

- full nutrition information as defined in current EU legislation even where this is not legally necessary
- Guideline Daily Amounts (GDAs) to provide a simple 'ready-reckoner'
- salt equivalence as well as the legally required sodium information.

A survey conducted by FDF showed that by the end of 2006:

- 97% of products, worth almost £3.3bn at retail value, would have full nutrition information on pack
- 58% of products, worth 31.5bn at retail value, would have the GDAs on pack
- 59% of products, worth £1.5bn at retail value, would have the salt equivalent as well as the legally required sodium information on pack.

It would seem simpler to not label sodium and simply label 'salt'. However this would not be strictly accurate as not all sodium in food is present as sodium chloride. For example milk and milk products contain a relatively high level of naturally occurring sodium that is not in the form of sodium chloride. Several food additives contain sodium, e.g. sodium nitrite preservative. Certain types of raw tomatoes contain naturally present monosodium glutamate, so that per 100 g, they contain quite a high level of sodium, none of which comes from sodium chloride. Bakery products contain a significant level of sodium bicarbonate and/or disodium diphosphate as raising agents.

As a guide, most manufacturers have worked out the salt equivalent by multiplying the sodium value by 2.5. For example a product which contains 400 mg of sodium per 100 g contains the salt equivalent of 1000 mg (or 1 g) of salt. Salt equivalent labelling is not ideal; the IGD undertook some qualitative consumer research in 1999 which showed that there was a lot of confusion about 'salt equivalence' with many consumers, particularly those with low education, believing that a salt replacement had been used. More consumer education about how to interpret the label may help to overcome this.

7.7.3   Accuracy of salt labelling

The accuracy of nutrition labelling comes in for criticism from time to time. It must be remembered that nutrients are components of products rather than ingredients, not necessarily being added as such but deriving both from direct addition (e.g., of sugar, salt or fat) and from their presence at varying levels in the various ingredients of a product. The scope for natural variation in the total quantity of nutrients in a product is, therefore, significant. Accordingly, the Directive, in requiring that declared figures be 'average values', states that: ' "Average value" means the value which best represents the amount of the nutrient which a given food contains, and reflects allowances for seasonal variability, patterns of consumption and other factors which may cause the actual value to vary.'

Added to these natural variations will be small variations in actual product composition (particularly for heterogeneous products) and the nature of product sampling and analysis, which mean that it is very difficult to get identical repeat analyses. Composition tables (such as McCance and Widdowson's *Composition of Foods* (FSA, 2002)) advise that the value is an average taken from a number of samples and that values may vary according to season, etc.

The current salt reduction work that is being undertaken by a number of manufacturers, to reduce the salt contents of prepared foods, also means that added sodium/salt is being reduced gradually over a period of time. This gradual reduction may only show up as a very small change in actual salt value, and so because of the cost of changing the label, this change is not always immediately reflected in the data on the label. The direct cost of a label change is around 2000–4000 Euro (EAS, 2004). This is largely for re-design and reprint and will increase if changes are timed such that large stocks of existing labels have to be destroyed.

7.8   Guideline daily amounts (GDAs)

The GDA labelling scheme is a 'ready-reckoner' tool that can help consumers compare how much of a certain key nutrient they are eating in a particular food with recommendations made by the UK government and nutrition experts.

Many of the major UK retailers have been labelling packs with GDAs since they were first developed by the Institute of Grocery Distribution (IGD) in 1998 (Institute of Grocery Distribution, 1998). IGD is an industry think-tank, which regularly conducts consumer research. In the summer of 2004 a decision was made by members of the food chain, under the auspices of IGD, to re-look at and extend GDAs so that they could be used by food manufacturers, retailers and the hospitality industry alike. Discussions between representatives of the food chain and some key academics led to the development of a range of GDA nutritional values for males, females, and children covering a range of ages. The technical rationale paper on how the GDA values were developed was distributed widely to the academic community and to industry members for comment, with an academic workshop, so that the scheme received 'buy-in' and approval. This GDA Technical Report can be found at: www.igd.com. The work has also been shared with the FSA who have approved the GDA values for salt and all the other nutrients (except for sugar which they are looking into themselves). FSA recommend that manufacturers and retailers use the GDA system on the back of packs (Food Standards Agency, 2005a).

GDA adult values are derived from the Estimated Average Requirements (EARs) for energy for healthy men and women aged between 19 and 50, of normal weight. The fat and saturates GDAs are based on the dietary reference values for these nutrients published by the Department of Health (Department of Health, 1991). Figures for salt are based on the 6 g per day recommendation made by COMA (Department of Health, 1994) then later by the Scientific Advisory Committee on Nutrition (SACN) in 2003 (SACN, 2003). Figures for children are similarly based on the COMA and SACN guidelines as appropriate.

An IGD GDA communication group, looked at how the GDA information should appear on the back of the pack. GDA format testing involved qualitative and quantitative consumer research. This research allowed a format recommendation to be made by IGD. Many food manufacturers and retailers are already labelling, or are about to label, back of packs with the GDA system. The final 'blue-print' for the GDA system of labelling was available early in 2006. Several leading food manufacturers and retailers are now considering a front of pack serving based scheme using actual and percentage GDA values for key nutrients.

7.8.1   Note on the development of the salt GDA

When IGD first set its GDA values in 1998, there was no value set for salt. However, later in 2000, an IGD working group set values for salt at 5g for women and 7 g for men. This was based on the COMA document on Nutritional Aspects of Cardiovascular Disease (COMA, 1994). In 2003, SACN produced its report on salt and made its recommendations for salt intake for adults and children (SACN, 2003). No distinction was made between male and female or boys and girls. As this is the most updated official guide on recommended intakes for salt, the present IGD group developing GDAs propose that these SACN values are set as the GDAs for salt.

IGD is also looking to set a GDA for sodium, but it will not recommend that this is used for labelling purposes. The setting of a sodium GDA would be useful as a benchmark for product development, taking account of all the sources of sodium in a food.

7.8.2   Status of GDAs within the EU

GDAs are intended as guidance to help consumers in their understanding of their recommended daily consumption of energy (calories), fat and saturates and a base against which the content of individual foods can be compared. Currently the GDA declaration is voluntary and is not covered by any EU or UK legislation. However, several European countries have also become interested in the concept of GDAs. The EU Commission has also expressed an interest, but it is too early yet to forecast whether they will become incorporated into European law. In the interim, the CIAA (Confederation of the Food and Drink Industries of the EU) has recommended that European food manufacturing members should use the 6g GDA for salt if using the GDA system.

Table 7.1 represents the best format as indicated by consumer testing. How-ever, due to size restrictions on a label, it may not be possible for all food labels to carry all this information. IGD have therefore set out a hierarchy of preferred

Table 7.1   Example of a back of pack label showing GDAs

One box: men, women and children

Typical values	Nutritional information		Guideline daily amounts		
	per 100 g serving	per 350 g serving	Women	Men	Children (5–10 years)
Energy – kJ	480 kJ	1680 kJ			
– kcal	115 kcal	405 kcal	2000	2500	1800
Protein	9.5 g	33.3 g	75 g	95 g	65 g
Carbohydrate	8.6 g	30.1 g	230 g	300 g	220 g
of which sugars	2.0 g	7.0 g	90 g	120 g	85 g
Fat	4.6 g	16.1 g	70 g	95 g	70 g
of which saturates	3.0 g	10.0 g	20 g	30 g	20 g
Fibre	1.5 g	5.3 g	24 g	24 g	15 g
Sodium*	0.3 g	1.1 g	2.5 g	2.5 g	1.6 g
*Equivalent as salt	0.8 g	2.8 g	6 g	6 g	4 g

formats for labels which have restricted space. They recommend that GDA tables carry a strap-line that explains how to use the table, who it is appropriate for and how some people's needs may vary.

7.8.3   Advantages of labelling with guideline daily amounts

- Consumers can see at a glance what their average dietary requirements are, and use this information to help them plan their meal, and see how it fits into a balanced diet.
- Consumers are already familiar with the concept of GDAs: research by IGD's consumer information group at the end of 2004 indicated that consumer awareness of GDAs is high, with 72% of consumers claiming to have seen them (Institute of Grocery Distribution, 2004a).
- The size of the pack may mean that the extent of GDA labelling may have to be limited in some cases. However, the pack could indicate where more detailed or specific information can be found if required (e.g. a GDA information website which could also give healthy eating advice).
- Unlike the 'signposting'* schemes based on thresholds and colours currently being looked at by the FSA (such as traffic light labelling, see page 146), GDAs take into account the portion size of the food. The categorising options being considered by FSA are based on *per 100 g* of food, so that the system

*   'Signposting': FSA wants to highlight the nutritional value of a food by using a simple front of pack indicator, which it is calling a 'signpost'. The most well known current example of such a 'signpost' is the proposed multiple traffic light, where each nutrient will be labelled red, amber or green according to the level of that nutrient in 100g of the product (see page 146). FSA is currently consulting on its preferred 'signposting' option of coloured multiple traffic lights with the hope that such a scheme will be adopted fully by the food industry.

quickly becomes unrealistic for foods where a portion size tends to be much smaller or larger than this.

- FSA's proposed 'traffic light scheme' does not take into account the frequency of consumption of various foods. GDAs allow people to see exactly how much of their daily intake a portion of a particular food will provide and they can quickly get a feel of how many calories, fat, salt, etc., they are consuming. Account must be taken of the fact that both portion size and frequency of consumption influence the overall amount of food consumed (Matthiessen *et al.*, 2003).
- Because the GDA label indicates the portion size, this can also act as a guide to a consumer and may help people to avoid serving themselves a larger portion size than is suggested!
- Although a GDA scheme, like any other new labelling scheme, will require backing by consumer education, it will actually provide opportunity to encourage consumers to eat a varied balanced diet. A subjective judgement scheme such as traffic lights would also require educational back-up, but such a scheme would not directly be teaching consumers about eating a variety of foods in the right proportions, appropriate to their needs.

7.9   Consumer use of nutrition labelling

FSA's fifth annual consumer attitudes survey (FSA, 2004) showed that nutrient content and labelling had replaced BSE as the biggest food concern among the British public. This survey also showed the amount of salt in food is now UK consumers' top concern, with fat and sugar content also making the top five. The number of people who said they were wary about the accuracy of nutrition labels rose from 34% in 2002 to 44% in 2004.

The last adult National Diet and Nutrition Survey (NDNS) showed that adults were consuming an average of 9.6 g of salt a day (NDNS, 2003). FSA hopes to get this level down to the SACN recommended 6 g a day by 2010. However, there has as yet been no evaluation of actual salt intakes on a national level to see whether the interest in eating less salt by the consumer has been transformed into action.

Many experts feel that such a high salt intake is contributing to excessive deaths due to cardiovascular disease, so FSA has been running a concerted salt campaign (www.salt.gov.uk) to encourage consumers to eat less salt. FSA has also been encouraging the food industry to use less salt in the food it produces and sells. The majority of the food industry has agreed, as a precautionary principle, to work at reducing salt levels in foods where feasible (i.e., from consumer taste acceptability and technical perspectives).

Research from the Food Standards Agency (FSA, 2005b) showed a rise in people making an effort to cut down on how much salt they eat, including taking account of the salt content of the food they bought. The FSA research showed that between August 2004 and January 2005 (which covers the period when an FSA public salt awareness campaign took place):

- there had been a 32% increase in people claiming to be making a special effort to cut down on salt (34% in August 2004, rising to 45% in January 2005)
- there had been a 31% increase in those who look at labelling to find out salt content (29% in August 2004, rising to 38% in January 2005)
- there had been a 27% increase in those who say that salt content would affect their decision to buy a product 'all of the time' (25% in August 2004, rising to 33% in January 2005).

In November 2004 IGD conducted a survey on the type of information consumers looked for when food shopping (IGD, 2004b). Results showed:

- The majority of consumers actively look for some information about products they buy when food shopping. Only 5% claim not to look for any information.
- The key information requirements are price and sell by date, which are relevant to approximately one third of consumers across different food types (fruit and vegetables, red meat and poultry, and processed foods, e.g. ready meals).
- Fewer consumers look for production and nutrition information, however specific information requirements are dependent on the food type.
- Only 11% of consumers looked for information about sodium/salt content on the back of pack, but this survey was conducted prior to the main impact of FSA's salt campaign. In comparison, for single nutrients, fat was the nutrient most consumers looked for (27%), followed by sugar and calories (15% each).
- Only 3% of consumers looked for low salt/sodium claims on the front of pack.

7.10   EUFIC (the European Food Information Council) findings (EUFIC, 2004)

Consumer focus groups recently conducted for EUFIC in France, Germany, Italy and the UK showed that consumers both understand the benefits of nutrition and are positive towards 'healthy and balanced eating'. However, whilst people know about certain nutrition basics, the terminology used on the label is not really understood. The research showed that consumers tend not to use the figures featured on nutrition labels as a tool for building a healthy diet.

Consumers believed the information on the labels to be accurate – but could not understand it. The role of the labels was not clear, nutrition information was often confused with the ingredients list, and in many cases people did not understand how to integrate the provided information into their daily dietary choices.

The research also showed that consumers were alienated by unclear language and would prefer to see terms they could relate to, which would help them determine what is important. The research also backed up findings that consumers do not understand the difference between sodium and salt.

The research highlighted several indicators about what consumers wanted from nutrition labelling:

- Labels should be readable, clear, attractive and well structured.
- The nutrition information should stand out.
- There needs to be directions to further help (websites, for example, where lifestyle guidance can be found).
- Consistency or uniformity of information is required across products.

EUFIC concluded that there are many things in the current labelling terminology that can be improved. However, as long as consumers lack a basic understanding of nutritional terms and requirements, the label information will be lost on them. There is an immediate need, therefore, for better nutrition education and improved nutrition knowledge. EUFIC suggests that this is now the big challenge for government, educators, health professionals and all operators in the food chain.

7.11   Non governmental organisations' views on nutrition labelling

Most consumer groups (such as the National Consumer Council and Which?) and several charities (such as the British Heart Foundation) are in support of signposts depicting traffic lights on the front of packs, such as those which are currently out for consultation by the FSA (FSA, 2005a). The traffic light model which is being proposed by the FSA has colour coding for four nutrients (which includes salt) (see Fig. 7.1). In order to assign the appropriate colour for each of the nutrients, threshold values have been set for each nutrient per 100 g of food. For example, the red traffic light threshold for salt in the FSA's scheme is proposed to be set at 1.5 g per 100 g. This would mean that a food with a salt

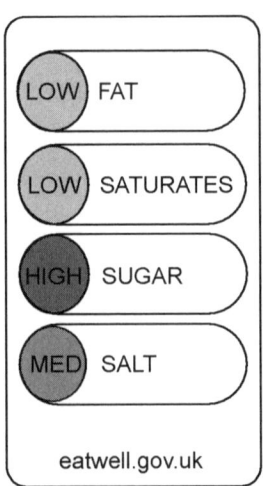

Fig. 7.1   The multiple traffic light.

content of 1.8 g per 100 g would have to carry a red traffic light for salt. It is proposed that the high criteria for nutrients is set according to the FSA definition of 'what's a lot' (see page 149), while the low definition is based on the levels in the current EU proposal on nutrition and health claims.

7.12 Why traffic light labelling is not accepted by the majority of food manufacturers

The majority of the food manufacturing industry is against such a subjective scheme as traffic light labelling for the following reasons:

- It only gives a very narrow 'picture' about the particular food. For example:
  - A food (such as a fortified breakfast cereal) may be a very good source of iron, or folic acid, suitable *inter alia* for women, yet it could end up carrying two out of four red traffic lights (for sugar and salt).
  - It would also not distinguish fat spreads compositionally designed to be more appropriate for people with high blood cholesterol – a substantial percentage of the population.
  - No account would be taken of the overall nutritional value of foods, including, for example, their micronutrients or fruit and vegetable and omega 3 content.
- Furthermore it provides little help on educating consumers about balanced eating, moderation or appropriate portion sizes.
- It is hard to develop a simple 'signpost' which can give appropriate information to an individual person. Thus such schemes may mislead consumers as they fail to give information about the adequacy of their own diet over time, or address the importance of physical activity. For example, a threshold scheme that classifies energy dense foods as 'less healthy', may not be sending the 'signpost message' appropriate for an active adolescent boy with a high energy requirement.
- If a food product were to bear just one red traffic light, a consumer's instant reaction would be that that food is not healthy: a red traffic light means 'STOP'. Indeed, FSA's provisional qualitative consumer research showed that some consumers *would* interpret a red traffic light as 'stop, don't eat it!' (Food Standards Agency, 2004) If this were to be the end result, it may have serious consequences for the availability of many of the food and drink products provided by the industry, yet *still* not encourage individuals to consume a balanced diet made up from a wide range of foods. It is far more likely that many of those who most need to take care of their diets would carry on as before and take no notice of any over-simplistic and potentially patronising 'signposting' scheme.
- If a food is to carry multiple traffic lights (i.e., a traffic light coding for each designated nutrient), it will probably carry the range of colours. In this case, how will a consumer know 'at a glance' whether that particular food is a 'healthy choice' or not?

- A simple front of pack traffic light scheme may deter the consumer from trying to understand the more objective and legally prescribed information on the back, which may include Guideline Daily Amounts. Consumers do not always need to absorb all the nutrition information about the food they buy at the point of purchase. They can plan their dietary intake by looking at a label in the home setting and thus take on board a much fuller picture about a particular food by reading the nutrition panel. In addition many food and drink products are linked to other sources of information such as company websites and consumer helplines.

- FSA has acknowledged that if a signposting scheme, based on categorising nutrients as red, amber or green, is to be used on front of pack, it would need to educate consumers about it. Indeed any labelling scheme will have to be underpinned by a very substantial education programme. The food industry feels it is more appropriate to use the opportunity to educate consumers on how to get the most from the label using the Guideline Daily Amount system. In this way consumers will also be able to learn how to build a diet appropriate for their own needs from the information on the label.

- Finally, much work is currently being done by the food industry to bring down levels of fat, salt and sugar. However, despite this some manufacturers for technical or consumer acceptance reasons just cannot bring down levels of certain nutrients beyond a certain point. When this point is checked against proposed schemes for assigning the traffic light colours, it has been found that, for some products, lowering nutrient levels does not take the product down from a red to an amber or from an amber to a green. Hence there is less incentive to make such changes as reformulation will not allow the product to improve its traffic light profile. An example of this would be bread; bread manufacturers have estimated that based on currently proposed schemes, it would be technically impossible to reduce the salt level of bread to the point where it would be able to carry a green traffic light for salt. Indeed the significant reductions that have taken place in bread salt levels have still not been able to move it from one traffic light category to a lower one. Factual, numerical information on nutrient content would, however, show the reduced levels in comparison with other products.

The food industry believes that many of the above shortfalls could be overcome by use of Guideline Daily Amounts (see page 143), backed by simple, factual consumer education.

7.13   FSA guidance on nutrition claims

In the past in the UK the Food Advisory Committee (FAC) has issued guidelines on nutrition claims. This body was abolished in December 2001 and some of its recommendations have been adopted by FSA. The items on salt from the 'FSA Guidance on Nutrition Claims in Labelling and Advertising' are listed below. In the UK manufacturers still tend to use these guidelines when making a claim:

- A recommendation that information in the nutrition panel on sodium levels should be accompanied by an equivalent salt figure.
- 'Low' salt/sodium should relate to no more than 40 mg (0.04 g) sodium/100 g or 100 ml of food.

FSA also suggest using the following 'rules of thumb' to judge 'what is a little and what is a lot' of sodium in ready-prepared foods:

- 0.5 g sodium or more per 100 g of food = a lot
- 0.1 g sodium or less per 100 g of food = a little.

This approach does not take account of the portion size of the food eaten. For example, certain foods, such as yeast extracts, are relatively high in sodium but are eaten in very small quantities, as an average serving is 4g. It has no legal status and is not used by industry in any significant way.

7.14   Codex guidelines on claims

The Codex Guidelines on Claims (CAC/GL 23-1997) establish general principles to be followed and leave the definition of specific claims to national regulations. Definitions are provided for a number of claims (nutrient content, comparative claims, nutrient function claims) as well as general requirements concerning consumer information in relation to claims.

Codex sets out a table of conditions for nutrient content claims. For sodium the following is suggested:

- low = not more than 0.12 g per 100 g
- very low = not more than 0.04 g per 100 g
- free = not more than 0.005 g per 100 g.

The values are the same as those suggested in the proposed EU Regulation on health and nutrition claims (see next section).

7.15   Proposed EU regulation on health and nutrition claims

There has never been harmonised EU legislation on claims, which is why the Commission introduced, in 1993, a proposal for a Regulation on nutrition and health claims. This is currently at an advanced stage of negotiation. This proposal includes the following criteria for claims on salt:

- *Low sodium/salt*: 'A claim that a food is low in sodium, and any claim likely to have the same meaning for the consumer, may only be made where the product contains no more than 0.12 g of sodium, or the equivalent value for salt per 100 g or per 100 ml'. 'In the case of foods naturally low in sodium, the term "naturally" may be used as a prefix to this claim.'
- *Very low sodium/salt*: 'A claim that a food is very low in sodium, and any claim likely to have the same meaning for the consumer, may only be made

where the product contains no more than 0.04 g of sodium, or the equivalent value for salt per 100 g or per 100 ml.' This would equate to 0.75 g of salt. In the case of foods naturally very low in sodium, the term "naturally" may be used as a prefix to this claim.'

- *Sodium-free or salt-free*: 'A claim that a food is sodium-free, and any claim likely to have the same meaning for the consumer, may only be made where the product contains no more than 0.005 g of sodium, or the equivalent value for salt per 100 g or per 100 ml.' This would equate to 0.0125 g of salt. In the case of foods naturally very low in sodium, the term "naturally" may be used as a prefix to this claim.'

7.16   Conclusions

Future EU claims legislation looks likely to refer to sodium/salt. This will provide harmonised quantitative criteria for the various sodium/salt claims. When the Nutrition Labelling Directive is revised, it will have to be made consistent with this. Sodium/salt (or possibly just salt) looks likely to be included in a core group of required nutrients for nutrition labelling, which might also be made compulsory, whether or not a nutrition claim is made. Sodium has, however, to be the reference element for analytical purposes. Accordingly, the salt value will be a 'salt equivalent' and an EU-agreed conversion factor will be needed (the accurate factor is close to 2.54, although 2.5 is generally used in the UK).

Consumers want information on food labels, but are not always confident at understanding it all. In order to reduce the confusion, there has been a general call from academics, the food industry and the EU Commission not to increase the overall amount of information on a label, including the nutrition information, but to make the information given as clear and understandable as possible.

As was stated at the beginning of this chapter, a label can inform but should never be used as a prime educational tool. An impact assessment undertaken for DG Sanco, European Commission, on the introduction of mandatory nutrition labelling in the European Union concluded:

> Nutrition labelling provides consumers with vital 'pegs' on which to
> hang their nutritional knowledge, but that knowledge needs to be gained
> and stimulated in other ways, be it consumer education or new
> marketing strategies.

There are other means by which manufacturers and retailers can provide additional information about the nutritional content of a product, including information on salt. Some suggestions are:

- Manufacturers' or retailers' telephone help-lines on nutritional issues.
- Referring to websites, which could give manufacturers' or retailers' product-specific information and link to particular, respected websites about salt to provide extensive information and education.

- Point of sale information. This could particularly tie in with events and campaigns such as blood pressure week or FSA's salt campaign.
- In store magazines/leaflets.
- In store nutritional advice.
- Advances in technology will open further routes to allow consumers to access information.

Consumer understanding of sodium/salt figures is necessary for practical value to be obtained from this information on food labels. There is evidently a need for further, and continuing consumer education on how to use nutrition labelling – a vital role for central government. This in turn will help motivate more consumers to read nutrition labels and facilitate their use in guiding dietary choices.

7.17   Sources of further information and advice

FDF
Food and Drink Federation, 6 Catherine St, London WC2B 5JJ
Tel: +44 (0)20 7836 2460
Website: www.fdf.org.uk

Foodfitness
Foodfitness, c/o Food and Drink Federation, 6 Catherine St, London WC2B 5JJ
Tel: +44 (0)20 7836 2460
E-mail: foodfitness@fdf.org.uk
Website: www.foodfitness.org.uk

A Guide to Nutrition Labelling: http://www.foodfitness.org.uk/guidetonutrition
labels.pdf
Salt and your health: publications@fdf.org.uk

IGD
IGD, Grange Lane, Letchmore Heath, Watford, Hertfordshire WD25 8GD
Tel: +44 (0) 1923 857141
E-mail: igd@igd.com
Website: www.igd.com/index.asp

BNF
British Nutrition Foundation, High Holborn House, 52–54 High Holborn, London WC1V 6RQ
Tel: +44 (0)20 7404 6504
E-mail: postbox@nutrition.org.uk
Website: www.nutrition.org.uk

EC
European Commission, DG Health and Consumer Protection, B-1049 Brussels
Website: www.europa.eu.int/comm/health/index.html

CODEX
http://www.codexalimentarius.net/web/index_en.jsp
or contact FSA

EUFIC
European Food Information Council, 19 rue Guimard, 1040 Brussels, Belgium
Tel: +32 2 50689 89
Website: www.eufic.org/gb/home/home.htm

FSA
Food Standards Agency, Aviation House, 125 Kingsway, London WC2B 6NH
Tel: +44 (0)20 7276 8000
Website: www.food.gov.uk

Specifically for salt:
info@salt.gov.uk
www.salt.gov.uk

Salt Manufacturers' Association
7 Bartley Road, Northenden, Manchester, UK M22 4BG
Tel: 07071 222425
E-mail: saltmanufacturers@msn.com
Website: http://www.saltinfo.com

7.18   References and further reading

BETTER REGULATION TASK FORCE REPORT (2004) *Make It Simple Make It Better*. London: Cabinet Office.

BLACK, A. and RAYNER, M. (1992) *Just read the label: understanding nutrition information in numeric, verbal and graphic forms*. London: HMSO.

COMA (1994) Report on Health and Social Subjects (46) Nutritional Aspects of Cardiovascular Disease. London: HMSO.

COMMISSION OF THE EUROPEAN COMMUNITIES (2003) Proposal for a Regulation of the European Parliament and of the Council on nutrition and health claims made on foods Brussels 16.7.2003 COM (2003) 424 final 2003/0165 (COD)

CO-OPERATIVE WHOLESALE SOCIETY LTD (2002) Lie of the label11. UK: Co-operative Wholesale Society.

DEPARTMENT OF HEALTH (1991) *Dietary Reference Values for Food Energy and Nutrients for the United Kingdom*. London: HMSO.

DEPARTMENT OF HEALTH (1994) *Nutritional Aspects of Cardiovascular Disease*. London: HMSO.

DEPARTMENT OF HEALTH (2004a) *Choosing a Better Diet: a food and health action plan.* London: Department of Health.

DEPARTMENT OF HEALTH (2004b) *Choosing Health: making healthier choices easier.* London: Department of Health.

EFRA SELECT COMMITTEE REPORT (2005) *Environment, Food and Rural Affairs Committee – Food Information – Seventh Report of Session 2004–2005.* London: TSO.

EUFIC CONSUMER WORK 2004: http://www.eufic.org/fr/heal/heal12_annex1.htm

EU Nutrition Labelling Directive1990 (90/496/EEC). *Official Journal of the European Communities* No L 276, 6.9.90, pp. 40–44.

EUROPEAN ADVISORY SERVICES (EAS) (2004) The introduction of Mandatory Nutrition Labelling in the European Union: Impact Assessment undertaken for DG Sanco, European Commission. SANCO/2004/D4/EAS/S12.378734. http://europa.eu.int/comm/food/food/labellingnutrition/nutritionlabel/impact_assessment.pdf

EUROPEAN HEART NETWORK (2003) A systemic review of the research on consumer understanding of nutrition labelling. Brussels.

FOOD AND AGRICULTURE ORGANISATION OF THE UNITED NATIONS AND THE WORLD HEALTH ORGANISATION (1998). Codex Alimentarius: Food Labelling Complete Texts: Codex Guidelines on Claims CAC/GL 23-1997.

FOOD AND AGRICULTURE ORGANISATION OF THE UNITED NATIONS AND THE WORLD HEALTH ORGANISATION (1998). Codex Alimentarius: Food Labelling Complete Texts: Codex Guidelines on Nutrition Labelling CAC/GL 2-1995.

FOOD STANDARDS AGENCY (2002) *McCance and Widdowson's The Composition of Foods*, 6th summary edition. Cambridge: Royal Society of Chemistry.

FOOD STANDARDS AGENCY (2003) *National Diet and Nutrition Survey of adults aged 19–64 years*, vol. 3. Food Standards Agency.

FOOD STANDARDS AGENCY (2005a). Consultation on a voluntary front of pack signpost labelling scheme for the UK, 16 November 2005. http://www.food.gov.uk/foodindustry/Consultations/consulteng/signpost2005eng

FOOD STANDARDS AGENCY (2005b) *Consumers choose less salt* (2 February) http://www.food.gov.uk/news/newsarchive/2005/feb/saltresearch.

FOOD STANDARDS AGENCY AND COI COMMUNICATIONS (2004) Consumer Attitudes to Food Standards Ref 0960/0205 Crown copyright UK.

FOOD AND DRINK FEDERATION (2005) *FDF Food and Health Manifesto*: Delivering on our Commitments https://www.fdf.org.uk/resources/final%2019%20sept%20manifesto_premier.pdf.

INSTITUTE OF GROCERY DISTRIBUTION (1998) *Voluntary nutrition labelling guidelines to benefit the consumer* (01.02): IGD http://www.igd.com/CIR.asp?menuid=36&cirid=78.

INSTITUTE OF GROCERY DISTRIBUTION (2004a) *Nutrition labelling – the consumers' choice* (15 November): http://www.igd.com/cir.asp?cirid=1303&search=1.

INSTITUTE OF GROCERY DISTRIBUTION (2004b) Food Labelling – Information Used by Shoppers (16 November) http://www.igd.com/cir.asp?cirid=1227&search=1.

MATTHIESSEN, J., FAGT, S., BILTOFT-JENSEN, A., BECK, A.M. and OVESEN, I. (2003) Size makes a difference. *Public Health Nutrition,* **6**(1): 65–72.

SACN (2003) *Salt and Health*. London: TSO. Published for the Food Standards Agency and Department of Health.

Part II

Strategies for salt reduction in food products

8

Technological functions of salt in food products

C. M. D. Man, London South Bank University, UK

8.1 Introduction

Salt (sodium chloride, NaCl) is a unique food ingredient that is used extensively in the home, in food service and in food manufacture. The practice of salting foods, which probably began shortly after the discovery of agriculture that led to the dawn of civilisation in Mesopotamia some 10,000 years ago, is one of the oldest ways of preserving foods (Desrosier, 1970). Pure salt is a transparent, colourless and crystalline powder with a specific gravity of 2.165. Its solubilities in water at 0°C and at 100°C are 35.7 g/100 g and 39.8 g/100 g respectively. It is hygroscopic; it absorbs moisture from damp atmospheres with a relative humidity above 75 per cent. It stimulates saltiness, one of the basic tastes, unmatched by any other chemicals. It has low toxicity, although excessive amounts of it have been known to cause fatal salt poisoning, particularly in very young children.

Today, besides its roles in human physiology, nutrition and health, salt performs a multi-purpose role in many manufactured foods and drinks (Brady, 2002). First and foremost, salt imparts a salty taste in a product, and it enhances the effect of other flavour constituents that are present in food. Second, in many food products it has specific properties that are important to the handling and processing of these products. Third, it acts as a preservative against microbial growth, primarily through its influence on water activity, and commonly in combination with other antimicrobial agents (Hutton, 2002). A summary of the many functions performed by salt in some food products is given in Table 8.1. In many of these products, it is often difficult to separate the functions of salt into distinct roles, which can be interrelated, and interdependent. For instance, in ambient-stable emulsified sauces (e.g., mayonnaise and salad dressings) and

Table 8.1 A summary of functions performed by salt in some food products (Brady, 2002; Hutton, 2002)

Food products	Functions performed by salt
Bread	• To impart flavour • To control yeast growth and fermentation rate • To assist product texture • To reduce spoilage
Breakfast cereals	• To impart flavour • To assist product texture
Margarine and spreads	• To impart flavour • To prevent spoilage • To control shallow-frying performance in some products
Sauces and pickles	• To impart flavour • To assist in preservation • To retain texture of pickled vegetables during storage prior to bottling • To inhibit clouding of vinegar in pickled products
Savoury snacks	• To impart flavour • To assist product texture in some expanded products • To act as a solid carrier of applied seasonings and flavours, enabling their accurate measurement and improving their flowability
Meat products	• To impart flavour • To assist in preservation • To increase water-holding capacity in some products, and increase meat binding in others
Cheese	• To impart flavour • To help reduce the metabolic activity of the starter-culture bacteria • To modify enzyme activity, thereby play an important part in the maturation of some cheeses

non-emulsified sauces (e.g., brown sauce and mint sauce), besides being a flavour modifier, salt is the second most important ingredient after vinegar in conferring the products' microbiological safety and stability (Jones, 2000). Attempts to reduce the level of salt in these products not only affect their taste and flavour but also their microbiological shelf lives, and in some cases, their texture and mouth feel. The following sections contain selected examples, which are intended to illustrate some of the many functions performed by salt in food.

8.2   Sensory effects of salt

Saltiness is one of the basic tastes imparted by salt, which is also used for its flavour modifying and enhancing effects in food and drinks. Classic salty taste is also given by lithium chloride (LiCl) and a few other inorganic salts. Chemically, it appears that cations cause salty tastes and anions modify them. That said, potassium and other alkaline earth cations produce both salty and bitter tastes. Among the anions commonly found in food and drinks, the chloride ion does not contribute a taste and is least inhibitory to the salty taste, while the citrate anion is more inhibitory than orthophosphate anions (Lindsay, 1996). The flavour-enhancing role of sodium chloride can be seen when sodium ions were eliminated from a mixture of amino acids, nucleotides, sugars, organic acids, and other compounds known to mimic the flavour of crab meat, the resultant mixture was totally lacking in crab-like character. Besides this tendency to enhance meaty flavours, salt is also known to decrease the sweetness of sugars, to give the richer, more rounded flavour that is desired in many confectionery products (Coultate, 2002). The sensory attributes of a food or drink tend to be perceived in the following order (Meilgaard et al., 2000):

- appearance
- odour/aroma/fragrance
- consistency and texture
- flavour (aromatics, chemical feelings, taste).

In practice, most or all of these attributes overlap in the process of perception, and it is therefore very difficult to evaluate them separately and independently.

In savoury snack products such as potato crisps and allied products, which are consumed primarily for pleasure and enjoyment rather than for their nutrition contents, the taste of salt appears to be one of the key sensory characteristics that determine consumer acceptance. A study conducted in the US looked at modelling consumer preference of commercial toasted white corn tortilla chip products, employing 80 consumers of age 18–35 and a sensory panel of nine trained assessors (Meullenet, 2003). Proportional odds models in conjunction with principal components were used for internal and external preference modelling of tortilla chip overall consumer acceptance. The study found that flavour was the most important attribute to consumer overall acceptance followed by the interaction of appearance by flavour and texture. It was shown that one flavour attribute (saltiness) and one texture attribute (crispness), and in that order of importance, contributed significantly to increase consumer overall acceptance. The authors suggested that flavour (and saltiness) was the most influential variable on consumers' purchase intent for tortilla chip products.

Analysed salt content was found to correlate with overall acceptance in reduced fat Edam and full fat Emmental-type cheeses in a Finnish study aimed to relate sensory attributes to consumer liking in cheeses with different fat content (Ritvanen et al., 2005). While results of this research cannot be generalised, the findings did throw some light on the sensory effects of salt in certain types of cheese. In Emmental-type cheeses, overall appeal of the product correlated significantly ($p < 0.05$) with the degree of sour and salt flavour. Comparing full fat and reduced fat Emmental cheeses, salt content was found to correlate significantly with pleasantness of flavour and mouth feel ($p < 0.05$) in the full fat sub-group only while saltiness correlated significantly with pleasantness of mouth feel in both reduced and full fat Emmental cheeses ($p < 0.05$). The authors maintained that findings such as these could help identify factors that would affect cheese choice.

As would be expected, sensory saltiness was found to be highly correlated with salt content in a study involving 14 different dry fermented lamb sausages in Norway (Helgesen and Naes, 1995). Moisture contents of the products ranged from 34.4 to 49.6%, salt contents from 4.1 to 7.0%, and water activity values from 0.81 to 0.92. Sausages with the highest intensity of juiciness were high in water content, had high water activity and low salt content. As salt increased, sausages became drier and harder. This effect on texture is undoubtedly attributable to the influence of salt on water activity. The salt level in processed meats has been declining over the years. Alternative salts have been used in their manufacture but potassium chloride, which leaves a bitter aftertaste, presents formulation problems (Reddy and Marth, 1991). The main reason for salt addition in many meat products, of course, is its preservative effect.

The level of salting is critical to obtaining a satisfactory product in sauerkraut production (Binstead et al., 1971). Besides facilitating and enabling desirable fermentation to take place, salt helps to maintain the crisp texture of the cabbage by withdrawing water from the vegetable, and inhibiting endogenous pectolytic enzymes, which otherwise cause the product to soften. Salt also contributes to the flavour of the product. Similarly, despite the development of non-fermentation techniques in the production of preserved cucumbers, the traditional lactic fermentation following pickling in brine does have a number of advantages over the more modern processes. Fermented cucumbers have flavour and texture characteristics, which the other products do not have. Traditional pickling in brine is a simple method, does not require special equipment, and is more economical with energy than techniques based on direct acidification followed either by pasteurisation or refrigeration (Adams and Moss, 2000).

Salt plays a flavour-enhancing role in product category such as ready-to-eat breakfast cereals as the taste of saltiness is not generally apparent to the palate. Table 8.2 gives declared salt contents of some popular breakfast cereals in the UK, which contain added salt. For a few of these, versions that have 'no added salt' actually exist. Oat-based products, too, do not contain any added salt. Given that the UK Food Standards Agency (FSA) regards 1.25 g salt or more per 100 g to be 'high in salt' (FSA, 2005), there appears to be scope for breakfast cereal manufacturers to reduce the salt content of some of their products. Whether or not this will happen, will depend on consumers' reactions through market research, as no manufacturer would want to see a drop in sale following a reduction in the use of added salt in its products (Brady, 2002).

In canned foods, salt is used mostly for flavouring purposes as an appropriate scheduled heat process, which either confers commercial sterility or ambient stability is central to the microbiological safety and stability of canned foods. There is therefore scope for reducing the salt content of canned products. In the UK, Asda Supermarket announced that all own-label canned vegetables would

Table 8.2   Salt contents of some popular breakfast cereals in the UK, which contain added salt

Breakfast cereal	salt (g per 100 g)
Corn flakes	1.8
Muesli	0.38
Whole wheat grain cereal	0.68
High-fibre wheat bran cereal	2.25
Bran enriched wheat flakes	2.0
Toasted crisp rice	1.65
Mixed whole grain rings	1.2
Mixed cereal flakes with dried fruits and nuts	1.5
Fruit and fibre cereal	1.5

have no added salt by the end of 2006 (*Food Quality News*, 2005). Consumers who use salt-free canned vegetables in preparing their own dishes will probably not tell the difference, or they will have to add salt 'to taste' if the vegetables are to be used on their own.

A comprehensive review of the sensory effects of salt is available (Hutton, 2002), which covers the following types of product in varying degrees of detail: soup; gravy; butter and spreads; bread, cakes and buns; biscuits; pastry; cheese; breakfast cereals; pasta and noodles; meat and meat products; savoury snacks; chocolate confectionery; ready meals; and fish products.

8.3   Processing and related properties of salt

8.3.1   Cheese

All varieties of cheese share a common basic technology in which starter cultures of lactic acid bacteria usually play a key role in the development of flavour, texture and keepability of the final products. Salt controls cheese ripening principally through its effects on water activity. The major effects of salt are (Guinee and Fox, 1993):

1. control of growth of the starter cultures and their activity;
2. control of the various enzyme activities in cheese;
3. syneresis of the curd resulting in whey expulsion and hence in a reduction in the moisture content of cheese, which also affect 1 and 2 above;
4. physical changes in cheese proteins, which influence cheese texture, protein solubility and probably protein conformation.

Inhibition of starter activity depends on the method of salting; the precise concentration of salt at which this occurs depends on the starter used, that is, species and its strain. For Emmental cheese, with a salt concentration of ~0.7% being the least heavily salted among major cheese varieties, salting is done by immersion of moulded cheese in brine solution. For Blue cheeses, which are among the most heavily salted varieties with a salt concentration of 3–5%, salting is done by rubbing of dry salt or a salt slurry to the surface of the moulded curds. In all major varieties, salt addition is made after curd formation, which plays a major role in regulating and controlling cheese microflora, as well as in regulating the pH of cheese. All this in turn influences cheese ripening, texture and flavour.

In many types of cheeses, including bacterially ripened and mould-ripened cheeses, the initial proteolysis is catalysed by residual coagulant, the rennet, which contains proteolytic enzymes chymosin, pepsin or microbial proteinases. The hydrolysis of α_{s1}-casein by these enzymes occurs over a very wide range of salt concentrations but is at an optimum around 6%. In contrast, proteolysis of β-casein by the same enzymes is strongly inhibited by 5% salt. α_{s1}-casein undergoes considerable proteolysis during ripening while β-casein remains unchanged until the later stage of ripening. A certain level of salt, i.e. ~5% is also necessary to prevent the development of bitterness in cheese (Guinee and Fox, 1993).

In making cheese from renneted or acidified milk, syneresis is an essential step (Walstra, 1993). Syneresis is important because, ultimately, it determines the moisture content of the final cheese. In practice, the rate of syneresis affects the method of processing, thereby the equipment and time needed, and the losses of fat and protein in the whey. And, in relation to other changes (e.g., acidification, proteolysis, inactivation of rennet enzymes) the rate of syneresis affects water content of the cheese, its composition and properties such as shelf life. While cutting undoubtedly initiates and encourages syneresis in the manufacture of hard, semi-hard and semi-soft cheese, and where it is employed, scalding enhances syneresis and whey expulsion, salt draws moisture from the curds by pseudo-osmosis. In reducing the moisture content of cheese via syneresis, salt also reduces the water activity, a_w, of cheese, which in the case of young cheeses particularly those containing more than 40% moisture, is determined almost entirely by salt content according to the equation, $a_w = 1 - 0.033m$, where m = molality of NaCl in cheese moisture (Guinee and Fox, 1993).

8.3.2   Meat and meat products

Besides its better known preservative effect, salt performs a variety of functions in meat and meat products. Salt has a tenderising action on meat. This in part is due to an enhanced water-holding capacity in meat. Work done many years ago at the UK ARC Meat Research Institute, Langford, Bristol, led researchers to conclude that (Offer *et al.*, 1984):

1. myofibrils swell in high concentrations (0.4–1.5 M or about 3–9%) of sodium chloride;
2. the degree of swelling is more than enough to explain the water uptake in meat processing;
3. the concentration of sodium chloride required for swelling is similar to that found for comminuted meat.

It was also found that pyrophosphate exerted a synergistic effect on salt such that its effect on the water-holding capacity of meat was enhanced.

Salt also increases meat binding; this is of great importance in the manufacture and quality of many meat products. Although different types of meat products such as emulsified (e.g., some sausages), particulate (e.g., hamburgers and meat balls), and formed meats (e.g., some ready-to-eat hams), vary greatly in their methods of manufacture and product texture, they do have one thing in common. This is their ability to bind their constituent meat components together to form a cohesive product. The mechanisms by which salt increases the binding ability of a protein matrix are:

- by increasing the amount of protein (myosin) extracted, which forms a salt and protein complex, like a kind of 'cement' binding pieces of meat together;
- by altering the ionic and pH environment such that the resultant heat set protein matrix forms a coherent three-dimensional structure (Schmidt and Trout, 1984).

In meat emulsions, salt performs an additional function in that it loosens the myofibrillar proteins and increases their ability to emulsify fat, especially at pH values near their iso-electric point. In these products, the ability of muscle proteins to hold fat is as important as the ability to hold water. Thus in comminuted and formed meat products salt is involved in three important interactions with the myofibrillar proteins: protein-water (water-holding), protein-protein (meat-binding) and protein-fat (fat-binding). The relative importance of these interactions obviously varies depending on the nature of the meat product. Besides these interactions, in cured meats, if nitrosomyoglobin, the pigment responsible for the attractive red colour, is converted to nitric oxide myohaemochromogen in which the globin portion of the molecule is denatured by salt, the stability of the red pigment is enhanced (Lawrie, 1974).

Not all the influences of salt in meat are desirable. It has an accelerating effect on the oxidation of fat. Consequently, cured meats are more liable than fresh to oxidative deterioration in the fat. It is believed that salt accelerates the reaction catalysed by lipoxidase present in the muscle, thereby promoting the development of oxidative rancidity (Lawrie, 1998).

8.3.3   Flour confectionery and other related baked goods

This category includes a huge group of largely heterogeneous products, which contain wheat flour as their main ingredient. From a physico-chemical point of view, products range from white bread in which full gluten development and starch gelatinisation take place, to short pie pastry where some starch gelatinisation occurs, and gluten development is avoided by the use of a weak flour with low/poor protein quality. Whatever the product, salt is always used in the formulation. In bread manufacture, salt has a profound effect on the quality of the final product. Salt influences the rate of yeast fermentation in bread. An insufficient amount of salt will result in excessive yeast fermentation manifested by outsize dough pieces at the end of proof and oversized bread with an open grain and poor texture (Williams, 1975).

Increased salt levels tend to reduce the rate of yeast fermentation due to the physical effect of high osmotic pressure and as the membrane permeability of yeast cells and hence their metabolic rate, begin to be affected. Generally the higher the level of salt, the longer is the proof time, and the taste of the finished good invariably gets saltier (Hutton, 2002). Salt also has a significant effect on the properties of wheat gluten, making it more stable and less extensible; the overall effect is less sticky dough, which is easier to handle. The biochemical basis for these effects is believed to be competition for hydrogen bonding sites in the gluten protein molecules, particularly the glutenin fraction, by the sodium ions (Matz, 1992). In fruit pie pastry, and quiche pastry, there is evidence to suggest that salt, at the low levels typically used (0.5–1.0% of dough weight), affect the machining characteristics of the dough, and the degree of shrinkage of the final product (Man, 2005). More work will need to be done on these products, however.

Similarly, short pastes used in the manufacture of meat pies, pasties and associated products are very sensitive to the amount of water used in the recipe. It is conceivable that salt can affect short pastes in the same way as a small amount of sugars, which is known to soften the paste and improves its handling properties (Hodge and James, 1981). In rotary moulded short-dough biscuits such as Lincoln biscuits, it was shown that a change in the level of salt from nil to twice normal slightly increased biscuit length and width, but caused a marked reduction in stack height. The net effect of these changes was to produce a harder, more dense biscuit of increased packet weight (Lawson et al., 1981). It is obvious that while salt has many important effects in flour confectionery and related products, they are often product and process-specific, and not necessarily always desirable.

8.3.4   Functional drinks

Intense exercising in high temperatures can result in dehydration, hyperthermia, heat shock or even death as a lot of water is lost through sweating and breathing. Remedy usually includes rehydration and energy replenishment. In recent years, functional drinks have appeared on the market, which claim to help athletes to hydrate and to regain energy rapidly. It is now known that carbohydrate, in the form of sucrose or glucose, is the most important nutrient for high-intensity performance, and, its content is the most important factor influencing the rate of gastric emptying of rehydration drinks. Most sports rehydration drinks, containing either sucrose or glucose, or both, are either slightly hypotonic or isotonic (i.e. ~300 mOsm/litre) as hypertonic solutions reduce the rate of net water absorption (Brouns and Kovacs, 1997). Indeed, the addition of sodium in the form of salt and sucrose and/or glucose to rehydration drinks is known to stimulate water absorption, influence hormones involved in the regulation of sodium and fluid retention, improve water retention during post-exercise rehydration, and supply carbohydrate energy, which will help to delay the onset of fatigue (Brouns, 1997). The inclusion of salt in sports drinks is for these physiological reasons rather than for it to perform a particular function in the drinks.

8.4   Preservative effects of salt

Salt is a very effective food preservative provided by nature. It exerts its preservative effects primarily by reducing the water activity of food, which, in practice, is used as a measure of the amount of water available for microbial growth. As the water activity of their environment is reduced, the number of groups of micro-organisms able to grow decreases. Table 8.3 gives a guide for the minimum water activities at which active microbial growth can occur (Adams and Moss, 2000). The water activity (a_w) of a substrate such as food is defined as the ratio of the partial pressure of water in the atmosphere in equilibrium with the substrate, P, compared with the partial pressure of the atmosphere in equilibrium with pure water at the same temperature, P_0. This is numerically equal to the equilibrium relative humidity (ERH) expressed as a fraction rather than as a percentage. Thus the relationship is given by:

$$a_w = \frac{P}{P_o} = \frac{1}{100} \, ERH \qquad (8.1)$$

Water activity is a colligative property and is therefore dependent on the number of soluble species (i.e., molecules or ions) in solution rather than their size. Salt which ionises completely in water into sodium and chloride ions, which collect water molecules about them through ion hydration making water unavailable for microbial growth, is therefore more effective at reducing the water activity of food than other ingredients like sucrose on a weight-for-weight basis. Besides this dehydration effect, the direct effect of the chloride ion, reduced oxygen tension and interference with the action of enzymes as water availability decreases all contribute to the preservative action of salt.

Table 8.3 Minimum water activities at which active microbial growth can occur (Adams and Moss, 2000)

Group of micro-organism	Minimum a_w
Most Gram-negative bacteria	0.97
Most Gram-positive bacteria	0.90
Most yeasts	0.88
Most filamentous fungi	0.80
Halophilic bacteria	0.75
Xerophilic fungi	0.61

8.4.1 Packaged cakes

For almost half a century, the baking industry in the UK has used the principle based on control of ERH to achieve desired mould-free shelf lives (MFSLs) for packaged cakes, sponges, pastries and related products. As sucrose is the main ingredient in these products, it has been used as a reference material and given a value of 1, assuming it is dissolved in water and goes fully into solution. In calculating ERHs of cakes and similar products for the estimation of their MFSLs, a coefficient called the 'sucrose equivalent' (SE) is used, which expresses an ingredient's effect on a_w (or ERH) compared with the effect that the same weight of sucrose would have. On this basis, salt has a SE of 11 (see Table 8.4); it is therefore 11 times more effective in reducing ERH than an equivalent amount of sucrose, assuming it is dissolved completely in the recipe. In the UK, typical levels of salt of 0.3–0.5 per cent of packaged cake and sponge recipes (Bent, 1997) equate roughly to 0.4–0.6 per cent of salt in the finished products. Compared with sucrose and other humectants such as glycerol, these salt contents make only a minor contribution towards the reduction of ERH, and the resultant MFSL. In practice, its use in the control and extension of MFSL of cakes is limited by its taste – saltiness, which is not a taste normally associated with flour confectionery, and which coming from 0.4–0.6 per cent of salt is precariously near its threshold of detection by the consuming public (Meilgaard et al., 2000).

8.4.2 Meat and meat products

Meat and meat products were probably among the first foods to be preserved with salt for ambient stability. Today, as a result of developments in food preservation such as canning, freezing, refrigeration and gas packaging, these products are rarely preserved with salt alone but in combination with other preservative agents and/or techniques. Also, in response to consumer preferences for less salty

Table 8.4 Approximate sucrose equivalents of some common cake ingredients (Bent, 1997)

Ingredient	Sucrose equivalent
Sucrose	1.0
Flour	0.2
Fat	0
Margarine, butter	0.2
Whole egg	0
Skimmed milk powder	1.2
Baking powder	3.0
Sodium chloride	11.0
42 DE glucose syrup solids	0.7
Sorbitol	2.0
Glycerol	4.0

products in recent years, average salt levels in products such as cooked ham, raw bacon and cooked corned beef have been halved between 1932 and 1979 in the US (Cerveny, 1980), and in the UK, salt contents of cured meats now typically fall between 2 and 5% compared with concentrations as high as 15% in traditional products (Wilson, 1981). Nevertheless, salt together with nitrite remain the essential ingredients in the preservation of cured meats, especially in the inhibition of spore-forming bacteria such as *Clostridium botulinum,* which will survive the heat process typically applied to many cured meats. The mechanism of the preservative action of nitrite is still poorly understood, partly due to the complexity of the interaction of a number of factors including salt content, pH, the heat process applied, and in some cases, other ingredients such as ascorbate and polyphosphates. The use of nitrite and nitrate in food has attracted much scientific as well as public attention since it was discovered in the 1950s that *N*-nitrosamines, formed by the reaction of nitrite with secondary amines, especially at low pH, can be carcinogenic (Adams and Moss, 2000). Although the use of nitrite and nitrate in food is controlled by legislation, there has been growing pressure for their levels in meat products to be reduced. In view of the complex interactions mentioned above, it is of paramount importance that any further attempts to reduce salt or nitrite, or both, in meat products do not result in microbiologically unsafe foods.

Similar developments are taking place in other parts of the world. In China, for example, experimental studies on the colour, flavour and taste have been conducted on the famous *jin-hua* ho tui. This is a traditional raw ham known for more than 800 years in China, and is much appreciated for its red muscle, white fat, characteristic flavour and taste, and the long ambient shelf life. It is a bone-in ham cured with salt only and ripened under carefully controlled environmental conditions. The salt contents of *jin-hua* ham can be as high as 15%. Recognising the fundamental requirement not to compromise micro-biological safety, attempts have been made to develop *jin-hua* ham with improved eating quality, which contains 5–7% salt, based on the intelligent application of hurdle technology principles. The combined method of preserving the ham that will be successful is likely to involve hygienic handling and processing, temperature control, and the careful control of a_w with salt (Leistner, 2000).

8.4.3   Pickled vegetables

Salt is pivotal to the production and preservation of many pickled vegetables. Pickled cabbage, gherkin, onion and olives are popular in Europe. Salt can be used on its own in the form of dry salt or saturated brine, although nowadays dry salting (2–3% w/w) is mainly used for sauerkraut production. It can also be used in a dilute form together with lactic acid, or vinegar. Raw vegetables destined for pickling have an extensive flora of micro-organisms, the majority of which are inhibited and prevented from causing spoilage when the vegetables are placed in a brine giving 8–11% equilibrium salt content overall (Ranken *et al.*,

1997). Thus salt allows desirable fermentation changes to take place and inhibits the growth of undesirable micro-organisms.

Heterofermenters such as *Leuconostoc mesenteroides* produce lactic acid, carbon dioxide and traces of alcohol and acetic acid. Acid accumulation eventually leads to fermentation by the more acid-tolerant homofermentative lactobacilli such as *Lactobacillus plantarum*, which converts sugars almost entirely to lactic acid, contributing further to the preservation of the vegetables. Thus salt plays a vital role in sorting the micro-organisms permitted to grow. The vegetables are usually freshened or debrined with potable water to about 5% salt in the freshened vegetables, before they are used to make pickles. Brined olives are not freshened but are packed either in the original brine, or repacked in brine containing 8–10% salt and 0.5% lactic acid (Ranken *et al.*, 1997). In pasteurised products, which have been semi-preserved in weaker brine or vinegar, salt is still one of the preservation factors, and it provides some protection against microbial spoilage after the product has been opened in the home.

Although pickling is an established way of preserving vegetables, it continues to find applications in many parts of the world. A study carried out at the University of Horticulture and Forestry, Nauni-Solan, India, found that blanched cauliflowers steeped in 10 and 15% salt solution containing 0.2% potassium metabisulphite remained acceptable for up to 180 days, after which the vegetable was freshened and prepared into pickle and pakora. The latter were ranked above acceptable by a panel of assessors using various quality attributes. Furthermore, the process is attractive because of its low cost and low energy consumption compared with refrigeration and freezing (Barwal *et al.*, 2005).

8.4.4   Ambient-stable sauces

These include the emulsified sauces such as mayonnaise and salad dressings, and the non-emulsified sauces such as tomato ketchup, brown sauce and horseradish sauce. These sauces have one thing in common, in that they rely in the main on acetic acid or vinegar, and salt and sugar to maintain their ambient stability. A code of practice – *the Code for the Production of Microbiologically Safe and Stable Emulsified and Non-Emulsified Sauces Containing Acetic Acid* issued by the European Trade Association, CIMSCEE (Comité des Industries des Mayonnaises et Sauces Condimentaires de la Communauté Economique Europeenne) (CIMSCEE, 1991) states that an intrinsically stable sauce is one where acetic acid tolerant organisms will not grow and that pathogens, e.g. *Salmonella* cannot survive in the sauce. The code specifies micro-organisms to be used in challenge testing of the sauce and recommends a twofold examination, covering both microbiological safety and stability. Furthermore, the code provides two formulae, which may be used to predict microbiological safety and stability respectively:

$$15.75(1 - \alpha) \text{ (total acetic acid\%)} + 3.08(\text{salt\%}) + (\text{hexose\%})$$
$$+ \ 0.5(\text{disaccharide\%}) + 40(4.0 - \text{pH}) = \Sigma_s \qquad (8.2)$$

For any sauce based on acetic acid, if the value of this formula (Σ_s) exceeds 63, safety from microbial pathogens is theoretically ensured.

$$15.75(1 - \alpha) \text{ (total acetic acid\%)} + 3.08(\text{salt\%}) + (\text{hexose\%})$$
$$+ 0.5(\text{disaccharide\%}) = \Sigma \qquad (8.3)$$

For any sauce based on acetic acid, if the value of this formula (Σ) exceeds 63, microbial spoilage should not occur.

Salt features in both formulae, and it clearly contributes significantly to the preservation of this type of sauce, although both formulae are not really applicable to the more modern types of condiment sauces, which may be of vastly different composition, pasteurised, chilled, or contain permitted preservatives.

8.4.5   Chilled foods

In the UK, since the 1980s, chilled foods have seen spectacular growth such that today there is a very wide range of chilled products sold in supermarkets. Parallel to this development and expansion in chilled foods, there have been a number of significant and fruitful developments in the area of predictive microbiology, both in the UK, and in other parts of the world notably in the US and Australia. The net result of these advances is that it is now possible to use software packages to predict the microbiological shelf lives of chilled foods. The two most widely known packages are the Food MicroModel (FMM), the product of a UK government-funded microbiological modelling programme, and the Pathogen Modeling Program (PMP), which was developed by the Eastern Regional Research Center (ERRC) of the United States Department of Agriculture (USDA). The FMM and PMP datasets, supplemented with additional information and data, have now been unified in a common database called ComBase – a free web-based database of food microbiology data, following a structure developed by the Institute of Food Research (IFR), Norwich, UK. ComBase is maintained by a consortium established in London on 5 May 2003, consisting of the UK Food Standards Agency, the IFR, the USDA Agricultural Research Service and its ERRC.

Chilled foods, by definition, are not shelf-stable. They usually have a short shelf life as the word 'chilled' seems somehow to have been perceived to mean 'fresh', an important selling point in the fiercely competitive food market. While salt is not primarily added to chilled foods for its preservative properties, as they have tended to be preserved by a combination of mild techniques such as hygienic handling and processing, pasteurisation, strict temperature control and gas packaging, salt at 3.5% in conjunction with an adequate chill temperature is known to be important for the preservation of vacuum-packed or modified atmosphere packed fish products against the risk of *Clostridium botulinum* (Betts, 1996). As a further illustration of the preservative properties of salt, the effect of salt contents on the growth of *Aeromonas hydrophila*, an emerging food-borne pathogen because of its ability to grow at chill temperatures, can be modelled using the PMP; the result is shown in Fig. 8.1.

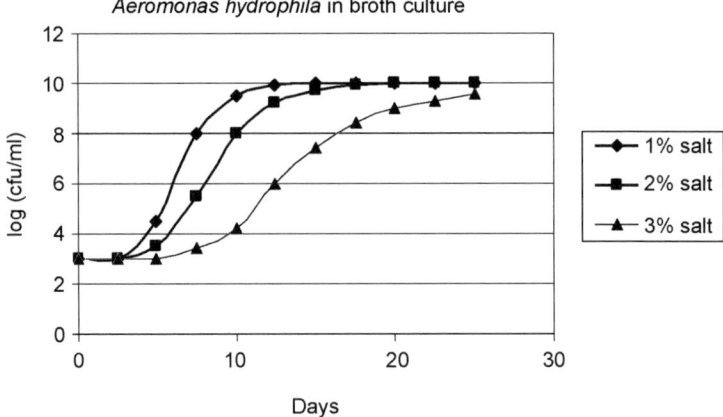

Fig. 8.1   The effect of salt contents on the growth of *Aeromonas hydrophila* (drawn from predictions by the PMP (Version 7.0, 2005)).

The preservative effects of salt are not universal, however. It has been demonstrated that salt stress proteins were induced in *Listeria monocytogenes* in response to salt stress (3.5% (w/v)) (Duché *et al.*, 2002), which could help to explain why the organism is frequently isolated from food containing high quantities of salt, such as cold-smoked salmon (Vogel *et al.*, 2001). The genetic basis of salt and alkaline tolerance of *Listeria monocytogenes* is now being investigated by the European *Listeria* Genome Consortium (Gardan *et al.*, 2003). Similarly, it was shown that at pH 5.0, salt (4–6 g per 100 g) in the presence of lactate (1.5 g per 100 g) promoted growth of the non-toxigenic *E. coli* O157:H7. These findings have potential implications for the use of acid and salt as a preservation treatment for the inhibition of *E. coli* O157:H7 in food such as chilled dairy products (Jordan and Davies, 2001).

8.5   Future trends

In response to the UK FSA's recommendations on an overall reduction in salt intake to reduce the effect on blood pressure, the UK Food and Drink Federation announced a few years ago an industry-wide programme to reduce salt – or more precisely sodium – in breakfast cereals, soups and sauces. The programme involves (IFST, 2003):

- a robust methodology for measuring sodium reduction; showing a 16% reduction in sodium achieved since 1998 in the breakfast cereal sector;
- an expectation that new products coming to market will continue the trend for products with lower sodium levels;
- a commitment to 10% sodium reduction for ambient soups and sauces by end of 2003;
- 'Operation Neptune' aimed at 30% reduction of salt in soups/sauces over 3 years.

It is clear from this review that salt is indeed a unique food ingredient, which performs a wide range of often product-specific functions in food. While some ingredients (e.g., potassium chloride) can perform a few of its functions, no ingredient can perform all the functions offered by salt. The food industry in the UK is expected to continue to co-operate with the Government's campaign to reduce the use of salt in the processed foods sector, and quite rightly so, as the lion's share of the nation's dietary sodium comes from processed foods (IFST, 2003). However, as more and more salt is being taken out of processed foods, future reductions will be more selective as no food company will want to put the acceptability of its products at risk or knowingly produce unsafe foods where their microbiological safety depends critically on the salt content. Its preservative effect aside, the role of salt in food processing, and in the creation of organoleptic quality of food requires more research on the basis of which further salt reduction in processed foods may be possible. It is therefore welcome news that a one-year feasibility project under the UK FoodLINK programmes, funded fully by the Department for Environment, Food and Rural Affairs (DEFRA) has begun at the University of Nottingham, UK. The project aims to assess the role of salt in the processing of starch-based foods, with emphasis on snacks, breakfast cereals, bread, soups and sauces (Farhat, 2005). It is hoped that further research into the non-preservative role of salt in food will follow.

8.6   Sources of further information and advice

Information about the technological functions of salt in the manufacture of food and drink products is scattered in both the primary and secondary literature. Two relatively recent and very useful papers concerned specifically with food products in the UK, one on 'the usage and functionality of salt' (Brady, 2002), and one on 'technological functions of salt' (Hutton, 2002), have been published in the *British Food Journal*. Other than these, readers will have to consult scientific journals or textbooks on specific food categories such as meat, dairy products and so on for detailed information about the technological functions of salt in these products.

8.7   References

ADAMS M R and MOSS M O (2000) *Food Microbiology*, 2nd edn, The Royal Society of Chemistry, London.

BARWAL V S, SHARMA R and SINGH R (2005) Preservation of cauliflower by hurdle technology. *Journal of Food Science & Technology*, 42(1), 26–31.

BENT A J (ed) (1997) Bennion and Bamford's *The Technology of Cake Making*, 6th edn, Blackie Academic & Professional, Glasgow.

BETTS G D (1996) *Code of practice for the manufacture of vacuum and modified atmosphere packaged chilled foods with particular regard to the risks of botulism,*

Guideline No. 11, Campden & Chorleywood Food Research Association, Chipping Campden.

BINSTEAD R, DEVEY J D and DAKIN J C (1971) *Pickle and Sauce Making*, 3rd edn, Food Trade Press, London.

BRADY M (2002) Sodium – Survey of the usage and functionality of salt as an ingredient in UK manufactured food products. *British Food Journal*, 104(2), 84–125.

BROUNS F (1997) Functional foods for athletes. *Trends in Food Science and Technology*, 8, 358–363.

BROUNS F and KOVACS E (1997) Functional drinks for athletes. *Trends in Food Science and Technology*, 8, 414–421.

CERVENY J G (1980) Effects of changes in the production and marketing of cured meats on the risk of botulism. *Food Technology*, 34(5), 240–243.

CIMSCEE (1991) *Code for the production of microbiologically safe and stable emulsified and non-emulsified sauces containing acetic acid.* Comité de Industries des Mayonnaises et Sauces Condimentaires de la Communauté Economique Européene, Brussels.

COULTATE,T P (2002) *Food – The Chemistry of its Components*, 4th edn, The Royal Society of Chemistry, London.

DESROSIER N W (1970) *The Technology of Food Preservation*, The AVI Publishing Company, Inc., Westport, CT.

DUCHÉ O, TRÉMOULET F, GLASER P AND LABADIE J (2002) Salt stress proteins induced in *Listeria monocytogenes*. *Applied and Environmental Microbiology*, 68(4), 1491–1498.

FARHAT I (2005) Salt, structure and flavour. *FoodLINK News*, issue 53, December 2005.

FOOD QUALITY NEWS.COM (2005) web news 05/09/2005 – www.foodqualitynews.com/news/ Asda cuts salt from entire tinned vegetable range.

FOOD STANDARDS AGENCY (2005) web page on salt – www.salt.gov.uk/index.shtml, visited in December 2005.

GARDAN R, COSSART P, THE EUROPEAN *LISTERIA* GENOME CONSORTIUM and LABADIE J (2003) Identification of *Listeria monocytogenes* genes involved in salt and alkaline-pH tolerance. *Applied and Environmental Microbiology*, 69(6), 3137–3134.

GUINEE T P and FOX P F (1993) Salt in cheese: physical, chemical and biological aspects. In: *Cheese: Chemistry, Physics and Microbiology, Volume 1 General Aspects* (ed. Fox, P F), Chapman and Hall, London, pp. 257–302.

HELGESEN H and NAES T (1995) Selection of dry fermented lamb sausages for consumer testing. *Food Quality and Preference*, 6, 109–120.

HODGE D G and JAMES C D (1981) *Pastry technology: factors affecting the consistency of short paste*. FMBRA Report No. 96, December 1981.

HUTTON T (2002) Technological functions of salt in the manufacturing of food and drink products. *British Food Journal*, 104(2), 126–152.

IFST (2003) *Salt – Information Statement*, Institute of Food Science and Technology, London.

JONES A A (2000) Ambient-stable sauces and pickles. In: *Shelf-life Evaluation of Foods* (Eds, Man, C M D and Jones, A A), 2nd edn, Aspen Publishers, Inc., Gaithersburg, MD, pp. 211–226.

JORDAN K N and DAVIES K W (2001) Sodium chloride enhances recovery and growth of acid-stressed E. coli O157:H7. *Letters in Applied Microbiology*, 32, 312–315.

LAWRIE R A (1974) *Meat Science*, 2nd edn, Pergamon Press, Oxford.

LAWRIE R A (1998) *Meat Science*, 6th edn, Woodhead Publishing, Cambridge.

LAWSON R, MILLER A R and THACKER D (1981) *Rotary moulded short-dough biscuits. Part II The effects of the level of ingredients on the properties of Lincoln biscuits.* FMBRA Report No. 93, July, 1981.

LEISTNER L (2000) Use of combined preservative factors in foods of developing countries. In: *The Microbiological Safety and Quality of Food* (eds. Lund, B M, Baird-Parker, T C and Gould, G W), Aspen Publishers, Inc., Gaithersburg, MD, pp. 294–314.

LINDSAY R C (1996) Flavors. In: *Food Chemistry* (ed. Fennema, O R) 3rd edn, pp. 723–765.

MAN C M D (2005) *Technological implications of reducing the sodium content in pie and quiche*: unpublished results, London South Bank University.

MATZ S A (1992) *Cookie and Cracker Technology*, Pan-Tech International, McAllen, Texas.

MEILGAARD M, CIVILLE G V and CARR B T (2000) *Sensory Evaluation Techniques*, 2nd edn, CRC Press, Inc., Boca Raton.

MEULLENET J F, XIONG R, HANKINS J A, DIAS P, ZIVANOVIC S, MONSOOR M A, BELLMAN-HORNER T, LIU Z and FROMM H (2003) Modeling preference of commercial toasted white corn tortilla chips using proportional odds models. *Food Quality and Preference*, 14, 603–614.

OFFER G, RESTALL D and TRINICK J (1984) Water-holding in meat. In: *Recent Advances in the Chemistry of Meat* (ed. Bailey, A J) The Royal Society of Chemistry, London, pp. 71–86.

RANKEN M D, KILL R C and BAKER C G J (1997) *Food Industries Manual*, 24th edn, Blackie Academic & Professional, Glasgow.

REDDY K A and MARTH E H (1991) Reducing the sodium content of foods: A review. *Journal of Food Protection*, 54(2), 138–150.

RITVANEN T, LAMPOLAHTI S, LILLEBERG L, TUPASELA T, ISONIEMI M, APPELBYE U, LYYTIKÄINEE T, EEROLA S and UUSI-RAUVA E (2005) Sensory evaluation, chemical composition and consumer acceptance of full fat and reduced fat cheeses in the Finnish market. *Food Quality and Preference*, 16, 479–492.

SCHMIDT G R and TROUT G R (1984) The Chemistry of Meat Binding. In: *Recent Advances in the Chemistry of Meat* (ed. Bailey, A J) The Royal Society of Chemistry, London, pp. 231–245.

VOGEL F B, HUSS H H, AHRENS B O P and GRAM L (2001) Elucidation of *Listeria monocytogenes* contamination routes in cold-smoked salmon processing plants detected by DNA-based typing methods. *Applied and Environmental Microbiology*, 67, 2586–2595.

WALSTRA P (1993) The Syneresis of Curd. In: *Cheese: Chemistry, Physics and Microbiology, Volume 1 General Aspects* (ed. Fox, P F), Chapman and Hall, London, pp. 141–191.

WILLIAMS A (1975) Problems associated with raw materials. In: *Breadmaking: The Modern Revolution.* (ed. Williams, A), Hutchinson Benham, London, pp. 163–176.

WILSON N R P (ed) (1981) *Meat And Meat Products – Factors Affecting Quality Control*, Applied Science Publishers, Barking.

9

Microbial issues in reducing salt in food products

G. Betts, L. Everis and R. Betts, CCFRA, UK

9.1 Introduction

Salt has been used for hundreds of years as a food preservative. There are many traditional food products like salted beef, salted cod and dried cured hams and fermented meats which rely on a high level of salt for preservative effects.

Fermented products such as soy sauce and pickled cucumbers also rely on the antimicrobial effects of salt, as do brined products such as pickled eggs and a range of pickled vegetables. Olives and salted cheeses such as feta-type cheese also rely heavily on the preservative effect of salt to achieve stability and safety.

In the products listed above, the presence of salt is apparent from the sensory characteristics, i.e. they taste salty, but many other foods such as cooked sliced meats, ready-to-eat recipe dishes (including pasta, fish and meat meals), sauces, soups, burgers, sausages and bread contain a lower level of salt – typically 1–3% – which may not taste overly salty but which add to the overall safety of the food product.

The contribution of salt to food preservation is discussed below.

9.1.1 Basis of food preservation
All foods will contain microorganisms which will attempt to grow in that food during storage. Growth of microorganisms will either cause food spoilage, i.e. an undesirable change in the quality (taste, smell or appearance) of the food, or food poisoning, i.e. illness in consumers due to presence of the microorganisms themselves or any toxins they have produced in the food.

The aim of food scientists is to develop a food formulation which will prevent the growth of undesirable microorganisms during the shelf-life of the food product. Most food-borne bacteria grow best in an environment which has a neutral pH, low salt (0.1%), and is moist and warm (25–37°C). By changing these conditions it is therefore possible to create an unfavourable environment where microbial growth is slowed down or stopped completely.

Alternatives to salt are described in more detail in Section 9.3 but it is worthwhile introducing the various antimicrobial factors at an early stage so that salt can be considered in context with the other commonly used factors.

The particular factors which can influence microbial growth are called 'intrinsic' or 'extrinsic' factors. Intrinsic factors are contained within the food itself and include the level of acidity or added salt. Table 9.1 shows a list of common intrinsic factors and their functions. As can be seen, intrinsic factors do not usually result in cell death but more often a slowing down of growth rate. Extrinsic factors are the environmental factors to which a food is exposed and include cooking temperature and packaging format (Table 9.2). Again, apart from the use of heat and other non-thermal physical processes, most extrinsic factors result in a slowing down of microbial growth rather than cell death.

Table 9.1   Examples of intrinsic factors controlling growth of microorganisms

Intrinsic factor	Effect on microorganisms
pH or acidity Either achieved by adding organic acids such as citric, acetic or formed naturally due to fermentation process	Diverts the cells' energy to removing the hydrogen ions in order to maintain homeostasis and thus reduces microbial growth. If the pH is low enough then cell death may occur
Water activity Removal of moisture by drying or evaporation or addition of solutes, e.g. salt and sugar	Lack of moisture will dry out the cell and reduce its growth. Microorganisms can survive in dry environments for a long time
Salt Lowers the water activity but has some antimicrobial action due to the ion	Cells will lose water in a high salt environment and divert energy to accumulate solutes in the cell to reach a balance with the conditions in the food. Growth will slow down or stop
Sugars Lowers the water activity. Less effective than salt and requires larger amounts	Effects are similar to salt but it requires a high level of sugar to achieve similar effects
Preservatives Addition of chemical preservatives such as sorbate, benzoate, propionate	Preservatives will divert the cells' energy to removing or counteracting the effects of preservatives in the cell. Microbial growth will slow down. Death may occur if levels of preservatives are high enough

Table 9.2   Examples of extrinsic factors controlling growth of microorganisms

Extrinsic factors	Effect on microorganisms
Heat treatment	
Application of heat during cooking	Dependent on times and temperatures achieved the heat treatment should destroy the microorganisms, i.e. those capable of growth in the product
Other non-thermal treatments	
Application of UV light, high pressure, ultrasound	Dependent on times and temperatures achieved the treatment should destroy the target microorganisms, i.e. those capable of growth in the product
Modified atmosphere	
Changing the atmospheric oxygen level, addition of CO_2	Removal of oxygen and addition of CO_2 will slow down/prevent the growth of aerobic organisms
Storage temperature	
Use of fast cooling and chill storage conditions	All microorganisms have a minimum temperature below which they cannot grow. Use of chill temperatures will slow down/prevent microbial growth

9.1.2   Mode of action of salt on microorganisms

The main effect that salt has on microorganisms is to reduce the water activity of the food so that it is unavailable for microorganisms to use. Salts (NaCl, KCl) and sugars (glucose and sucrose) have a similar effect but the amount of each solute required to reduce the water activity will vary.

Reducing the water activity (a_w) of the environment can have a dramatic effect on microorganisms. It can increase the lag phase, i.e. the period before microbial growth starts occurring when cells are adapting to their environment, and can slow down the growth rate once it starts. Each microorganism will have a minimum a_w below which is cannot grow and this limiting a_w value will depend on the type of solute used. When a microorganism is placed into a low a_w environment, the water from the inside of the cells will move across the microbial cell membranes towards the outside where there is a higher concentration of the solute. This loss of water results in plasmolysis of the cell and the rigid structure of the cell is lost, so the cell appears somewhat like a deflated balloon. Plasmolysed cells are unable to grow and need to restore the correct water levels before they can restart growth.

There are two mechanisms that can be used to return the cells to a balanced state:

(i)   the external solute, i.e. in this case, salt, could enter the cell until it was at a similar level inside and outside the cell. However, this would need the

microorganism to have a suitable transport mechanism for taking the salt from the food into the cell. It would also mean that the enzyme systems in the inside of the microorganisms would have to still be able to work in the presence of high levels of salt, which they are generally unable to do.
(ii) alternatively the cell could accumulate or produce a different solute from the one exerting the osmotic stress, which will allow the osmotic potential to be balanced and will allow the cytoplasmic enzymes to function correctly. Such solutes are called compatible solutes and include glutamate, carnitine and betaine.

In most bacteria, it is the second mechanism which is used to counteract the majority of salts and sugars used in food manufacture, except for glycerol, where the first mechanism is used. It is easy to see why counteracting large levels of salts and sugars uses up the microorganisms' energy and thus prevents them from growing. The ultimate response of the microorganism will depend on how much salt is present. If the levels are too high the microorganisms will not be able to match the internal level of solute to that of the environment and growth will not start. If the level of salt is within the growth limits of the microorganism then it will eventually manage to balance the internal solute level to that of the food and return to its normal size as it takes water back in. Growth will then be able to start, although it may be slower than in non-stressed cells.

9.1.3   Maximum salt levels permitting growth of microorganisms

Table 9.3 contains the minimum water activity levels permitting growth of a range of food-borne pathogens and spoilage organisms. It is important to note that each factor is considered independently and there will be interactions between these factors. For example, Betts et al. (2000) looked at the synergistic effect of salt, temperature and yeasts on growth of spoilage yeasts. At a good growth temperature (22°C) the lag time was 36 hours in broth containing 10% salt. As the temperature became more inhibitory, the lag time increased to 63 hours at 15°C, 234 hours at 8°C and 767 hours at 4°C.

Similar data is available for combinations of salt and pH, salt and preservatives and salt and nitrite. These will be discussed later.

9.2   Replacement of salt with other compounds used to increase osmotic pressure

9.2.1   Taste

One of the main issues to arise from replacing salt in foods is the change in taste. Salt serves several functions in foods. It has an effect on the structure of the food, it plays an antimicrobial role and it contributes to the overall flavour of the food.

If the salt level is reduced then other factors will need to be considered to maintain the desired taste. The majority of other solutes, i.e. sugars, do not give

Table 9.3   Minimum growth conditions for food-borne microorganisms

Type of microorganism	Minimum pH for growth	Minimum a_w for growth	Anaerobic growth (e.g., in vacuum pack)	Minimum growth temp °C
Salmonella	3.8	0.92–0.95	Yes	4–5.2
Staphylococcus aureus	4.0	0.83	Yes	7
Bacillus cereus (spores/heat resistant)	4.9	0.93–0.95	Yes	4
Clostridium botulinum				
proteolytic A, B, F	4.6	0.94	Yes	10
non-proteolytic B, E, F	5.0	0.97	Yes	3.3
Listeria monocytogenes	4.3	0.92	Yes	−0.4
Escherichia coli	4.4	0.935	Yes	*ca.* 7–8
Psychrotrophic spoilage				
(*Pseudomonas*)	5.0	0.97	No	0
(*Enterobacter aerogenes*)			Yes	
(Lactic acid bacteria)	3.5	0.90	Yes	4
Micrococci			No	
Yeasts	1.5	0.62	Yes	Pink yeast −34
Clostridium perfringens	4.5	0.93–0.95	Yes	
Vibrio parahaemolyticus	4.9	0.94	Yes	5
Yersinia enterocolitica	4.4	0.96	Yes	−1.3
Aeromonas hydrophila	<4.5	0.97	Yes	−0.1

Adapted from Betts *et al.* (2004)

a similar salty taste. Other salts, such as potassium chloride, could be used but there is relatively little data available on their antimicrobial properties. One of the modes of action of salt is to lower the water activity of the environment surrounding the microorganisms. If salt levels are reduced then the water activity will be higher and thus will increase the potential for growth of pathogens. Lactate, either as sodium or potassium lactate has been investigated as an alternative to salt. Sodium lactate works in a similar way to salt in that it lowers the a_w of the product but also at lower pH values the acid molecules are undisassociated and the lactic acid is able to diffuse across the bacterial membranes and cause further inhibition than that seen by reduced a_w alone.

Generally speaking, lower concentrations of sodium lactate are needed to prevent growth compared to salt (Houtsma *et al.*, 1996) possibly because lactates have a greater effect on reducing the a_w of a food than salt i.e. 3% sodium lactate will achieve a lower water activity than 3% sodium chloride. Inclusion of lactate in the formulation of cured meats has been shown to inhibit growth of *C. botulinum* and *L. monocytogenes*.

Other solutes such as sucrose and glucose will make the food sweet and would also have textural changes owing to the amount required to achieve an equivalent a_w to salt.

Table 9.4   Weight (g/l) of NaCl and sucrose required to reduce water activity

% NaCl	a_w	% Sucrose
1.7	0.99	15.5
3.4	0.98	26.1
6.6	0.96	39.7
9.4	0.94	48.2
11.9	0.92	54.4
14.2	0.90	58.5
16.3	0.88	62.8
18.2	0.86	65.6

9.2.2   Amount of solute required

In order to reduce the water activity of a solution using sugars, a much larger amount of the sugar needs to be added compared to salt.

For example, Table 9.4 shows the levels (%) of salt and sucrose required to achieve an a_w between 0.99 and 0.86. The amount of sucrose (g/l) required to achieve an equivalent a_w as salt can be up to eight-fold higher. This would have an impact on the taste, colour and texture of the product.

9.3   Methods to reduce salt without compromising microbial safety

Preservation techniques which are designed to prevent microbial growth in foods include low temperature storage, reduction of water activity (a_w), reduction of pH, modification of oxidation/reduction (E_h), addition of competitive microorganisms and addition of preservatives (Marechal *et al.*, 1999). It is unusual to use only one of these factors to achieve product stability. Combinations of preservatives can be more effective than just using one and the use of combinations is referred to as the hurdle effect (Leistner, 1978; 1995a; Leistner and Gorris, 1995). By using a number of different means of inhibition, it is possible to apply each individual hurdle in a reduced intensity and result in food products that are safe, have adequate shelf-life and may be more acceptable to consumers (Leistner, 1992; 1995b). The combined hurdles may have an additive or even a synergistic effect, allowing combinations that achieve microbial stability and safety to be chosen.

If the salt level of a product is reduced then the shelf-life of that product is also likely to be reduced as it will not be so antimicrobial. If the shelf-life needs to remain the same then levels of the other factors will have to be increased to replace the antimicrobial effect of the reduced salt.

9.3.1   Natural antimicrobials

Many natural food ingredients which are traditionally added to achieve a desired flavour, also have the potential to control microbial growth. This is known to be true for vegetable extracts, mustard, onion, garlic, horseradish and a range of other herb and spice ingredients including extracts and essential oils from the plants, which have been shown to inhibit the growth of a range of microorganisms. Natural antimicrobial compounds and their possible modes of action have been reviewed extensively (Nychas, 1995; CAST, 1998), but because they contain a variety of compounds from different chemical classes, it is not possible to identify a single mechanism by which all of these compounds act on microorganisms.

Essential oils are natural mixtures of aromatic compounds present in plants that are extracted by steam or solvent distillation, in yields of between 0.01% and 2.0% (calculated on the weight of fresh plant distilled). These compounds have been shown to have both fungistatic and bactericidal activity to suppress infection by plant pathogenic microorganisms. The flavour and fragrance industry and natural product industries use these compounds routinely. Such compounds that are approved for use in foods and combine antimicrobial activity with low toxicity have great potential as natural food preservatives (Smid and Gorris, 1999).

There has been much interest in recent years in the use of natural anti-microbials to help in food preservation as an alternative to the use of traditional chemical preservatives, which at present are used to achieve sufficiently long shelf-life and safety. Chemical preservatives such as sorbate and benzoate have long been used as reliable preservative factors to control microbial growth, but such compounds do not satisfy the demand of some consumers for natural and healthy foods that contain 'no artificial agents'. In order to meet such criteria, food manufacturers are searching for more natural alternatives to the use of these chemicals, such as naturally occurring antimicrobial compounds, alone or in combination with already existing preservative mechanisms (Smid and Gorris, 1999). The active components found in herbs and spices (e.g., thymol from thyme and oregano, cinnamaldehyde from cinnamon, and eugenol from clove) have been shown to have a wide spectrum of antimicrobial activity (Martini *et al.*, 1996; Friedman *et al.*, 2000; Lambert *et al.*, 2001).

Much work has been performed on establishing the potential of such natural antimicrobials to replace or reduce the use of chemical food preservatives (Deans and Ritchie, 1987; Friedman *et al.*, 2002; Ismaiel and Pierson, 1990; Nychas, 1995; Smith-Palmer *et al.*, 1998; Hammer *et al.*, 1999; Dorman and Deans, 2000; Hsieh *et al.*, 2001). Most of this work has been performed in laboratory broths, and only a few have been done in real foods. It would appear that the effectiveness of antimicrobial compounds is decreased when used in real foods, or a higher level of the compound needs to be added to achieve an antimicrobial effect. The type of food product has been shown to affect the antimicrobial effects observed. This has been suggested to be due to the composition of the food, which immobilise and inactivate the components (Farbood *et al.*, 1976; Shelef, 1983; Shelef *et al.*, 1984), particularly in high-fat foods.

9.3.2   pH reduction

Microorganisms have defined optimum ranges of external pH required for growth and survival, and so acidification is often quite effective in controlling microbial growth. Organisms are, in general, more sensitive to changes in the internal cytoplasmic pH than to changes in external pH, although significant changes in either will lead to loss of viability.

There are two main types of acid; 'strong' acids which are non-permeant, i.e. unable to enter the microbial cell and 'weak' acids which are able to enter the microbial cell and exert their action within the microorganism. The type of acid used will determine the minimum pH level at which microorganisms can grow. It is the 'weak' or 'organic' acids that are used in food production and these will be discussed in Section 9.3.3. Some strong acids such as phosphoric acid are used in the beverage industry. As strong non-permeant acids do not affect the pH of the cytoplasm to the same extent as weak permeable acids, relatively large changes in external pH are required for effective preservation, which may be detrimental to the foods' sensory properties. Therefore, in many cases, acidification alone is too detrimental to the sensory quality of foods to be acceptable as the only means of controlling microbial growth.

In general for growth and survival, bacteria require pH values that are between 4 and 8, whereas the yeasts and moulds are able to grow and survive at a wider range of pH, in some instances, the range can be between pH 2 and 11 (Wheeler *et al.*, 1991). However, microorganisms may survive in conditions of low pH and although growth may have stopped, the cells may still be metabolically active. The inside of a bacterial cell is at a pH close to neutral and needs to be maintained at this level for the organism to grow. Therefore growth is often slowed down or prevented in acidic environments, as microorganisms use increasing amounts of energy in an attempt to counteract the effects of the acid molecules and maintain the correct pH inside the cell.

Many of the studies which have been reported in the literature on growth of food pathogens and spoilage organisms in different pH environments will have used strong acids. As most foods will contain the weak acids as acidulants, this may affect the minimum pH values for growth. Generally speaking, weak acids are more effective than strong acids at inhibitory microbial growth so the pH required to prevent growth is higher for weak acids than strong acids.

9.3.3   Organic or weak acids

Organic acids can occur naturally in many fruits and vegetables and have been widely used to maintain microbial stability in low pH foods including fruit juices, beverages, wines, pickled vegetables, mayonnaise, and salad dressings (Pilkington and Rose, 1988; Sofos and Busta, 1981; Restaino *et al.*, 1982; Seymour, 1998). Spoilage of such foods is most often caused by yeasts, moulds and lactic acid bacteria, as environmental conditions in these foods generally inhibit the growth of bacteria (Beuchat, 1982).

The main types of acid used in food manufacture include acetic, citric, malic, tartaric and lactic. A characteristic of these acids is that they exist in two forms (dissociated and undissociated) dependent on the pH values. This is shown below for acetic acid.

Low pH, e.g. 4.5 to 5.5 found in foods		**High pH, e.g. 7.0 inside microbial cells**
CH_3COOH	$\longleftarrow\longrightarrow$	$CH_3COO^- + H^+$
Undissociated molecule		Dissociated molecule

This is important for food preservation because only the undissociated form of the acid is able to enter the bacterial cell. Charged molecules such as the dissociated form are not able to cross the bacterial membranes. All organic acids have a particular pH (the P_ka value at which half of their molecules are undissociated. These are shown in Section 9.3.4. The acids work most effectively as the pH approaches or falls below their P_ka value, as more acid is in the undissociated state and thus able to enter the microbial cell.

The choice of acid will depend on the desired flavour in the final product. Acetic acid is the most antimicrobial of the organic acids but has a strong vinegary odour and taste. Tomato-based products naturally contain citric acid and this is therefore the acidulant that is most often added to these products to further reduce the pH. Reduced pH works well with salt as a preservation system. Choosing the right level of each factor is critical for product safety and is best done using predictive models or challenge tests (see Section 9.4).

9.3.4   Chemical preservatives

There are a number of chemical preservatives than can be used in food products to inhibit growth of microorganisms.

The salts of organic acids are often added to low pH foods to reduce the growth of yeasts and moulds. For example, sodium benzoate and potassium sorbate are used for fruit products, pickles, mayonnaises and dried fruit. The active ingredient of many of these preservatives is the organic acid, e.g. benzoic or sorbic acid. Preservatives are generally used for lower pH foods because they work better in acidic environments at or near their P_ka values (Table 9.5).

Sorbate
Sorbic acid and potassium sorbate are widely used throughout the food industry for the preservation of cheese, in bakery products, vegetable-based products (pickles, olives, fresh salads), fruit based products (dried fruits, fruit juices), beverages and some other products such as smoked fish, margarine and mayonnaises.

Sorbate is more inhibitory to yeasts and moulds than bacteria. The yeast species inhibited by sorbates include *Brettanomyces*, *Candida*, *Cryptococcus*, *Saccharomyces*, *Zygosaccharomyces* and *Debaromyces*. Mould species include *Alternaria*, *Aspergillus*, *Cladosporium*, *Penicillium* and *Fusarium*. Bacteria that

Table 9.5 P_ka values for organic acids commonly used in food preservation

Acid	P_ka[1]
Acetic	4.75
Benzoic	4.20
Citric	3.14 (4.77, 6.39)[2]
Lactic	3.08
Malic	3.40 (5.11)[2]
Proprionic	4.87
Sorbic	4.80
Tartaric	2.98 (4.34)[2]

[1] P_ka – the pH value at which 50% of the acid is in its undissociated form.
[2] These acid molecules have more than one acid group. Figures in brackets refer to dissociation value of second or third acid group.

are inhibited by sorbates include *Bacillus*, *Clostridium*, *Enterobacter*, *Pseudomonas*, *Salmonella* and *Serratia* (Sofos and Busta, 1983).

Sorbates are selective in their antimicrobial activity; they are more effective against catalase positive organisms that catalase negative, and aerobes rather than anaerobes (Sofos and Busta, 1983) which means that they are extremely useful in preservation of fermented foods. This is because growth of pathogens such as *S. Typhimurium*, *E. coli*, and staphylococci will be inhibited, but lactic acid bacteria will still be able to grow. The antimicrobial activity of sorbate is dependent on pH, and is greatest at lower pH values, with 98% of sorbic acid being undissociated at pH 3 compared to just 37% at pH 5.

Benzoate
Sodium benzoate is commonly used in products where pH is low, for example mayonnaises, pickled vegetables, fruit products and drinks. It is commonly combined with potassium sorbate in mayonnaise-type products. This is because the mixture of the two preservatives is more effective than either of them individually and also sorbate is tasteless (Lueck, 1980). Benzoate exerts its primary antimicrobial action upon yeasts and moulds, including the aflatoxin-producing moulds. Many bacteria are also inhibited by the presence of benzoate. However, clostridia and the lactic acid bacteria are resistant (Lueck, 1980).

The antimicrobial action of benzoic acid is due to interactions with cell membrane enzymes. These enzymes are involved in oxidative phosphorylation and acetic acid metabolism. Benzoic acid can also affect the cell wall. There are many cases of yeasts being isolated from benzoate-containing products (Chipley, 1983), and also *Z. bailii* has been reported to grow in high concentrations of benzoic acid. This resistance is thought only to occur when there is sufficient energy to pump benzoate out of the cell.

Propionate
The main forms used are sodium and calcium propionate. Propionate is used in cheese production to prevent mould growth on the cheese surface. It is also used

widely with bakery products, and like sorbate and benzoate its action is pH dependent. However, it is able to work at higher pH values which makes it suitable for bakery use. Its mode of action is also similar to that of sorbate and benzoate, because it accumulates within the cell and acts upon enzymes (Lueck, 1980).

Gram negative bacteria are also inhibited and so is *B. mesentericus* which can cause rope spoilage of bread. The action of propionate against yeasts is weak and therefore the yeasts used in the baking process are not inhibited.

9.3.5   Nitrite

Nitrite is a food preservative typically used in cured and fermented meat products along with salt. In these types of products, the safety is reliant on a careful balance of these two compounds.

Nitrite was originally added to meat products to give the characteristic pink coloration of cured meats and to inhibit growth of spore forming bacteria, particularly *Clostridium botulinum*. It is usually added as a potassium salt (KNO_2, E249) or a sodium salt ($NaNO_2$, E250) with a legal maximum in-put value of 150 mg/kg or a residual value of 50 mg/kg. It has been extensively studied since the early 1950s and minimum inhibitory concentrations against a range of organisms have been established (Tompkin, 1993). Nitrite has limited effect against yeasts (Nielsen, 1983a) and many of the microbial groups associated with meat products such as *Enterobacteriaceae*, *Pseudomonas* and *Lactobacillus* (Singhal and Kulkarni, 2000).

Some reports (Nielson 1983a,b) show that Enterobacteriaceae, *Broch. thermosphacta*, and *Moraxella* spp. were inhibited by up to 200 ppm sodium nitrite in Bologna-type sausages, whereas yeasts, and LAB were only marginally inhibited and hence tended to predominate in these products.

These findings contrast with those of Gibson and Roberts (1986), who reported growth of *E. coli* and *Salmonella* at 10–35°C was not prevented by most combinations of salt (1–6% w/v), pH (5.6, 6.2 and 6.8) and sodium nitrite (0–400 ppm) tested. Inhibition only occurred under extreme conditions of pH (5.6), temperature (10°C) and nitrite (400 ppm). Sameshima *et al.* (1998) have shown that the growth rate of *Lac. viridescens* and *Enterococcus faecalis* was up to three times slower in vacuum-packaged pork sausage containing 200 ppm sodium nitrite and stored at 10°C compared to controls containing no nitrite.

Despite many years of research on nitrite the precise mode of action is still not certain and there are considered to be many possible target sites of action in the microbial cell. A good review of the biochemical basis for nitrite inhibition of *C. botulinum* in cured meats is given by Benedict (1980). The primary modes of action (Singhal and Kulkarni, 2000; Surekha and Reddy, 2000) include; inhibition of respiration by inactivation of key enzymes; release of nitrous acid and nitric oxides; formation of S-nitroso compounds by reaction of nitrite with ham proteins.

Like many food preservatives, nitrite works better under acidic conditions which favour the production of undisassociated nitrous acid and thus permit its entry into the bacterial cell. The synergistic effects of reduced pH and nitrite will be considered in more detail later.

Combined effect of salt and nitrite on Clostridium botulinum

Clostridium botulinum is fairly sensitive to changes in salt level. In cured meat products, the levels of both salt and nitrite, which are needed to achieve product stability, have been established over many years. It is apparent that nitrite acts synergistically with salt to inhibit *C. botulinum*. The amount required to achieve inhibition varies between different foods but in products with a low salt content and prolonged shelf-life, addition of between 50 and 150 mg/kg nitrite is necessary (Anon, 2003).

Much of the work done on inhibition of *C. botulinum* by nitrite has concentrated on proteolytic strains. At a pH value of 6.0, the probability of toxin production was 59% when the input level was 100 mg/kg $NaNO_2$ but only 1% when the level was 300 mg/kg (Robinson *et al.*, 1982). The amount of in-going nitrite needed to inhibit *C. botulinum* varies considerably dependent on the pH of the product, the heat treatment it receives, the presence of ascorbate and the amount of spores present.

Nitrite levels in cured meats fall during storage and the level of residual nitrite is usually comparable in a range of products irrespective of in-going amount. Sofos *et al.* (1980) found that the residual level of nitrite fell to between 1 and 3 ppm after 25 days storage irrespective of whether 40, 80 or 120 ppm was added.

It is now generally considered to be the residual amount of nitrite which is important with respect to *C. botulinum* control (Anon, 2003). However, there is evidence to show that the starting levels are important as well as the final residual level. Christiansen (1980) found that spores germinated readily at in-going amounts of 50 and 156 mg/kg once residual levels fell below 10 mg/kg.

It would appear that the protein type affects the rate of depletion of nitrite and also the antibotulinal effect. Sofos *et al.* (1979) found that 80 mg/kg nitrite was less effective in chicken products than those made with beef or pork. They proposed that it may be due to higher levels of iron in the mechanically deboned meat which made the nitrite less effective. They concluded that a level of 80 mg/kg nitrite was ineffective on its own and that higher levels were necessary. For example, they found that a level of 156 mg/kg nitrite was significantly inhibitory against *C. botulinum*.

In summary, it would appear to be a combination of initial amount of nitrite present and the level of residual nitrite that is important with respect to inhibition of *C. botulinum*, although it should be noted that most studies quote the required inhibitory values as in-going amounts rather than residual amounts. There is no clear link between in-going and residual amounts that would allow a defined residual amount to be calculated for any given in-going amount. The residual levels of nitrite become depleted during storage to eventually reach the same level, however, it would appear that for any given residual level there is a greater anti-botulinal effect when the in-going amount was higher.

The amount of in-going nitrite required to achieve inhibition ranged from 50 mg/kg to 200 mg/kg dependent on pH, storage temperature, salt level and inclusion of other antimicrobials such as sorbic acid.

Effect of nitrite and salt on other pathogens
Gibson and Roberts (1986) looked at the combined effects of salt and nitrite on *Clostridium perfringens* at different temperatures and pH values. At storage temperatures of 20 to 35°C, they found that growth of this organism was prevented by the levels of curing salts used commercially (up to 200 mg/kg) provided the pH was 6.2 or below. This was usually accompanied by a salt level of 3%. At temperatures below 20°C, this organism grew slowly without the addition of nitrite or salt.

Nitrite has been shown to inhibit *Listeria monocytogenes* in conditions found in cured meat products (Buchanan *et al.*, 1989). The inhibitory effects were temperature and pH dependent. For example, the time for a 4-log increase in numbers of *L. monocytogenes* was 8 hours at 37°C and 287 hours at 5°C when the salt level was 0.5%, the nitrite was 100 mg/kg and the pH was 6.0. When the salt level was increased to 4.5%, the times for a 4-log increase were 11 and 479 hours for 37°C and 5°C respectively.

Further studies (McClure *et al.*, 1991) showed that at 20°C and below no growth of *L. monocytogenes* was detected within 21 days at nitrite concentrations of 50 mg/kg when the pH was 5.3 or below. When the pH was at 6.0 and above, nitrite had little effect even at 400 mg/kg and temperatures down to 10°C.

The growth of *Bacillus cereus* was also affected by combinations of pH, salt and nitrite (Benedict *et al.*, 1993). For example at 12°C, pH 6.75 and 50 mg/kg nitrite, the lag time was 136 hours with 3.5% salt and 81 hours with 1.5% added salt. Under adverse conditions of temperature or pH, the addition of salt or nitrite completely inhibited growth of *B. cereus*.

Similar data were found for *Shigella flexneri* (Zaika *et al.*, 1994) where under adverse conditions of temperature or pH, the addition of salt or nitrite was sufficient to prevent growth of this organism. At 15°C, pH 7.0, 2.5% salt, there was a lag time of 274 hours without nitrite and no growth at all with 200 mg/kg nitrite. At 19°C, pH 6.5 and 0.5% salt growth occurred rapidly for 50, 100 and 200 mg/kg. Growth was only inhibited in the presence of 1000 mg/kg nitrite. However, when the salt was increased to 2.5%, growth only occurred with 50 mg/kg nitrite but was inhibited at levels of 100 mg/kg or higher.

Yersinia enterocolitica was not inhibited by up to 200 mg/kg nitrite at low temperatures although when the pH was low, and salt was included, the growth rate was slower with the addition of nitrite. At 5°C and pH 5.5 the lag time was 79 hours with 0.5% salt but was 242 hours with 2% salt and 50 mg/kg nitrite (Bhaduri *et al.*, 1994).

With respect to *Salmonella* at 10–35°C, growth was not prevented by most combinations of salt (1–6% w/v), pH (5.6, 6.2 and 6.8) and sodium nitrite (0–400 ppm) tested. Inhibition only occurred under extreme conditions of pH (5.6), temperature (10°C) and nitrite (400 ppm) (Gibson and Roberts, 1986).

With *Staphylococcus aureus* the bacteriostatic effect of nitrite was also dependent on pH and oxygen availability. However, this organism is highly tolerant to increased salt levels and at the levels present in cured meat products, growth of *Staph. aureus* occurred at up to 200 mg/kg nitrite, albeit slowly at refrigeration temperatures (Buchanan *et al.*, 1993).

Salt is an important food preservative. Many traditional products, particularly cured meat products such as sliced meats and bacon, have used salt as a primary preservative for many years. If salt levels are reduced then the impact of this on food safety should to be assessed on a case-by-case basis using one of the techniques described in Section 9.4.

9.3.6 Heat treatment

With the exception of raw products, most foods will have some heat process during manufacture. The heat treatment is used as one of the hurdles in an overall preservation strategy. The choice of heat treatment will depend on:

(i)   the product characteristics
(ii) target microorganisms likely to be present, survive and grow under the product characteristics
(iii) heat resistance characteristics of the target organisms.

There are three main categories of heat treatment used to stabilise foods:

(i)   Pasteurisation for short shelf-life products. This is designed to inactivate vegetative microorganisms. Typically a process of 70°C/2 minutes or equivalent (z value of 7.5°C) is given to chilled food products which will achieve 6-log reduction in *L. monocytogenes* and *Salmonella* but it is also sufficient to inactivate most Enterobacteriaceae, *Pseudomonas* and yeasts that could spoil chilled foods.
(ii) Pasteurisation for long-life products. This is designed to inactivate bacterial spores. A process of 90°C/10 minutes or equivalent (z value of 9°C) is often given to chilled food products which are vacuum packed or modified atmosphere packed and have a shelf-life of greater than 10 days. This process is designed to achieve a 6-log reduction of psychrotrophic strains of *C. botulinum* but will also inactivate vegetative spoilage organisms. A process of 95°C/5 minutes or 95°C/10 minutes or equivalent (z value of 8.3°C) is given to acidic ambient stable products. It is designed to inactivate acid tolerant sporeformers which could grow and spoil the product if present after heat treatment, i.e. *Clostridium butyricum, Bacillus polymyxa.*
(iii) Sterilisation to achieve commercial sterility in canned goods. The treatment will depend on the spore formers present but will achieve the equivalent to 3 minutes at 121°C (F_o3) as a minimum.

If the salt level of a food is changed then this may well change the product characteristics and reduce its inherent antimicrobial properties. In turn this may widen the range of organisms able to grow in the product. If these additional microorganisms have greater heat resistance characteristics then the heat treatment may need to be increased accordingly.

9.3.7   Non-thermal processing options

There are a number of physical treatments that can be used to achieve inactivation of microorganisms instead of heat treatments. These include:

High pressure

The use of high pressures to inactivate microorganisms is not new. A wide range of pathogens and spoilage organisms are inactivated by high pressure and it also appears that there is a link between cell shape, size and cell wall structure, and the effectiveness of pressure treatment (Hoover *et al.*, 1989). For example, yeast cells are more sensitive that bacteria and inactivation begins at 200 MPa. Gram negative rods are the next sensitive, where greater than 350 MPa may be required to cause injury or death (e.g., *Escherichia coli* and *Pseudomonas aeruginosa*). The Gram positive bacteria are more resistant to pressure than Gram negative bacteria, requiring greater than 400 MPa (e.g., *Listeria monocytogenes*) and Gram positive cocci such as *Staphylococcus aureus* may require pressures greater than 450 MPa to cause inactivation. Finally vegetative cells are more susceptible than spores which require greater than 600 MPa, usually in cycles (6 cycles of 600 MPa at 70°C), to achieve spore inactivation (Smelt, 1998). A suggested explanation for the lower pressure resistance of Gram negative microorganisms is that the more complex cell membrane has greater susceptibility to environmental changes brought about by pressurisation.

High pressure is used for liquid and fruit products, jams and pâtés and ethnic products like guacamole. It is used for heat sensitive products where a pasteurisation or sterilisation effect is required but where the application of heat may damage the product.

Ultrasound

Ultrasonication is known to disrupt biological structures and between 1930 and 1970 much research was done. For example, in the 1960s research was focused on the mechanism of ultrasound interaction with microbial cells (Hughes and Nyborg, 1962). This involved investigating the cavitation phenomenon and associated shear disruption, localised heating and free radical formation. By 1975 it was shown that brief exposure to ultrasound caused a thinning of cell walls, attributed to the freeing of the cytoplasmic membrane from the cell wall (Alliger, 1975). In 1987, the combined effect of ultrasound with heat was found to be effective at causing inactivation of the vegetative bacterium, *Staphylococcus aureus* (Ordonez *et al.*, 1987). This was followed by a report where the combined application of ultrasound and heat was shown to produce a decimal reduction time decrease of up to 99.9% in *Bacillus subtilis* spores, in the temperature range 70°C to 90°C using a 20 KHz, 150 W ultrasound reactor (Garcia *et al.*, 1989).

The antimicrobial efficacy of ultrasound was examined using *E. coli*, *S. aureus*, *B. subtilis* and *P. aeruginosa* by exposing them to an ultrasound frequency of 26 KHz at a temperature of 39°C (Scherba *et al.*, 1991). Their findings showed that the intensity of ultrasound had a direct effect, with high intensity ultrasound causing the greatest degree of cell inactivation.

Further research was carried out in 1992, when Wrigley looked at the effect of ultrasound treatment of *S. Typhimurium* in a number of different systems including a broth, skim milk and egg. This research demonstrated that ultrasound treatment alone produced a 2–3-log reduction in the number of bacteria present. This effect was enhanced if the menstruum was simultaneously heated at 50°C.

The effect of sonication for *Salmonella* decontamination of chicken skins has also been investigated (Lillard, 1993). The findings showed that a combination of chlorine treatment with ultrasound resulted in a greater number of *Salmonella* being killed, compared with a chlorine only treatment.

The research done to date involving ultrasound has important implications for the food industry. Present sterilisation and pasteurisation methods involve high temperature processing which can give rise to undesirable flavours and textures in some foods. It is envisaged that the use of ultrasound treatments could lead to a decrease in process times and/or temperatures and a subsequent improvement in product quality and convenience to the consumer.

Pulsed light
Light in the form of ultraviolet (UV) light has been used for many years to sterilise hospital equipment, water and other liquids. It has also been used to sterilise food packaging and the surface of food products such as meats or baked goods. There are three different types of UV light.

Near-ultraviolet (near-UV) radiation (also called (UV-A) can be considered to lie in the wavelength range of 320–400 nm. The long-wavelength limit represents the beginning of the visible spectrum, whilst the short-wavelength limit corresponds roughly to the point below which proteins and nucleic acids begin to absorb significantly; this region is called the 'mid-UV' region or UV-B (320–290 nm) and can cause sunburn and skin cancer. The final type is UV-C which is in the wavelength 190–280 nm.

More recently, a different type of light energy has been investigated for use i.e. pulsed light technologies. The main difference between the germicidal UV lamps reviewed so far and pulsed light technology is that the latter uses intense pulses of broad spectrum light that include all the wavelengths of natural sunlight. Pulsed light is produced using a technology that increases light power (intensity/time) many times by accumulating electrical energy in a capacitor over relatively long times (fractions of a second) and releasing the stored energy in much shorter times (thousandths of a second). Broad spectrum pulsed light has been reported to be able to achieve a high level of inactivation of a range of microorganisms, for example two pulses at low treatment level ($0.75 \, J/cm^2$ per flash) gave a 7-log reduction of *Staphylococcus aureus* on Petri dish whilst one single flash at 4 J/cm^2 was enough to give a 7-log reduction of *Escherichia coli* and *Bacillus subtilis* spores.

Hard crusted white bread rolls inoculated with mould spores were exposed to two pulses of light (0.5 ms duration time, 16 J/cm^2 per flash and 1 Hz). Pulsed light was able to eliminate mould spores without burning the food product. Baked cake surfaces inoculated with mould spores were packed in clear plastic containers. After treatment with three pulses of 16 J/cm^2 per pulse at five-second intervals, they were stored at room temperature and checked for mould growth. Untreated caked exhibited mould growth within three days, whilst the average number of days before visible mould growth appeared on the treated cakes was 10.

9.3.8   Modified atmosphere

The use of modified atmospheres can have a large effect on the shelf-life of a product. Day (1992) gives a good review of modified atmosphere packaged (MAP) goods.

Chilled foods are often stored in a mixture of gases where oxygen is excluded in order to minimise growth of aerobic spoilage organisms such as *Pseudomonas* species. Typical gas mixtures will contain carbon dioxide (CO_2) at a level of 25 to 40% and Nitrogen (N_2) at 60% to 75%. The use of such a gas mixture will change the microbial population within a food. Whilst the growth of *Pseudomonas* will be minimised, facultative spoilage organisms such as lactic acid bacteria, Enterobacteriaceae and yeasts (all of which can grow with or without oxygen) may dominate.

There is also the potential for growth of anaerobic pathogens such as *Clostridium botulinum* in MAP foods. Due to these risks, it is currently recommended in the UK that the shelf-life of chilled MAP or vacuum packaged foods should be restricted to 10 days or less if they do not contain sufficient controlling factors, i.e. pH of 5.0 or less, a_w of 0.97 or less, aqueous salt level of 3.5% or greater; heat treatment equivalent to 90°C for 10 minutes (Betts, 1996). This applies to all VP/MAP foods stored between 3°C and 8°C. The reduction of salt from VP/MAP foods has large implications for their safety and shelf-life (see Section 9.4).

For some foods, e.g. raw meat, a gas mixture containing elevated levels of oxygen is used in order to prolong the visual quality of the product. For ambient-stable snack products such as crisps and nuts, an atmosphere excluding oxygen, e.g. 100% N_2 or even argon, may be used to minimise rancidity due to fat oxidation and this prolongs the shelf-life (Day, 2001).

9.4   Techniques to assess the effect of salt reduction on the safety and quality of food

The Food Standards Agency have proposed a series of reductions in salt levels for a range of manufactured food types. The target reduction in salt levels ranges from 10 to 68% dependent on product type and the average weight of such products likely to be consumed. The biggest risk to microbial safety and stability will be for chilled products where the salt level plays a key part in product stability. For canned or dried goods, the salt reduction may well affect the quality of the product but will have less of an impact on the microorganisms, as other factors, e.g. heat, will be the critical control points in these products. Table 9.6 shows the target level of reduction in salt levels for chilled and perishable foods.

It should be noted that in publications on salt in foods, the terms salt level and sodium level can sometimes appear synonymous. The two terms, however, are very different. Salt is sodium chloride and therefore is made up of both sodium and chlorine ions, whilst sodium is solely sodium. By law, food labels have to contain levels of sodium (not salt); therefore, to establish the level of salt in a food from a sodium level, a small calculation must be made. It is usually taken that multiplying the sodium level by 2.5 will give the corresponding level of salt (sodium chloride) (FSA, 2003). It should also be noted that in some foods sodium chloride (salt) is not the only source of sodium, e.g. baked products will often contain baking powder (sodium bicarbonate).

The figures in Table 9.6 are all taken from those published on the FSA website. NaCl concentrations are calculated from Na levels, given by FSA in their October 2003 table using the multiplier of 2.5. For example, the current salt level in bacon is given by FSA as 1491 mg/100 g Na; this is equivalent to 1.491% Na, which is converted to % salt (NaCl) by multiplying by 2.5, which gives 3.73% NaCl.

9.4.1 Shelf-life trials

The aim of food manufacturers is to produce food products that meet customer expectations in terms of safety and quality. This is achieved by good product and process design and assignment of appropriate shelf-life and storage conditions. There are many aspects of product quality which may determine how long it should be stored, such as biochemical changes, e.g. rancidity of fats; sensory changes, e.g. loss of texture or colour; and microbiological changes. Each of these needs to be considered during shelf-life assessment.

Samples of the product need to be stored in conditions representative of those likely to be seen during distribution and sale and examined at regular intervals for important factors such as levels of microorganisms.

There are many different factors that will limit the shelf-life of a food product such as:

• raw materials
• hygiene
• product formulation
• distribution
• processing
• storage

Table 9.6   FSA salt reduction model target sodium reduction levels for chilled foods

Food group	Current sodium level (mg/100g)	Proposed sodium level (mg/100g)	Current salt level (% NaCl)	Proposed salt level (% NaCl)	% reduction
Pizza	600	300	1.50	0.75	50
Bought sandwiches	500	350	1.25	0.87	30
Milk and cream	43	43	0.10	0.10	0
Processed milk products	82	82	0.20	0.20	0
Fromage frais and yoghurt	61	61	0.15	0.15	0
Ice cream and diary desserts	64	64	0.16	0.16	0
Cheese	700	500	1.75	1.25	29
Quiche, manufactured	550	250	1.37	0.62	55
Other processed egg products	410	300	1.02	0.75	27
Fat spreads	726	400	1.18	1.00	45
Bacon and ham	1491	750	3.72	1.87	50
Homemade meat dishes	163	163	0.40	0.47	0
Meat roll/sliced meat (excluding bacon/ham)	848	450	2.12	1.12	47
Other processed meat	823	450	2.05	1.12	45
Burgers, kebabs	503	300	1.25	0.75	40
Sausages	962	550	2.40	1.37	43
Meat pies	465	300	1.16	0.75	35
Other processed fish products	896	650	2.24	1.62	27
Processed vegetable-based products	393	260	0.98	0.65	34
Ready meals, meat-based	400	250	1.00	0.62	38
Meal centre meat-based (e.g., coated meat)	461	350	1.15	0.87	24
Ready meals, fish-based	300	200	0.75	0.50	33
Meal centre fish-based (e.g., fish fingers)	430	250	1.07	0.62	42
Ready meals, vegetable-based	300	200	0.75	0.50	33
Meal centre vegetable-based (coated veg products)	260	200	0.65	0.50	23
Ready meals, pasta-based	326	250	0.81	0.62	23
Meal centre pasta-based (stuffed pasta)	128	100	0.32	0.25	22
Take away meat-based	386	250	0.96	0.62	35
Take away fish	248	200	0.62	0.50	19
Take away vegetable/ potato	115	200	0.28	0.50	−74

- packaging, including gas atmosphere
- consumer handling.

The shelf-life of food products should be determined in a logical sequence of events:

(i) The kitchen/pilot-scale assessment where the product and process characteristics are defined and a target shelf-life decided.
(ii) The factory-scale trials where the majority of product testing is done on samples produced under routine manufacturing conditions and the shelf-life of the product is assigned.
(iii) Full-scale production where any changes to the shelf-life are monitored.

This is described in detail in CCFRA Guideline No. 46 (Betts *et al.*, 2004). It is recommended that if there are any changes to ingredients, compositions or manufacturing processes that the shelf-life of a food product should be re-evaluated. This may be able to be done using predictive models but may also require the use of new shelf-life studies. Any changes to the level of salt in a recipe should instigate a review of the shelf-life.

9.4.2   Challenge tests

Shelf-life evaluation as described above is designed to determine how long a product remains within the designated quality parameters during normal production and storage conditions.

In this case, only the growth of those microorganisms that are likely to be present in or able to cross-contaminate the batch of product will be assessed. Ideally, under good manufacturing conditions, there will be minimal chance of food pathogens, e.g. *Salmonella*, being present in the product and the effect of the product formulation on growth of such pathogens will, therefore, not be evaluated during shelf-life studies. Reliance should not be placed on shelf-life evaluation to establish the safety of the product with respect to microbiological pathogens and other control measures such as HACCP and separation of high/ low risk operations should be the primary tools for identification and control of microbiological safety hazards. As part of the HACCP study it may be considered appropriate to carry out a challenge test study. Challenge testing is designed to determine whether the product formulation and storage conditions would control growth of pathogens during the designated shelf-life, if they were present in a food.

Challenge testing involves deliberate inoculation of the product with relevant microorganisms that have the potential to survive or grow within the product during normal storage conditions. Careful consideration needs to be given to the planning and interpretation of results from challenge tests.

It may be possible to demonstrate that a pathogen can grow in a product, however, the challenge test does not imply that the pathogens would be present in the product, nor does it attempt to quantify the likelihood of this occurring. Routine microbiological analysis and HACCP can help answer this question.

However, challenge tests can allow the risk of food poisoning to be evaluated if contamination occurred.

The organisms used in challenge studies should have been identified as being likely to be present in the product from ingredients or from cross-contamination post-cooking/preparation. The levels of organisms inoculated into the product must be consistent with those likely to occur in practice. Further information on challenge testing is given in CCFRA Technical Manual 20, Guidelines for Microbiological Challenge Testing (Rose, 1987).

9.4.3   Predictive modelling

Predictive models have been developed extensively over the past few years and are powerful tools for looking at the effect of environmental conditions on growth of pathogens and spoilage organisms.

Predictive microbiological models are developed from laboratory data obtained under a defined set of experimental conditions or to predict the likely responses under new combinations of those conditions of conditions not previously tested. For example, data describing the effect of temperature on an organism at 5, 10 and 15°C can be used to predict the likely growth at 8°C. The same principle applies to any test parameter such as salt, pH or preservatives. The power of this tool for new product development is very apparent. Modifications of new or existing recipes can be tried on the computer before embarking on expensive laboratory experiments or pilot-scale production runs.

There are currently a number of publicly available modelling systems that can be assessed by the food industry. For food pathogens there is the USDA Pathogen Modelling Program (PMP), Growth Predictor (which is based on the data which was formerly available as FoodMicromodel) and for spoilage organisms there is the Campden and Chorleywood Food Research Association (CCFRA) *FORECAST* Service.

Because nitrite and salt interact with each other and with other environmental factors such as pH and storage temperature, it is not easy to define critical combinations of these factors once you move away from the generally accepted levels of these factors required to prevent growth, e.g. 150 mg/kg nitrite and 3.5% salt for psychrotrophic *C. botulinum*.

Predictive models can be used to assess the likely change in growth characteristics caused by changing the product formulation. The only limitation is that many of the models available for food pathogens are based on salt, pH and temperature only and do not include nitrite. With regard to cured meats, the CCFRA meat spoilage model does contain nitrite as a factor but very few pathogen models currently contain nitrite as a factor. Specific pathogen models which included all of the relevant factors would be useful for assessing the safety of reformulated cured meats.

Two models which do contain nitrite are *L. monocytogenes* and *Salmonella*. The effect of changing levels of salt and nitrite can be predicted for these organisms using these models.

Table 9.7 Effect of nitrite (0–200 ppm) and salt (0.5 to 3.5%) on growth of *L. monocytogenes* at pH 6.0 and 8.0°C

Salt		Nitrite (ppm or mg/kg)							
% w/v		200	150	125	100	75	50	25	0
0.5	Lag	105	90	50	65	70	55	45	40
	T3	400	320	280	250	230	200	170	150
1.0	Lag	110	90	70	60	55	45	40	40
	T3	375	305	270	240	210	190	170	150
1.5	Lag	100	80	70	60	55	50	40	40
	T3	360	300	270	240	210	190	165	145
2.0	Lag	105	80	70	60	55	50	45	40
	T3	370	300	270	240	215	190	190	150
2.5	Lag	100	80	70	60	55	50	45	40
	T3	370	310	280	250	220	190	190	150
3.0	Lag	110	80	70	60	60	50	45	40
	T3	380	320	290	260	230	210	190	160
3.5	Lag	110	90	70	60	60	55	50	40
	T3	410	340	300	270	240	220	195	170

T3 = Time for a 3-fold increase in number (hours)
Lag = Lag time (hours)

Table 9.7 shows the lag time and the time for a 3 log increase in levels of *L. monocytogenes* at a constant pH of 6.0, a temperature of 8°C and a range of salt and nitrite levels. It would appear from Table 9.7 that nitrite has a greater effect on the growth of *L. monocytogenes* than salt level. For any given nitrite level there is relatively little additional effect caused by increasing the salt level from 0.5 to 3.5%. This may expected due to the high resistance of this organism to salt.

At any given salt level, the lag time is approximately 2.5-fold higher at 200 mg/kg nitrite than 0 mg/kg nitrite and it takes more than twice as long to achieve a 3-log increase in numbers.

Table 9.8 shows the lag times for *Salmonella* under the same conditions of nitrite and salt. The pattern is similar as for *L. monocytogenes*. At any given salt level, there is approximately a 3-fold increase in lag time caused by raising the nitrite level from 0 mg/kg. When the nitrite level is 100 mg/kg or lower then there is also some effect on the lag time by decreasing the salt concentration. For *Salmonella*, it would appear that the effect of high levels of nitrite is greater than any changes in salt levels. Yet when the nitrite level is reduced and becomes less inhibitory then the growth characteristics are also influenced by changing salt levels.

Such data demonstrates why simultaneous changes in both salt and nitrite may lead to increased growth potential of pathogenic organisms.

Table 9.8 Effect of nitrite (0–200 ppm) and salt (0.5 to 3.5%) on growth of *Salmonella* at pH 6.0 and 8.0°C

Salt		Nitrite (ppm or mg/kg)							
% w/v		200	150	125	100	75	50	25	0
0.5	Lag	260	200	170	140	110	90	70	55
	T3	1150	930	750	650	540	430	330	250
1.0	Lag	240	200	160	140	110	90	70	55
	T3	1020	850	720	630	520	420	310	250
1.5	Lag	240	200	160	140	110	90	80	60
	T3	1000	850	740	630	520	420	340	260
2.0	Lag	240	200	160	140	110	100	80	60
	T3	1000	850	740	640	550	450	360	280
2.5	Lag	240	200	170	140	120	120	90	70
	T3	1020	870	800	680	590	490	400	320
3.0	Lag	240	200	170	160	140	130	100	80
	T3	1100	960	840	750	650	540	450	350
3.5	Lag	250	220	200	170	160	140	110	90
	T3	1150	1040	920	830	720	600	500	400

T3 = Time for a 3-fold increase in number (hours)
Lag = Lag time (hours)

9.5   References

ALLIGER, H. (1975). Ultrasonic disruption. *American Laboratory*, 10, 75–85.

ANON. (2003). The effects of nitrites/nitrates on the microbiological safety of meat products: opinion of the scientific panel on biological hazards on the request from the commission related to the effects of nitrites/nitrates on the microbiological safety of meat products. *The EFSA Journal*, 14, 1–31.

BENEDICT, R.C. (1980). Biochemical basis for nitrite-inhibition of *Clostridium botulinum* in cured meat. *Journal of Food Protection*, 43 (11), 877–891.

BENEDICT, R.C., PARTRIDGE, T., WELLS, D. and BUCHANAN, R.L. (1993). *Bacillus cereus*: aerobic growth kinetics. *Journal of Food Protection*, 56 (3), 211–214.

BETTS, G.D. (1996). *Code of practice for modified atmosphere and vacuum packaged goods.* Guideline No. 11. Campden and Chorleywood Food Research Association, Glos., UK.

BETTS, G.D., LINTON, P. and BETTERIDGE, R.J. (2000). Synergistic effects of sodium chloride, temperature and pH on growth of spoilage yeasts: a research note. *Food Micro.*, 17 (1), 47–52.

BETTS, G.D., BROWN, H.M. and EVANS, L.K. (2004). *Evaluation of product shelf life for chilled foods.* Guideline No. 46. Campden and Chorleywood Food Research Association, Glos., UK.

BEUCHAT, L.R. (1982). Thermal inactivation of yeasts in fruit juices supplemented with food preservatives and sucrose. *Journal of Food Science*, 47, 1679–1682.

BHADURI, S., TURNER-JONES, C.O., BUCHANAN, R.L. and PHILLIPS, J.G. (1994). Response surface model of the effect of pH, sodium chloride and sodium nitrite on growth of

Yersinia enterocolitica at low temperatures. *International Journal of Food Microbiology*, 23, 333–343.

BUCHANAN, R.L., STAHL, H.G. and WHITING, R.C. (1989). Effects and interations of temperature, pH, atmosphere, sodium chloride and sodium nitrite on the growth of *Listeria monocytogenes*. *Journal of Food Protection*, 52 (12), 844–851.

BUCHANAN, R.L., SMITH, J.L., McCOLGAN, C., MARMER, B.S., GOLDEN, M. and BELL, B. (1993). Response surface models for the effects of temperature, pH, sodium chloride and sodium nitrite on the aerobic and anaerobic growth of *Staphylococcus aureus* 196E. *Journal of Food Safety*, 13, 159–175.

CAST (1998). *Naturally occurring antimicrobials in foods*. Council for Agricultural Science and Technology, Task Force report, 132 (April).

CHIPLEY, J.R. (1983). In *Sodium Benzoate and Benzoic Acid Antimicrobials in Foods*. Ed. Branen, A.L. and Davidson, P.M. Marcel Dekker. Chapter 2, 11–35.

CHRISTIANSEN, L.N. (1980). Factors influencing botulinal inhibition by nitrite. *Institute of Food Technologists*, May, 237–239.

DAY, B.F.P. (1992). *Guidelines for the good manufacturing and handling of modified atmosphere packed foods*. Technical Manual No. 34. Campden and Chorleywood Food Research Association, Glos., UK.

DAY, B.F.P. (2001). *Fresh prepared produce. GMP for high-oxygen and non-sulphide dipping*. Guideline No. 31. Campden and Chorleywood Food Research Association, Glos., UK.

DEANS, S.G. and RITCHIE, G. (1987). Antibacterial properties of plant essential oils. *International Journal of Food Microbiology*, 5, 165–180.

DORMAN, H.J.D. and DEANS, S.G. (2000). Antimicrobial agents from plants: antibacterial activity of plant volatile oils. *Journal of Applied Microbiology*, 88, 308–316.

FARBOOD, M.I., MacNEIL, J.H. and OSTOVAR, K. (1976). Effect of rosemary spice extractive on growth of microorganisms in meats. *Journal of Milk and Food Technology*, 39 (10), 675–679.

FRIEDMAN, M., HENIKA, P.R. and MANDRELL, R.E. (2002). Bactericidal activities of plant essential oils and some of their isolated constituents against *Campylobacter jejuni*, *Escherichia coli*, *Listeria monocytogenes* and *Salmonella Enterica*. *Journal of Food Protection*, 65 (10), 1545–1560.

FRIEDMAN, M., KOZUKUE, N. and HARDEN, L.A. (2000). Cinnamaldehyde content in foods determined by gas chromatography-mass spectrometry. *Journal of Agricultural Food Chemistry*, 48, 5702–5709.

FSA (2003). UK Salt Intakes: Modelling Salt Reductions. www.food.gov.uk/multimedia/spreadsheets/saltmodel.xls.

GARCIA, M.L., BURGOS, J., SANZ, B. and ORDONEZ, J.A. (1989). Effect of heat and ultrasonic waves on the survival of two strains of *Bacillus subtilis*. *Journal of Applied Bacteriology*, 67, 619–628.

GIBSON, A.M. and ROBERTS, T.A. (1986). The effect of pH, sodium chloride, sodium nitrite and storage temperature on the growth of *Clostridium perfringens* and faecal streptococci in laboratory media. *International Journal of Food Microbiology*, 3, 195–210.

HAMMER, K.A., CARSON, C.F. and RILE, T.V. (1999). Antimicrobial activity of essential oils and other plant extracts. *Journal of Applied Microbiology*, 86, 985–990.

HOOVER, D.G., METRICK, C., PAPINEAU, A.M., FARKAS, D.F. and KNORR, D. (1989). Biological effects of high hydrostatic pressure on food microorganisms. *Food Technology*, 43 (3), 99–107.

HOUTSMA, P., DE WIT, J.C. and ROMBOUTS, F.M. (1996). Minimum inhibitory concentration (MIC) of sodium lactate and sodium chloride for spoilage organisms and pathogens at different pH values and temperatures. *Journal of Food Protection*, 59 (12), 1300–1304.

HSIEH, P.C., MAU, J.L. and HUAG, S.H. (2001). Antimicrobial effect of various combinations of plant extracts. *Food Microbiology*, 18, 35–43.

HUGHES, D.E. and NYBORG, W.L. (1962). Cell disruption by ultrasound. *Science*, 138, 108–114.

ISMAIEL, A.A. and PIERSON, M.D. (1990). Inhibition of germination, outgrowth and vegetative growth of *Clostridium botulinum* 67B by spice oils. *Journal of Food Protection*, 53 (9), 755–758.

LAMBERT, R.J.W., SKANDAMIS, P.N., COOTE, P.J. and NYCHAS, G.J.E. (2001). A study of the minimum inhibitory concentration and mode of action of oregano essential oil, thymol and carvacrol. *Journal of Applied Microbiology*, 91 (3), 453–462.

LEISTNER, L. (1978). Hurdle effect and energy saving. In *Food Quality and Nutrition*. (ed. Downey, W.K.). Applied Science, London.

LEISTNER, L. (1992). Food preservation by combined methods. *Food Res. Internat.*, 25, 151-158.

LEISTNER, L. (1995a). Stable and safe fermented sausages worldwide. In *Fermented Meats*. (ed. G. Campbell-Platt and P.E. Cook). Blackie Academic & Professional, Glasgow, 160–175.

LEISTNER, L. (1995b). Principles and applications of hurdle technology. In *New Methods of Food Preservation* (ed. Gould, G.W.). Blackie Academic and Professional, Glasgow.

LEISTNER, L. and GORRIS, L.G.M. (1995). Food preservation by hurdle technology. *Trends Food Science Technology*, 6, 41–44.

LILLARD, H.S. (1993). Bactericidal effect of chlorine on attached salmonellae with and without sonication. *Journal of Food Protection*, 56 (8), 716–717.

LUECK, E. (1980). *Antimicrobial Food Additives*. Springer-Verlag, Berlin.

MARECHAL, P.A., MARTINEZ DE MARNANON, I., POIRIER, I. and GERVAIS, P. (1999). The importance of the kinetics of application of physical stresses on the viability of microorganisms: significance for minimal food processing. *Trends Food Science Technology*, 10, 15–20.

MARTINI, H., WEIDENBÖRNER, M., ADAMS, S. and KUNZ, B. (1996). Eugenol and carvacrol: the main fungicidal compounds in clove and savory. *Italian Journal of Food Science*, 1, 63–67.

McCLURE, P.K., KELLY, T.M. and ROBERTS, T.A. (1991). The effects of temperature, pH, sodium chloride and sodium nitrite on the growth of *Listeria monocytogenes*. *International Journal of Food Microbiology*, 14, 77–82.

NIELSEN, H.J.S. (1983a). Composition of bacterial flora in sliced vacuum packed bologna-type sausage as influenced by nitrite. *Journal of Food Technology*, 18, 371–385.

NIELSEN, H.J.S. (1983b). Influence of nitrite addition and gas permeability of packaging film in a sliced vacuum-packed whole meat product under refrigerated storage. *Journal of Food Technology*, 18, 573–585.

NYCHAS, G.J.E. (1995). Natural antimicrobials from plants. In *New Methods of Food Preservation*, G.W. Gould (ed.) pp. 58–89. Blackie Academic & Professional, Glasgow.

ORDONEZ, J.A., AGUILERA, M.A., GARCIA, M.L. and SANZ, B. (1987). Effect of combined ultrasonic and heat treatment (thermoultrasonication) on the survival of a strain of *Staphylococcus aureus*. *Journal of Dairy Research*, 54 (1), 61–67.

PILKINGTON, B.J. and ROSE, A.H. (1988). Reactions of *Saccharomyces cerevisiae* and *Zygosaccharomyces bailli* to sulphite. *Journal of General Microbiology*, 134, 2823–2830.

RESTAINO, L., LENOVICH, L.M. and BILLS, S. (1982). Effect of acids and sorbate combinations on the growth of four osmophilic yeasts. *Journal of Food Protection*, 45, 1138–1142.

ROBINSON, A., GIBSON, A.M. and ROBERTS, T.A. (1982). Factors controlling the growth of *Clostridium botulinum* types A and B in pasteurised meats v. prediction of toxin production. *Journal of Food Technology*, 17, 727–744.

ROSE, S.A. (1987). Guidelines for Microbiological Challenge Testing. CCFRA Technical Manual No. 20. CCFRA, Chipping Campden, UK.

SAMESHIMA, T., TAKESHITA, K., MIKI, T., ARIHARA, K., HOH, M. and KINDON, Y. (1998). Effect of sodium nitrite and sodium lactate on the growth rate of lactic acid spoilage bacteria isolated from cured meat products. *Japanese Journal of Food Microbiology*, 13 (4), 159–164.

SCHERBA, G., WEIGEL, R.M. and O'BRIEN, W.D. (1991). Quantitative assessment of the germicidal efficacy of ultrasonic energy. *Applied and Environmental Microbiology*, 57 (7), 2079–2084.

SEYMOUR, I.J. (1998). The weak acid preservative stress response in *S. cerevisiae*. Ph.D thesis. University of Nottingham, UK.

SHELEF, L.A. (1983). Antimicrobial effects of spices. *Journal of Food Safety*, 6, 29–44.

SHELEF, L.A., JYOTHI, E.K. and BULGARELLI, M.A. (1984). Growth of enteropathogenic and spoilage bacteria in sage-containing broth and foods. *Journal of Food Science*, 49, 737–740, 809.

SINGHAL, R.S. and KULKARNI, P.R. (2000). Permitted preservatives. Nitrite and Nitrate. In *Encylopaedia of Food Microbiology*. Volume 3. Ed. R.K. Robinson, C.A Batt and P.D. Patel. Academic Press, London, UK, 1762–1769.

SMELT, J.P.P.M. (1998). Recent advances in the microbiology of high pressure processing. *Trends in Food Science and Technology*, 9, 152–158.

SMID, E.J. and GORRIS, L.G.M. (1999). Natural antimicrobials for food preservation. In *Handbook of Food Preservation* (ed. Rahman, M.S.), Marcel Dekker Inc., New York, 285–308.

SMITH-PALMER, A., STEWART, J. and FYFE, L. (1998). Antimicrobial properties of plant essential oils and essences against five important foodborne pathogens. *Letters in Applied Microbiology*, 26, 118–122.

SOFOS, J.N. and BUSTA, F.F. (1981). Antimicrobial activity of sorbate. *Journal of Food Protection*, 44 (8), 614–622.

SOFOS, J.N. and BUSTA, F.F. (1983). Sorbates. In *Antimicrobials in Foods* (ed. Branen, A.L. and Davidson, P.M.). Marcel Dekker, Chapter 6, 141–175.

SOFOS, J.N., BUSTA, F.F. and ALLEN, C.E. (1979). *Clostridium botulinum* control by sodium nitrite and potassium sorbate in various meat and soy protein formulations. *Journal of Food Science*, 44, 1162–1666.

SOFOS, J.N., BUSTA, F.F. and ALLEN, C.E. (1980). Influence of pH on *Clostridium botulinum* by sodium nitrite and sorbic acid in chicken emulsions. *Journal of Food Science*, 45, 7–13.

SUREKHA, M. and REDDY, S.M. (2000). Preservatives. Classification and Properties. In *Encyclopaedia of Food Microbiology*. Volume 3. (ed. R.K. Robinson, C.A. Batt and P.D. Patel). Academic Press, London, UK, 1710–1717.

TOMPKIN, R.B. (1993). Nitrite. In *Antimicrobials in Foods*. (ed. P.M. Davidson and A.L.

Branen). Marcel Dekker, Inc., New York, 191–262.

WHEELER, K.A., HURDMAN, B.F. and PITT, J.I. (1991). Influence of pH on the growth of some toxigenic species of *Aspergillus*, *Penicillium* and *Fusarium*. *International Journal of Food Microbiology*, 12, 141–150.

ZAIKA, L.L., MOULDEN, E., WEIMER, L., PHILLIPS, J.G. and BUCHANAN, R.L. (1994). Model for the combined effects of temperature, initial pH, sodium chloride and sodium nitrite concentrations on anaerobic growth of *Shigella flexneri*. *International Journal of Food Microbiology*, 23, 345–358.

10

Sensory issues in reducing salt in food products

D. Kilcast and C. den Ridder, Leatherhead Food International, UK

10.1   Introduction

Salt in our diet is essential to a healthy life, but in common with many other dietary components, too much can be harmful. The main risk to health arises through increased blood pressure (hypertension) and is a consequence of the sodium content of common salt (sodium chloride). Currently the average daily intake of salt in the UK population is around 9 g/day, against a target set in 1994 in the report by the Committee on Medical Aspects of Food and Nutrition Policy (COMA) of 6 g/day. Concerns have been increasing that salt levels in the diets of children are particularly high. Consequently, food manufacturers and retailers in the UK are coming under intense pressure to reduce salt levels in processed foods, which contribute on average 75% of dietary salt. The target is to reduce the salt level in processed foods by 30% in three years.

10.2   The role of salt in food

Salt has three major functions in food: preservation, flavour and processability. The classic historical use of salt is as a preservative, lowering the water activity to prevent microbial growth (Kurlanski, 2002). Salt continues to have this role in chilled foods, meat and fish products, cheese, pickled vegetables, sauces and bakery products. Salt contributes not only its own taste to many foods, but also contributes more generally to overall flavour contributions from other components. In addition, salt can reduce the perception of other stimuli, such as bitter compounds. Third, salt has specific processing functions in different food categories. In the bakery sector, salt has important effects on gluten development, reducing the sticky texture. In meat and fish products, salt improves water-holding capacity and increases binding in comminuted meat products. The activity of microorganisms and enzymes in cheese maturation is influenced by salt, as is the water activity. In some foods, salt can fulfil all these preservative, flavour and process functions.

10.3   The basis of flavour

Flavour is not a single entity, and is generally regarded as a combination of three types of chemosensory responses: the basic tastes, aroma volatiles and trigeminal stimuli (Keast *et al.*, 2004).

10.3.1   The taste involatiles

The foundations of any flavour are the involatile chemicals that are dissolved in saliva and detected in the mouth on cellular aggregates called taste buds, and universally given the name basic tastes (Shallenberger, 1993). Unfortunately a misunderstanding has been widely propagated that the basic tastes are perceived only on certain locations on the tongue. This is not true: although some regions are more sensitive than others, tastes can be perceived over much of the tongue, and also on other soft oral surfaces, such as the palate and the throat. We also now recognise that we need to expand the four basic tastes that have long been recognised (sweet, salt, sour, bitter) to include umami, the sensation from monosodium glutamate (MSG) and other flavour enhancers. The importance of the contributions of these stimuli to flavour should not be underestimated. For example, bitterness in the right context (such as coffee and beer) is essential to enjoyment, and yet in the wrong context is a source of consumer complaints. We now know that sensitivity to bitterness has a genetic basis, and three populations have been identified (non-tasters, tasters and supertasters) (Bartoshuk, 2000). In Caucasian populations, consumers can be classified as non-tasters, tasters and supertasters in a ratio of approximately 25:50:25. This gives product developers an uncomfortable segmentation to deal with, and the delivery of optimum bitterness levels becomes even more complicated when we realise that the segmentation is not necessarily the same in different ethnic groupings, as the proportion of supertasters is known to be higher in other populations, for example Indonesian.

Other sensations that are sometimes classified as basic tastes, and which are frequently encountered in salt reduction programmes, are metallic and astringent. The metallic note is classically associated with the taste of metal ions, for example from iron and copper (Yang and Lawless, 2005), but can also result from the response to by-products of fat breakdown in foods. Astringency is not strictly a taste, but a drying sensation that arises through precipitation of proteins present in saliva by phenolic compounds present in the food, and the consequent loss of lubrication (de Wijk and Prinz, 2005).

10.3.2   Aroma volatiles

The basic tastes provide the foundation for the aroma volatiles that give interest and variety to foods. In contrast to the limited number of basic tastes, there are many thousands of chemical species that deliver an odour response in the nasal cavity, and these are essential in building flavour variety. We are also learning much more about how these volatile chemicals are perceived, particularly how they reach the receptors in the nose when we are consuming foods (Taylor and Hort, 2004).

10.3.3   Trigeminal stimuli

Trigeminal stimuli have received less attention than other stimuli, but are important in delivering excitement to foods and drinks. These are stimuli that trigger the trigeminal nerve that is responsible for the pain response to heat and cold (Green, 2004). The chemicals that cause this are present in ingredients such as pepper, horseradish and chilli, but this response is also produced from alcohol and the carbonic acid that is formed on carbonation of soft drinks. In an age in which consumers are experiencing sensations from foods encountered in foreign travels, it is not surprising that this stimulus is finding itself employed in unusual circumstances – for example in the use of chilli in fruit-based soft drinks, and also in alcoholic beverages. As such, trigeminal stimuli can be considered as adding excitement to flavour.

10.4   Contribution of salt to flavour

Salt has its own characteristic taste that we are all familiar with, and has been long accepted (Shallenberger, 1993) as one of the four basic tastes (sweet, salt, acid, bitter), recently extended to five with the addition of umami. Other chemical species can also deliver a salty taste, but only sodium chloride gives what is generally recognised as a pure salty taste (i.e. absence of any other tastes). It is now thought that either the halide anion (Cl^-, Br^-, I^-) or the metal cation (Li^+, Na^+, K^+, Mg^{2+}) can deliver saltiness, but that the combined sodium and chloride ions deliver the clean saltiness that is needed.

One of the most important functions of salt, in common with the other basic tastes, is to form the foundation for building the overall flavour response and is the consequence of responses to the basic tastes, to volatile odour molecules and to the chemical species that generate an irritant response and delivering excitement to flavour (Fig. 10.1).

A related function of salt lies in its ability to enhance the flavour of products. Without salt, foods such as soup, bread and biscuits are perceived as bland and unappetising. Although this enhancement is most common in savoury foods, salt is also important in delivering flavour in sweet products such as chocolate. An additional function of salt is to contribute to a fuller mouthfeel (Hutton, 2000). It is uncertain how crucial perceived saltiness, fullness, or overall balance is in relation to the acceptance of different low-salt foods.

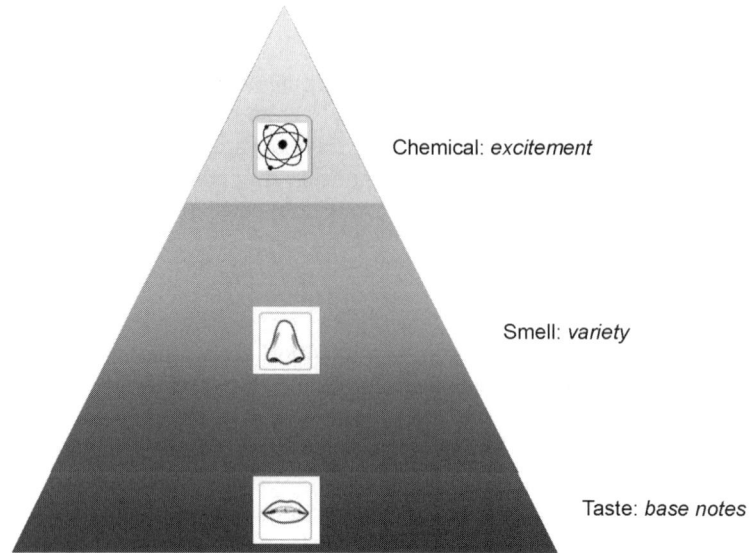

Chemical: *excitement*

Smell: *variety*

Taste: *base notes*

Fig. 10.1   Contribution of the basic tastes to flavour.

Together with changes in perceived saltiness, the most notable sensory change resulting from a reduction in salt is often an increase in bitterness, which can result from either a loss of bitterness inhibition by salt, or bitterness inherent in some salt replacers. Perceived bitterness is considered to be a negative and undesirable attribute in many products, and is another factor to be taken into consideration.

Removal of salt as a fundamental building block will have a profound effect on overall flavour, and this is basis for the difficulties that the industry faces in reducing salt, even if the other functions of salt can be compensated.

10.5   Challenges in reducing salt

Sensory challenges arising from salt reduction relate not only to maintaining an acceptable salt perception but also relate to the additional sensory properties mentioned above. Ideally, salt reduction should not adversely change the characteristic flavour and mouthfeel of a product.

The reduction of salt affects food in different ways. Sensory characteristics of one set of reduced salt products are not necessarily reflected if the same strategy is applied to other products as a consequence of the very different environment in which the reduced salt system is perceived. For this reason, conducting evalua-tions of salty taste in aqueous solution is useful only for initial screening purposes. In addition, individual recipes will require specific salt reduction strategies.

In the remainder of this chapter, only the sensory aspects of salt reduction will be considered. However, maintaining microbiological stability and safety is an essential requirement for any salt reduction programme, and aspects related to processability must also be considered.

10.6   Main approaches to salt reduction

A number of approaches can be followed in developing reduced sodium food. These can be applied independently, but can also be applied in combination. One important consideration is that many retailers with a 'clean label' policy do not welcome any salt enhancer or replacer that contravenes this policy. In addition, there are substantial cost implications for some materials (for example, peptides) that are likely to limit their application.

10.6.1   Reduction of overall salt content: reduction by stealth

This refers to the reduction of salt content in a product formulation, which can be undetected in many cases if the reductions are small and are carried out stepwise. Anecdotal evidence suggests the slow and gradual reduction of salt can pass consumers unnoticed as the palate adjusts to the revised sensory profile if changes are small enough (Bertino *et al.*, 1982; Anonymous, 1998; Wheelock and Hobbiss, 1999). This approach is occasionally referred to as reduction by stealth and when gradually continued over time, large reductions can be achieved. This approach has been used successfully by a number of manufacturers in the UK (FSA, 2006). In the period between 1998 and 2005 there has been a 33% reduction in the salt content of cereals and since the late 1980s a 25% reduction in the salt content of bread. Reductions by stealth have also been made in some major branded products, with reductions of 11–18% across the Heinz product range, and a 33% reduction in Kraft processed cheese products.

However, a level will almost inevitably be reached below which loss in appealing flavour will be noticed by consumers, with a significant risk of reduced sales. In addition, this process is unlikely in itself to achieve the significant reductions in salt content of processed foods sought by the UK government in the short term.

10.6.2   Compensation

The contribution of salt to overall flavour can be compensated for by the use of substitutes that deliver the required flavour by other means. This can be achieved by a number of means, including an increase in the quantity of the main components that are responsible for flavour delivery, the use of herbs and spices, and for other specific products the use of ingredients such as lemon, onion/garlic, vinegar or other acids. Examples of this can be found in Chapter 17.

10.6.3   Saltiness enhancement

A saltiness enhancer can be defined as a material that increases the perception of saltiness of a salt-tasting compound without having any significant saltiness itself. Numerous compositions have been formulated in which sodium has been partially replaced by other substances such as other mineral salts and amino acids, which are reported to enhance the salty taste of sodium chloride. At present salt enhancers are not widely used in thermally processed food products (Bonorden *et al.*, 2003).

Research related to salt enhancers has been well documented. The main focus of taste-enhancing compounds has generally been upon compounds that are already present in foods. Recently, taste-enhancing compounds that are generated from precursors during processing have gained attention. It should be noted that the distinction between saltiness enhancers and salt replacers (see Section 10.6.4) is often blurred, and compounds classified as enhancers can often have a salty character. There is also a view (Kemp and Beauchamp, 1994) that the term flavour potentiation should be used to describe an increase in intensity, with the term flavour enhancer reserved to describe an increase in pleasantness. However, in this chapter the more common usage of enhancement to cover intensity will be followed. Some of the most commonly investigated flavour and salt enhancers are listed below.

Glycine and glycine monoethyl ester
Glycine has been used in a number of reduced salt products (Matsumoto *et al.*, 2000; Omura *et al.*, 2001). Its function appears to be a combination of reducing water activity as well as acting as a salt enhancer for various types of sausages (Gelabert *et al.,* 2003; Gou *et al.*, 1996). Kuramitsu *et al.* (1997) and Segawa *et al.* (1995) have investigated the use of glycine ester in soy sauce. It should be noted that some confusion between glycine and glycine ester is evident in published papers.

The results obtained from the addition of glycine at 20% to sausages produced a slight reduction of acid and salty taste. Results obtained from work on soy sauce indicate that glycine ester elicits adverse taste characteristics, namely sour/acid, at higher concentrations, suggesting that its use at low concentrations is to provide a salty taste.

L-lysine and L-arginine
L-lysine, an amino acid, is made from fermented corn starch, and elicits a salty but also an astringent response (described variously as harsh, acrid taste, irritating the throat). A number of commercially available salt enhancers are based upon the effects of L-lysine and L-arginine. These amino acids have been used in combination with ornithine, citric acid, succinic acid, potassium chloride, and ammonium chloride. Some mention has been made of a stringent feeling in the stomach after intake of L-arginine (Mosciano, 1999; Kimoto and Morishige, 2002).

Guerrero *et al.* (1995) describe a salt-taste-enhancing composition, which can be used in a wide variety of foods and beverages. The composition is prepared by enzymically hydrolysing a protein in the absence of added sodium. The resulting hydrolysate contains peptides and free amino acids, particularly free lysine and free arginine, in combination with an ammonium salt. When the composition is dehydrated and added to a food or beverage containing a reduced amount of sodium chloride, the salty taste is enhanced and, in some cases, the flavour of the food or beverage is enhanced. The composition has a nutritional value as it contains significant amounts of dietary protein.

Lactates

Three main types of lactate salts have been reported as saltiness enhancers – potassium, sodium and calcium lactate. Potassium and sodium lactates are widely used in the meat and poultry industries as antimicrobial agents. Sodium and potassium lactate have been used in reduced salt meat products and found to maintain a certain salty level (Price, 1973). Their application for this purpose is, however, not widespread. Calcium is often added to foods as a salt of organic anions such as lactate (Lawless *et al.*, 2003b). However, calcium lactate has a considerable sour component. Small quantities of volatile lactic acid produced by interaction of the lactate with water are easily detectable in a neutral carrier (Tordoff, 1996).

Mycoscent

Mycoscent (www.mycoscent.com) is a by-product from the production of a mycoprotein (marketed as Quorn) and described as a flavouring preparation. The range is claimed to act as a flavour enhancer, acting synergistically with dairy and savoury flavours and is suggested for use as a salt enhancer/substitute, with the potential for 50% sodium reduction in savoury products.

Trehalose

Trehalose is a non-reducing disaccharide. Commercially known as ASCEND^{TM}, it has been used as a flavour enhancer in ready to eat meat and poultry products. From studies carried out by the manufacturers, Cargill, ASCEND^{TM} can effectively reduce or eliminate attributes such as metallic, bitter and astringent (www.cargill.com, 2004). European Patent Application EP0813820 (Toshio *et al.*, 1997) describes the use of trehalose to enhance the saltiness of sodium chloride without imparting any unsatisfactory taste and flavour to food products.

L-ornithine

Several dipeptides derived from L-ornithine have been described by Tada *et al.* (1984) as being salty. Amongst them L-ornithyltaurine monohydrochloride (Orn-Tau·HCl) and ornithyl-β-alanine monohydrochloride (H-Orn-β-Ala-OH·HCl) have been claimed to exhibit the clearest and strongest salty taste without being accompanied by a bitter after taste. However, in 1987, Huynh-ba re-examined this work and reported the dipeptides not to be salty. They concluded that the saltiness of Orn-Tau·HCl claimed earlier, probably resulted from sodium chloride present as an artefact of the method of preparation

(Huynh-ba and Philipposian, 1987). Their taste panel ($n = 17$) tested a 0.5% solution and judged the solution to be slightly sour, bitter and metallic. This result was in complete disagreement with what had been claimed (Tada *et al.*, 1984). The original research has been repeated by Seki (Seki *et al.*, 1990). They confirmed the earlier identified saltiness of Orn-Tau·HCl and supplied a possible explanation for the aforementioned disagreement. Nakamura *et al.* (1996) reported that the highest saltiness level is delivered by Orn-Tau·1.2HCl, which also shows enhancement, but with an accompanying sourness.

O-aminoacyl sugars

Ornithyl-β-alanine (OBA) has been found to produce saltiness as well as give an enhancement to the saltiness of sodium chloride (Tamura *et al.*, 1989; 1993). However, this is an extremely expensive ingredient to synthesise, and Tamura *et al.* (1993) also point out that a saltiness 20 times that of sodium chloride is needed to avoid excessive intake of amino acids or peptides.

Glutamates

Monosodium glutamate (MSG) has long been used in several Asian cuisines for its taste, known as 'umami'. There is no English word synonymous with umami, (it roughly translates to 'savoury deliciousness' (Prescott, 2001)) and the closest related terms are savoury, meat or broth like (Ninomiya, 2001). The characteristic taste of umami is also imparted by 5′-ribonucleotides such as inosine-5′-monophosphate (IMP) and guanosine-5′-monophosphate (GMP) (respectively present in dried skipjack and shiitake mushrooms). Taste synergism between glutamate and nucleotides has been reported (Yamaguchi and Ninomiya, 2000) and markedly enhances the umami taste (Halpern, 2000). Several studies of interactions between glutamic acid, glutamates, IMP, GMP, sodium chloride, and human perception of foods have been summarised in reviews (Yamaguchi and Takahashi, 1984; Bellisle *et al.*, 1991; Bellisle, 1998).

One umami-imparting compound, monosodium glutamate, generally receives more attention. Whilst MSG does not have a pleasant taste by itself (Halpern, 2000), at low concentrations it can enhance the taste of other compounds. It has been often suggested that MSG could be used as a means of reducing sodium chloride levels while maintaining acceptable flavour (Okiyama and Beauchamp, 1998). A number of studies have indicated that MSG can enhance the taste of salt in a number of products. In principle use of MSG could maintain flavour characteristics at lower salt concentrations, but its disadvantage is that it contains sodium itself. In some cases more sodium was added to the product salted with MSG or MSG plus salt than with salt alone (Pangborn and Braddock, 1989).

MSG does not supply the same saltiness, fullness, or overall balance that sodium chloride does and it has also been reported to impart additional bitter/metallic notes (Gillette, 1985). Research published by Yamaguchi (1991), however, contradicts these findings and states that addition of umami leads to an increase in body and even thickness. The mechanism behind this is unclear, but it may be that we associate greater body with more savoury foods (Yamaguchi, 1991; Prescott, 2001).

The taste characteristics of conventionally salted soup and low salt soup with added MSG were investigated by Ball *et al.* (2002). Results showed that even though low salt soups with MSG were rated as being less salty, they were more preferred because of their overall flavour characteristics. When MSG is used to enhance the taste of products the pleasantness of the taste is often described as being meatier and more brothy. Meatiness and brothiness are anecdotally used to describe umami. MSG does not generally elicit descriptors of the four primary tastes when described by untrained panellists (Okiyama and Beauchamp, 1998).

A small amount of research has been done on the flavour characteristics of calcium diglutamate (CDG), sometimes also referred to as calcium glutamate (Prescott, 2001). Being free of sodium, it could potentially achieve equivalent taste characteristics at lower sodium concentrations than MSG (Bellisle *et al.*, 1992). However, little research has been published on the taste characteristics of CDG in relation to foods and it appears to be little used as an additive (Ball *et al.*, 2002).

The research that is available suggests that umami substances are capable of enhancing other flavours of food and increasing pleasantness of reduced salt foods (Roininen *et al.*, 1996). MSG and other glutamates have, however, been linked with a variety of health conditions including hyperactivity, sickness and migraine and so are avoided by some consumers. Although there is minimal scientific evidence to support these claims, many manufacturers and retailers have a strict policy against glutamate use.

Alapyridaine
Alapyridaine (N-(1-Carboxyethyl)-6-hydroxymethyl-pyridinium-3-ol) is a compound discovered in heated glucose/alanine solutions (Frank *et al.*, 2001). Alapyridaine naturally occurs in beef bouillon and does not exhibit a taste on its own. Besides enhancing sweetness and umami tastes, the presence of alapyridaine also influences salt perception (Ottinger and Hofmann, 2003).

Soldo *et al.* (2003) investigated binary and ternary combinations of sodium chloride, L-arginine, and alapyridaine. Threshold detection of an aqueous solution of sodium chloride was found to be significantly decreased in the presence of equimolar amounts of either alapyridaine or L-arginine. Combinations with alapyridaine scored significant higher than L-arginine. The most intense salty taste was found for ternary mixtures of sodium chloride, L-arginine, and alapyridaine. Currently, however, alapyridaine is not commercially available.

10.6.4   Salt replacement
Combinations of ingredients for use as salt replacers have been based on the replacement of the Na^+ cation by potassium, ammonium, calcium and lithium and by anions such as phosphate and glutamates (British Nutrition Foundation

(BNF) 1994). Salts such as lithium chloride, and ammonium chloride, although providing a salty taste are considered unsuitable due to their poor stability, smell and, in the case of lithium, toxicity. Amino acids and peptides have been described as providing a mixture of tastes; however, the majority are described as being bitter, sweet or sour, with saltiness rarely being cited as a relevant attribute.

Potassium chloride

Potassium chloride (KCl) is the most common choice as a feasible salt replacer (Renqvist, 1919; McBurney and Lucas, 1966; Dzendolet and Meiselman, 1967; Frank and Michelsen, 1970; Murphy *et al.*, 1981; Klaauw and Smith, 1995; Rosett *et al.*, 1995). Its acceptability is, however, limited due to its pronounced bitter/chemical/metallic taste and aftertaste. Currently, sodium reduction is achieved by a straight substitution of a proportion of the sodium chloride with potassium chloride. Replacement of 30% sodium chloride by potassium chloride is generally feasible for many products, and 50% substitution is possible for specific products. However, masking the undesirable sensory attributes associated with potassium chloride remains a problem. This is exacerbated by the wide distribution of sensitivities to bitterness, which is known to have a genetic basis (Bartoshuk, 2000). Consequently, even low levels of bitterness will be detected by the sensitive end of the population.

In recent years, efforts have been made to eliminate the undesirable bitter and metallic tastes associated with the introduction of other salts. Most commonly, potassium chloride is combined with MSG, ammonium chloride, magnesium sulphate and amino acids.

There are currently a number of commercial salt replacers available, made up with various combinations of sodium chloride and potassium chloride (e.g., LoSalt, Morton and Pansalt).

Other potassium salts

The application of potassium glutamate as a substitute for salt has been investigated in split pea soup. The potassium glutamate offered little more than the untreated product did, and in no way approached the flavour of salt (Gillette, 1985). Potassium sulphate is distinguished from others by having significant scores for the basic tastes of sweetness, bitterness, sourness and saltiness. Saltiness, sourness and bitterness increased with increasing concentration, while sweetness diminished. The mixed taste appears to be a combination of the fundamental properties of individual ions, and clearly several salts are capable of these four basic tastes (Shallenberger, 1993). Potassium lactate, as previously described, can be used to reduce the amount of salt present while maintaining a certain level of perceived saltiness.

Calcium chloride

The taste of calcium chloride is predominantly described as bitter, sour and sweet at 1 mM but bitter, salty and sour at 100 mM. Concentrations generally used in the reported literature range from 1 mM to 100 M (Klaauw and Smith, 1995; Tordoff, 1996; Lawless *et al.*, 2003b).

Outside of the classical basic tastes, the three oral properties – metallic taste, astringency and irritation – are also ascribed to calcium chloride (the irritation comprises the sensation similar to hot salsa, hot peppers or Tabasco-like sensation). The nature of the metallic taste is unclear and may be primarily a retronasal smell sensation. More work is required, with panellists trained with reference standards such as alum for astringency, capsaicin for irritation, and ferrous sulphate for metallic taste (Lawless *et al.*, 2003a).

Sodium chloride enhanced the salty taste of calcium chloride and suppressed any other tastes, especially bitterness. Sucrose and citric acid also had a generally suppressive effect on calcium chloride. These effects have the potential of enhancing the palatability of calcium salts as fortifying agents (Shallenberger, 1993; Lawless *et al.*, 2003a,b).

Magnesium sulphate
Magnesium sulphate provides both a bitter and a salty taste, depending on its concentration (Delwiche *et al.*, 1999; Lawless *et al.*, 2003a). At low levels it is associated with a salty taste compared to high levels where it is perceived as being bitter (Shallenberger, 1993). For this reason it may have the potential to be used as a salt replacer. In contrast to this is the research published by Breslin and Beauchamp (1995), stating that magnesium sulphate has a suppressing effect on the saltiness at intermediate concentrations of NaCl. At higher concentrations magnesium sulphate had no influence on the saltiness of NaCl.

Magnesium sulphate has been reported in a number of different patents relating to salt replacement such as WO8500958 (Rood, 1984), EU377119 (Kurppa, 1988) and UK 2396793 (Wilson, 2004), where it has been used in conjunction with other salts to reduce the level of sodium present. The reduced sodium salt, Icelandia Life sea salt, contains magnesium sulphate at 17%. It has also been suggested that the addition of magnesium sulphate into salt mixtures has potential health benefits, such as reducing blood pressure.

Metal ions and sea salt taste
It is often claimed in the media that the unique taste of speciality salts, including sea salt, is due to the presence of metal ions. Although the cumulative effects of a number of different metal ion contaminants cannot be discounted as a factor, experiments on the effect of metal ions have been inconclusive (Steingarten, 2002). Analysis of sea salts showed that levels of potassium and magnesium can be associated with bitter notes, but the presence of impurities does not necessarily correlate with poor quality. Oshima Blue sea salt from Japan, the most expensive salt in the world, has substantial quantities of potassium, magnesium, sulphur and calcium (Fig. 10.2) but is perceived as having a smooth, salty taste. An Icelandic sea salt comprising 41% sodium chloride, 41% potassium chloride, and 17% magnesium chloride is currently being marketed on its normal salt taste.

Fig. 10.2   Photomicrograph of Oshima Blue sea salt showing high impurity levels (courtesy of Leatherhead Food International).

10.6.5   Bitterness inhibitors

The use of potential salt replacers such as potassium chloride is limited primarily by associated non-salty tastes, the principal problem being bitterness. This is a particular problem as a result of the trimodal nature of bitterness response, since the supertaster population will be sensitive to even minute traces of bitterness. The reduction of salt and perceived bitterness in food, has been investigated by both the food and pharmaceutical industries for many years (Kurihara *et al.*, 1994; Kurihara and Nirasawa, 1994). A number of compounds have been proposed to inhibit bitter taste (Pangborn, 1960, 1962; Haga *et al.*, 1984; Godshall, 1988), yet a suitable bitterness inhibitor for general use has not been identified.

When two taste compounds are mixed, there is potential for one compound to interfere with taste receptor cells or taste transduction mechanisms associated with the other compound. For example, this type of peripheral interaction occurs between sodium salts and certain bitter compounds (Keast and Breslin, 2002). Sodium salts and bitter compounds generally interact so that bitterness is suppressed to some variable degree and saltiness is unaffected (Bartoshuk *et al.*, 1988; Breslin, 1996; Breslin and Beauchamp, 1997a,b).

The use of sweeteners has been previously demonstrated to be effective in reducing bitterness. Used at low concentrations, in conjunction with other compounds, the perceived sweet aftertaste is removed, while masking the perceived bitterness. Sucrose is commercially the most commonly used sweetener, but other sweeteners have been investigated, and in view of the long persistence of bitterness that can occur, the use of intense sweeteners such as thaumatin also characterised by long persistence, has been investigated exhaustively.

Thaumatin, an intensely sweet protein marketed under the trade name Talin is isolated from the fruit of Katamfe (*Thaumatococcus daniellii*, Benth). It is normally found in the form of a pale brown hydroscopic powder which has a sweetness level some 2000–3000 times that of sucrose, when measured on a weight basis equivalent to 8% sucrose (Shallenberger, 1993). Initially limited for use as a sweetener and flavour enhancer in chewing gum and cigarettes, it has been used in savoury products, where it is used to improve the umami effect.

Thaumatin has the ability to mask bitterness both in the taste and aftertaste of a product, giving the mixture a more sugar-like profile. It is often used to increase and round off the natural flavour profile of products containing mint, such as mouthwash, toothpaste and breath fresheners, as well as being used as a masking agent in pharmaceutical preparations. Often the active ingredients in these types of products provide a bitter or astringent note that is unacceptable to consumers. In the majority of applications, thaumatin can be used at levels below 10 ppm providing the best flavour modification or even flavouring (Birch, 2000).

2,4-dihydroxybenzoic acid (DHB) and its salts have been identified as potential bitter inhibitors which do not affect sweetness. It has also been reported to be effective at eliminating the undesirable metallic aftertaste often associated with saccharin. Work carried out has mainly focused on its addition to salt mixtures used to season baked potatoes, popcorn and other edibles. Although it has a good taste profile, a problem associated with it is that it is prone to decarboxylation at low pH (Kurtz and Fuller, 1997).

10.6.6   Increasing saltiness delivery

Maximum saltiness of topically applied salt is usually not achieved as the largest salt crystals do not dissolve fast enough to reach the sodium receptors in the mouth before swallowing. It seems reasonable to hypothesise that the efficiency of a given amount of salt could be increased if the ions comprising salt could reach the receptors more quickly.

Sodium chloride elicits a salty taste only when in solution. A rapid dissolution rate could intensify saltiness in some foods and thus reduce the levels required. The dissolution rate of sodium chloride in the mouth is partly determined by the exposed surface area, is a function of crystal size and crystal form.

Smaller crystal sizes and a low bulk density will result in a large surface area; for example, due to a long flat shape, flakes have a lower bulk density, and thus a faster rate of dissolution. Crystal shape and crystal density will also influence the rate of dissolution; for example, dendritic salt has voids throughout the crystal, thus drastically increasing exposed surface area, which again increases the dissolution rate (Bravieri, 1983).

Differences in salt crystal form and ensuing differences in salt dissolution rate are likely to be the main reasons for the perceived higher quality of many sea salts in comparison to common table salt, which is inevitably in a cubic crystal form. For example, Maldon salt from the east coast of England has an unusual hollow pyramidal shape that helps confer a high dissolution rate. This is likely to be a more important factor in perceived salt quality than the presence of metal ions in trace quantities.

In a collaborative research project carried out at Leatherhead Food International, the effect of a wide range of salt crystal sizes on the rate of saltiness

Table 10.1   Different forms of salt types evaluated

Type	Supplier	Process	Structure	Particle size (μ) range (majority)
Table salt	Sainsbury's	Vacuum dried	Cubic crystalline	200–500 (400–500)
Dendritic salt	Morton Salt	Vacuum dried	Dendritic	50–300 (200–300)
Alberger fine prepared flour salt	Cargill Salt	Alberger process	Open structure?	40–300 (50–80)
Premium fine prepared flour salt	Cargill salt	Vacuum dried and ground	Cubic crystalline	20–200 (50–100)
Microfine salt	Salt Union	Vacuum dried and ground	Cubic crystalline	10–100 (20–40)
Microsized 95 extra fine salt	Cargill Salt and pulverised	Vacuum dried	Crystalline?	5–30 (10–20)
100% salt	LFI	Freeze dried	Cubic/glass?	5–10μ particles aggregated together
50% salt/50% polydextrose	LFI	Freeze dried	Cubic/glass?	20–500μ pieces, easily broken down.
10% salt/90% polydextrose	LFI	Boiled and deposited	Glass	Large fragments of glassy material

perception has been investigated in conjunction with other approaches to salt reduction. A range of salt types with different sizes and crystal forms was obtained from commercial salt producers. In addition, salt was prepared in a novel form by various types of processing. The salts tested are shown in Table 10.1, and photomicrographs of their physical form are shown in Fig. 10.3.

Evaluation of the rate at which the perception of saltiness was perceived when applied to an unflavoured potato snack was measured by a trained sensory panel using time-intensity assessment, in which the perception of saltiness was recorded as a function of time in the mouth. The tests confirmed that the salts with the finest crystal sizes gave a more rapid release of saltiness than the larger crystal sizes, although there were additional differences in the maximum level of saltiness attained (den Ridder and Kilcast, 2005). As there is no general agreement on how the form of the time-intensity curve relates to consumer liking, the importance of these findings remains to be determined.

Whilst approaches based on salt crystals can only see practical applications in foods with salt in the solid form, such as surface-coated snacks, improved delivery in foods containing salt in solution would also be attractive commercially. In another part of the LFI collaborative project, the increase in sensory saltiness of emulsion-based systems was investigated by modifying the emulsion structure. In particular, the aim was to establish whether multiple emulsion systems could be prepared and, if so, whether such systems could have potential in modifying the sensory perception of products.

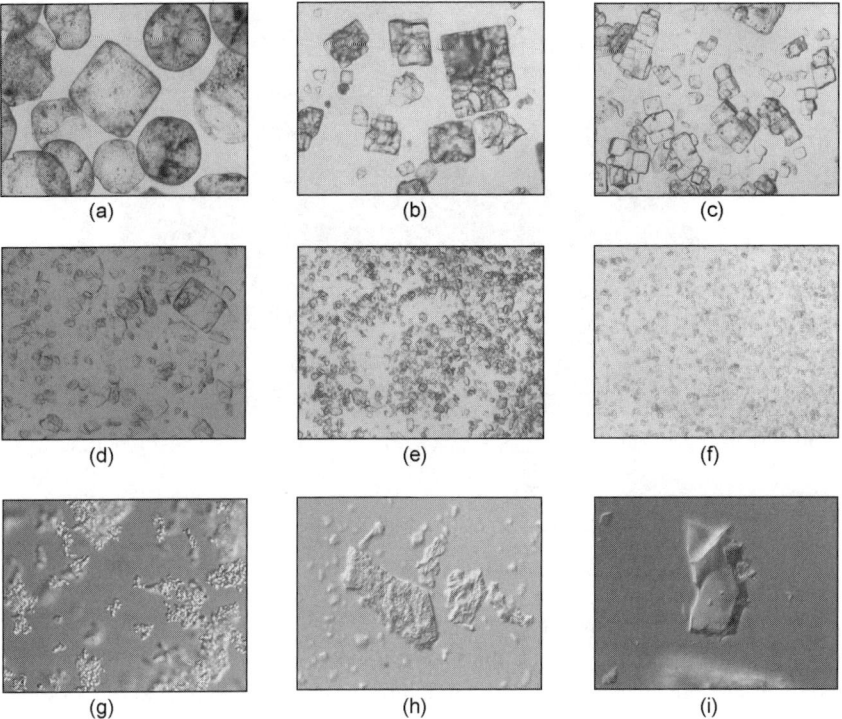

Fig. 10.3   Photomicrographs of salts tested (courtesy of Leatherhead Food International). (a) Sainsbury's Table Salt; (b) Morton Dendritic Salt; (c) Alberger Salt; (d) Premier Fine; (e) Microfine Salt; (f) Microsized 95 Extra Fine Salt; (g) LFI 100% Salt (Freeze Dried); (h) LFI 50% Salt/50% Polydextrose (Freeze Dried); (i) LFI 10% Salt/90% Polydextrose (Glass).

Simple oil-in-water (o/w) emulsion containing salt would have all the salt dissolved in the external aqueous phase. A simple water-in-oil (w/o) emulsion would have all the salt dissolved in the internal aqueous phase. The sensory perception of salt from these two systems is different, since in the o/w emulsion the salt is directly in contact with the palate, while in the w/o emulsion, it is the oil that is in contact with the palate and not the aqueous phase.

However, if a water-in-oil-in-water (w/o/w) double emulsion could be produced, then in theory, the salt could be either dissolved in the internal aqueous phase, the external aqueous phase or both. Thus, it would be anticipated that for a double emulsion system it should be possible to attain a range of sensory saltiness, depending on the relative amounts of salt dissolved in the internal and external aqueous phases. In the limit, maximum saltiness would be obtained when all the salt is dissolved in the external aqueous phase. Considering the difference between an o/w emulsion containing salt and a w/o/w emulsion containing the same level of salt (but only in the external aqueous phase), it follows that the w/o/w emulsion will have a higher concentration of salt in the external aqueous phase and would therefore be expected to be perceived as more salty.

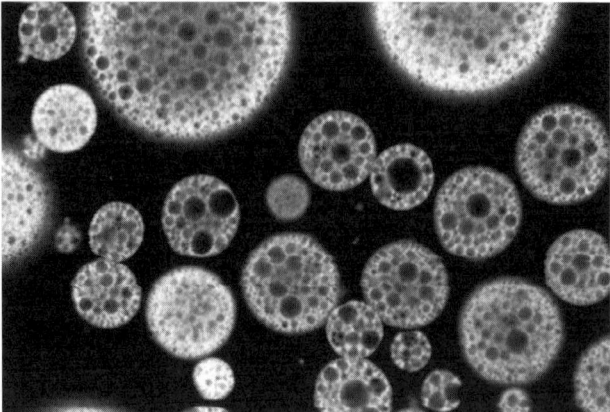

Fig. 10.4   Confocal scanning laser image of w/o/w double emulsion (courtesy of Leatherhead Food International).

It was found possible to form w/o/w emulsions in which salt can be incorporated into either the external or internal aqueous phase. An example of the double emulsion structure prepared for the experiments is shown in Fig. 10.4. The taste characteristics depend on the location of the salt. In order to maximise stability, it appears that the osmotic pressure of the internal and external aqueous phases of the w/o/w emulsion should be as equal as possible, which will require formulation modifications.

10.7   Conclusions and future trends

Saltiness is an important sensory attribute of many foods, and salt itself contributes even more to the characteristic flavour of many food types. Whilst ensuring an adequate salt dietary intake is vital to health, the intake in the UK has become too high, with substantial health risks. Changing the salt content of a consumer population that has adapted to a high salt diet will not be easy, and will entail a number of strategies, one of the most important of which will be consumer education. Otherwise, a strategy will almost certainly be needed combining a 'stealth' approach of gradual salt reduction, with a more technological approach toward delivering salt perception at lower salt content. Considerable efforts are now being made to utilise the recent progress in understanding the mechanisms underlying taste perception to find new materials that can deliver salty taste or suppress unwanted tastes such as bitterness. However, most options are likely to have cost implications for the food manufacturer, as any strategy involving enhancement or replacement will require the incorporation of a more expensive ingredient. Against this, though, are the substantial benefits to health that will accrue from an innovative approach.

10.8   Sources of further information and advice

The chemistry of taste perception is covered in detail in:

SHALLENBERGER R S (1993). *Taste Chemistry*. New York: Blackie Academic & Professional.

The historical importance of salt can be found in a readable form in:

KURLANSKI M (2002). *Salt. A World History*. London: Random House.

Accounts of recent developments in flavour perception can be found in a number of excellent chapters in:

TAYLOR A J and ROBERTS D D (eds) (2004). *Flavour Perception*. Oxford: Blackwell.

10.9   References

ANONYMOUS (1998). Salt Science. *Health Which?*, 12, 24–26.

BALL P, WOODWARD D, BEARD T, GHOOBRIDGE A and FERRIER M (2002). Calcium diglutamate improves taste characteristics of lower-salt soups. *European Journal of Clinical Nutrition*, **56**, 519–523.

BARTOSHUK L M (2000). Comparing sensory experiences across individuals: recent psychophysical advances illuminate genetic variation in taste perception. *Chemical Senses*, **25**, 447–460.

BARTOSHUK L M, RIFKIN B, MARKS L E and HOOPER J E (1988). Bitterness of KCl and benzoate: related to genetic status for sensitivity to PTC/PROP. *Chemical Senses*, **13(4)**, 517–528.

BELLISLE F (1998). Nutritional effects of umami in the human diet. *Food Review International*, **14**, 309–319.

BELLISLE F, MONNEUSE M O, CHABERT M, LARUE-ACHAGIOTIS C, LANTEAUME M T and LOUIS-SYLVESTRE J (1991). Monosodium glutamate as palatability enhancer in the European diet. *Physiology & Behavior*, **49**, 869–873.

BELLISLE F, DARTOIS A M and BROYER M (1992). Two studies on the acceptability of calcium glutamate as a potassium-free sodium substitute in children. *Journal of Renal Nutrition*, **2**, 42–46.

BERTINO M, BEAUCHAMP G K and ENGLEMAN K (1982). Long term reduction in dietary sodium alter the taste of salt. *American Journal of Clinical Nutrition*, **36**, 1134–1144.

BIRCH G (2000). *LFRA Ingredients Handbook: Sweeteners*, 2nd edn. Leatherhead: Leatherhead Publishing.

BONORDEN W R, GIORDANO D A and LEE B L (2003). Salt flavor enhancing compositions, food products including such compositions, and methods for preparing such products. In *US6 541 050; Campbell Soup Co.*: United States Patent.

BRAVIER E R (1983). Techniques for sodium reduction and salt substitution in commercial processing. In *Research & Dev. Assn*: Norfolk, VA.

BRESLIN P A S (1996). Interactions among salty, sour and bitter compounds. *Trends in Food Science and Technology*, **7(12)**, 390–399.

BRESLIN P A S and BEAUCHAMP G K (1995). Suppression of bitterness by sodium: variation among bitter taste stimuli. *Chemical Senses*, **20(6)**, 609–623.

BRESLIN P A S and BEAUCHAMP G K (1997a). Salt enhances flavour by suppressing bitterness. *Nature*, **387(6633)**, 563.

BRESLIN P A S and BEAUCHAMP G K (1997b). Suppression of bitterness by sodium: implications for flavor enhancement. In Roy G (ed.), *Modifying Bitterness: Mechanism, Ingredients and Applications*. Lancaster, PA: Technomic Publishers, pp. 179–213.

BRITISH NUTRITION FOUNDATION (BNF) (1994). *Salt in the diet*. British Nutrition Foundation Research Report.

DE WIJK R A and PRINZ J F (2005). The role of friction in perceived oral texture. *Food Quality and Preference*, **16(2)**, 121–129.

DELWICHE J F, HALPERN B P and DESIMONE J A (1999). Anion Size of Sodium Salts and Simple Taste Reaction Times. *Physiology & Behavior*, **66(1)**, 27–32.

DEN RIDDER and KILCAST D (2005). Unpublished results.

DZENDOLET E and MEISELMAN H L (1967). Gustatory quality changes as a function of solution concentration. *Perception & Psychophysics*, **2**, 29–33.

FSA (FOOD STANDARDS AGENCY) (2006). www.food.gov.uk/news/newsarchive/2006/mar/salt targets.

FRANK O, OTTINGER H and HOFMANN T (2001). Characterization of an intense bitter-tasting 1H,4H-quinolizinium-7-olate by application of the taste dilution analysis, a novel bioassay for the screening and identification of taste-active compounds in foods. *Journal of Agricultural and Food Chemistry*, **49(1)**, 231–238.

FRANK R L and MICHELSEN O (1970). Sodium-potassium chloride mixtures as table salt. *Third Symposium on Salt*, **2**, 135–139.

GELABERT J, GOU P, GUERRERO L and ARNAU J (2003). Effect of sodium chloride replacement on some characteristics of fermented sausages. *Meat Science*, **65(2)**, 833–839.

GILLETTE M (1985). Flavour effects of sodium chloride. *Food Technology*, **39(6)**, 47–52.

GODSHALL M A (1988). The role of carbohydrates in flavour development. *Food Technology*, **11**, 71–76.

GOU P, GUERRERO L, GELABERT J and ARNAU J (1996). Potassium chloride, potassium lactate and glycine as sodium chloride substitutes in fermented sausages and in dry-cured pork loin. *Meat Science*, **42(1)**, 37–48.

GREEN B G (2004). Oral chemesthesis: an integral component of flavour. In Taylor A and Roberts D (eds), *Flavor Perception*. Oxford: Blackwell, pp. 151–171.

GUERRERO A, KWON S S Y and VADEHRA D V (1995). Compositions to enhanced taste of salt used in reduced amounts. *US Patent 5711985*: United States.

HAGA F, KOMINE H, KONDO E and KUWANO N (1984). Study on effect of seasoning with potassium salts by sensory test. *Japanese Journal of Nutrition*, **42(2)**, 225–234.

HALPERN B P (2000). Glutamate and the flavor of foods. *American Society of Nutritional Sciences*, **4S**, 910S–914S.

HUTTON T (2000). Technological functions of salt in the manufacturing of food and drink products. *Food & Drink Federation*, 1–16.

HUYNH-BA T and PHILIPPOSIAN G (1987). Alleged salty taste of L-ornithyltaurine monohydrochloride. *Journal of Agricultural and Food Chemistry*, **35(1)**, 165–168.

KEAST R S J and BRESLIN P A S (2002). An overview of binary taste–taste interactions. *Food Quality and Preference*, **14**, 111–124.

KEAST R S J, DALTON P and BRESLIN P A S (2004). Flavor interactions at the sensory level. In Taylor A and Roberts D (eds), *Flavor Perception*. Oxford: Blackwell, pp. 228–255.

KEMP S E and BEAUCHAMP G K (1994). Flavor modification by sodium chloride and monosodium glutamate. *Journal of Food Science*, **59**, 682–686.

KIMOTO E and MORISHIGE F (2002). Arginine/ascorbic acid mixed powder as an oral supplement. *US Patent 2002091156*: United States.

KLAAUW N J V D and SMITH D V (1995). Taste quality profiles for fifteen organic and inorganic salts. *Physiology & Behavior*, **58(2)**, 295–306.

KURAMITSU R, SEGAWA D, NAKAMURA K, MURAMATSU S and OKAI H (1997). Further studies on the preparation of low sodium chloride-containing soy sauce by using ornithyl-taurine hydrochloride and its related compounds. *Bioscience, Biotechnology, and Biochemistry*, **61(7)**, 1163–1167.

KURIHARA K, KATSURAGI Y, MATSUOKA I, KASHIWAYANAGI M, KUMAZAWA T and SHOJI T (1994). Receptor mechanisms of bitter substances. *Physiology & Behavior*, **56(6)**, 1125–1132.

KURIHARA Y and NIRASAWA S (1994). Sweet, antisweet and sweetness-inducing substances. *Trends in Food Science and Technology*, **5(2)**, 37–42.

KURLANSKI M (2002). *Salt. A World History*. London: Random House.

KURPPA L (1988). Agent for reducing adverse effects of table salt. *European Patent Application Nmr EU0 377 119*: Finland.

KURTZ R J and FULLER W D (1997). Development of a low-sodium salt: a model for bitterness inhibition. In Roy G (ed.), *Modifying Bitterness: Mechanism, Ingredients and Applications*. Lancaster, PA: Technomic Publishers, pp. 215–226.

LAWLESS H T, RAPACKI F, HORNE J and HAYES A (2003a). The taste of calcium and magnesium salts and anionic modifications. *Food Quality and Preference*, **14(4)**, 319–325.

LAWLESS H T, RAPACKI F, HORNE J, HAYES A and WANG G (2003b). The taste of calcium chloride in mixtures with NaCl, sucrose and citric acid. *Food Quality and Preference*, **15(1)**, 83–89.

MATSUMOTO Y, FUKUSHI H and HIRAKI J (2000). Food Preservative. In *Japanese Patent Application; Chisso Corp. JP2000-270821*: Japan.

MCBURNEY D H and LUCAS J A (1966). Gustatory cross adaptation between salts. *Psychonomic Science*, **4(8)**, 301–302.

MOSCIANO G (1999). The creative flavorist. *Perfumer and Flavorist*, **24(6)**, 10–13.

MURPHY C, CARDELLO A V and BRAND J G (1981). Tastes of fifteen halide salts following water and NaCl: Anion and cation effects. *Physiology & Behavior*, **26**, 1083–1095.

NAKAMURA K, KURAMITU R, KATAOKA S, SEGAWA D, TAHARA K, TAMURA M and OKAI H (1996). Convenient synthesis of L-ornithyltaurine-HCL and the effect on saltiness in a food material. *Journal of Agricultural and Food Chemistry*, **44(9)**, 2481–2485.

NINOMIYA K (2001). An overview of recent research on MSG: sensory applications and safety. *Food Australia*, **53(12)**, 546–549.

OKIYAMA A and BEAUCHAMP G K (1998). Taste dimensions of monosodium glutamate (MSG) in a food system: role of glutamate in young American subjects. *Physiology & Behavior*, **65**, 177–181.

OMURA K, SUZUKI T, FURUBE K and YOSHITAKE S (2001). Seasoning preparation for food. In *Japanese Patent Application; Eisai Co. Ltd. JP2001-178393*: Japan.

OTTINGER H and HOFMANN T (2003). Identification of the taste enhancer alapyridaine in beef broth and evaluation of its sensory impact by taste reconstitution experiments. *Journal of Agricultural and Food Chemistry*, **51(23)**, 6791–6796.

PANGBORN R M (1960). Taste Interrelationships. *Food Research*, **25**, 245–256.

PANGBORN R M (1962). Taste Interrelationships III: Suprathreshold Solutions of Sucrose and Sodium Chloride. *Journal of Food Science*, **27**, 495–500.

PANGBORN R M and BRADDOCK K S (1989). *Ad libitum* preferences for salt in chicken broth. *Food Quality and Preference*, **1(2)**, 47–52.

PRESCOTT J (2001). Taste hedonics and the role of umami. *Food Australia*, **53(12)**, 550–554.

PRICE S (1973). Phosphodiesterase in tongue epithelium: activation by bitter taste stimuli.

Nature, **1(241)**, 54–55.

RENQVIST Y (1919). Uber den Geschmack. *Skandinavisches Archiv für Physiologie*, **38**, 97–201.

ROININEN K, LÄHTEENMÄKI L and TUORILLA H (1996). Effect of umami taste on pleasantness of low-salt soups during repeated testing. *Physiology & Behavior*, **60(3)**, 953–958.

ROOD R P (1984). Salt substitute. *US Patent 4473597*: United States.

ROSETT T R, WU Z, SCHMIDT S J, ENNIS D M and KLEIN B P (1995). KCl, CaCl$_2$, Na$^+$, and salt taste of gum systems. *Journal of Food Science*, **60(4)**, 849–867.

SEGAWA D, NAKAMURA K, KURAMITSU R, MURAMATSU S, SANO Y, UZUKA Y, TAMURA M and OKAI H (1995). Preparation of low sodium chloride containing soy sauce using amino acid based saltiness enhancers. *Bioscience, Biotechnology and Biochemistry*, **59(1)**, 35–39.

SEKI T, KAWASAKI Y, TAMURA M, TADA M and OKAI H (1990). Further study on the salty peptide ornithyl-beta-alanine. Some effects of pH and additive ions on the saltiness. *Journal of Agricultural and Food Chemistry*, **38(1)**, 25–29.

SHALLENBERGER R S (1993). *Taste Chemistry*. New York: Blackie Academic & Professional.

SOLDO T, BLANK I and HOFMANN T (2003). (+)-(S)-Alapyridaine – A general taste enhancer? *Chemical Senses*, **28**, 371–379.

STEINGARTEN J (2002). Salt Chic, in *It must've been something that I ate*, 47–58, Headline.

TADA M, SHINODA I and OKAI H (1984). L-ornithyltaurine, a new salty peptide. *Journal of Agricultural and Food Chemistry*, **32(5)**, 992–996.

TAMURA M, SEKI T, KAWASAKI Y, TADA M, KIKUCHI E and OKAI H (1989). An enhancing effect on the saltiness of sodium chloride of added amino acids and their esters. *Agricultural and Biological Chemistry*, **53**, 1625–1633.

TAMURA M, NAKAMURA K, KINOMURA K and OKAI H (1993). Relationship between taste and structure of O-aminoacyl sugars containing basic amino acids. *Bioscience, Biotechnology and Biochemistry*, **57(1)**, 20–23.

TAYLOR A J and HORT J (2004). Measuring proximal stimuli involved in flavour perception. In Taylor A and Roberts D (eds), *Flavor Perception*. Oxford: Blackwell, pp. 1–38.

TORDOFF M G (1996). Some basic psychophysics of calcium salt solutions. *Chemical Senses*, **21(4)**, 417–424.

TOSHIO M, SATOSHI I and YUKIO U (1997). Method for enhancing the salty-taste and/or delicious-taste of food products. *EP0813820*: Japan.

WHEELOCK V and HOBBISS A (1999), *All you ever wanted to know about salt but were afraid to ask*. Skipton, Yorkshire: Verner Wheelock Associates.

WILSON L (2004). Salt composition *UK Patent Application GB2 396 793*.

YAMAGUCHI S (1991). Basic properties of umami and effects on humans. *Physiology & Behavior*, **49**, 833–841.

YAMAGUCHI S and NINOMIYA K (2000). Umami and Food Palatability. *American Journal of Nutritional Sciences*, **4S**, 921S–926S.

YAMAGUCHI S and TAKAHASHI C (1984). Hedonic functions of monosodium glutamate and four basic substances used at various concentration levels in single and complex systems. *Agricultural and Biological Chemistry*, **48**, 1077–1081.

YANG H H-L and LAWLESS HT (2005). Descriptive analysis of divalent salts. *Journal of Sensory Studies*, **20(2)**, 97–113.

11

The use of bitter blockers to replace salt in food products

R. McGregor, Linguagen, USA

11.1 Introduction: the potential for taste modifiers in healthy food products

This chapter outlines how the elucidation of the biochemical mechanisms involved in taste is enabling the use of modern biotechnology approaches to identify compounds that improve the taste of food. The chapter begins with an overview of the drivers behind the push to reduce salt in food products. The second section reviews the science of taste. The next section discusses how biopharmaceutical discovery methods are being used to identify novel taste modifying compounds, and how these compounds can be used by the food industry to produce healthier, better tasting products. This section includes a case study of a bitter blocker discovered using biotechnology methods, including examples of how it can be used to decrease the sodium content in food products without the decrease in palatability that leads to consumer rejection. The chapter closes with a discussion of future trends in the industry, including how the further elucidation of taste signaling will enable the discovery of the next generation of taste modifiers, including new salt substitutes.

11.2 Why replace salt in foods?

It is now widely accepted that for a significant proportion of the population, those known as the salt-sensitive subpopulation, increasing intake of salt (sodium chloride) leads to increased blood pressure (Jones, 2004). It is a common perception with the public that a significant quantity of salt in the diet is inherently present, added during cooking or shaken on at the table. In fact, 75% of the salt in the typical Western diet is added during processing. Therefore, it falls to food manufacturers to identify ways of reducing salt in order to attain the dietary levels of sodium now being recommended by health authorities in many countries.

There is heated debate as to whether to target those people most at risk from consuming high levels of sodium, the salt-sensitive subpopulation, or to use a population-wide approach to reducing dietary salt intake. The salt-sensitive population consists primarily of individuals who are hypertensive, older, black or who have kidney disease.

The argument for population-wide sodium reduction is predicated on the perceived difficulty in identifying those individuals who are salt sensitive (Obarzanek et al., 2003), although, as with many aspects of dietary sodium, this view is far from universally held (Egan, 2003).

A number of studies have investigated the effect of sodium on blood pressure. Proposed mechanisms for a linkage include salt modulation of arterial wall gene expression (Ying and Sanders, 2002), denervation of arterial baroreceptors (DiBona and Sawin, 2002), perturbation of the rennin-angiotensin system (He et al., 2001; Harrison-Bernard et al., 2003), abnormalities in nitric oxide production/release (Barron et al., 2001), decreased activity of afferent renal nerves (Kopp et al., 2003), and increased activity of α_{1D}-adrenergic receptors (Tanoue et al., 2002).

With current technology, it has been virtually impossible for food manufacturers to reduce sodium content and still market a product the consumer is willing to purchase. The elucidation at the molecular level of the signaling pathways involved in taste perception has opened up new opportunities to successfully reduce salt.

11.3   The science of taste perception

The perception of taste occurs predominantly in specialized structures called taste buds, clustered into taste papillae on the surface of the tongue. Taste buds contain groups of elongated taste receptor cells linked to the peripheral nervous system. The apical surface of these cells contains protein receptors and channels, which bind to, or allow entry of, tastants that come into contact with the taste bud in the mouth cavity. The purpose of taste receptor cells is to transduce the physical interaction of tastants at the apical surface into release of neurotransmitter from the cell's basolateral surface. The basolateral surface is exposed to the peripheral nervous system and the resulting increase in neurotransmitter concentration triggers activation of nerve cells resulting in a nerve impulse to the brain. For a review of the basics of taste, read Smith and Margolskee (2001).

Five taste modalities are currently recognized: sweet, salty, sour, bitter, and umami (savory) (Smith and Margolskee, 2001; Kim et al., 2004). Etiologically speaking, recognition of these various tastes has evolved to enable humans and animals to discern important information about the quality of food. Sweet, salty and umami taste typically are associated with foods that contain nutrients important for well being. Sweet tasting foods are typically high in carbo-hydrates, salty food contains important minerals, and umami taste is coupled to the presence of amino acids. Sour and bitter taste perception is characteristically a protective mechanism against ingesting substances that may be deleterious to the body, such as spoiled food or poisons respectively. Other tastes, such as the taste of polysaccharides, may be accepted as distinct taste modalities in the future (Sclafani, 2004).

The past 15 years has seen great strides in the elucidation of the mechanisms by which the physical interaction of tastants at the exposed surface of taste receptor cells is transduced into neurotransmitter release from the basolateral surface of these cells. In the early 1990s, the laboratory of Robert F. Margolskee reported the identification of a protein involved in taste (McLaughlin *et al.*, 1992). This protein, called gustducin, interacts with receptor proteins in the apical membrane of taste cells and was the first taste protein to be characterized at the molecular level. Gustducin is a member of a family of proteins called G proteins, and hence the receptor proteins it interacts with are known as G protein coupled receptors (GPCRs). A number of G proteins have subsequently been identified in the taste system (McLaughlin *et al.*, 1994). Upon tastant binding, a change in conformation of the receptor leads to activation of G protein. G protein activation in turn results in the switching on of intracellular proteins known as effector enzymes. These effector enzymes modulate the concentration of molecules within the cell called second messengers. In the resting state, taste cells are polarized due to the active maintenance of concentration gradients of ions across the cell membrane. As second messenger levels change, various ion channels open, both in the cell membrane, and in intracellular membranes, resulting in the depolarization of the taste cell as ions flow down their respective concentration gradients. Depolarization results in release of neurotransmitter from the taste cell into the synaptic cleft and subsequent activation and depolarization of the adjacent neuron. This electrical signal is transmitted through the nervous system to the brain where it is interpreted as taste.

There are at least 25 GPCRs proposed to be involved in bitter taste (Adler *et al.*, 2000; Matsunami *et al.*, 2000). These GPCRs, known as the T2Rs, allow for the detection of the wide variety of structural classes of compounds that are bitter. This large family of receptors also speaks to the etiological importance of bitter taste detection. If only one receptor was involved in bitter taste, then a non-functioning mutation of this receptor would likely result in death as there would be no first line detection apparatus for many poisons.

It appears that not all bitter tastants exert their bitter taste by activating T2R receptors. It has been reported that certain amphiphilic substances, including H_1-receptor antagonists, can directly activate G proteins (Naim *et al.*, 1994; Burde *et al.*, 1996). It is possible that the GPCR-independent mechanism of G protein activation represents a bitter sensing mechanism for many bitter tastants present at high concentrations. Other bitter compounds, such as the methyl xanthines (e.g., caffeine in coffee, theophylline in tea and theobromine in cocoa) are known to interact with phosphodiesterase (PDE), which is one of the effector enzymes present in taste cells. Studies have shown that in response to caffeine and theophylline, taste cell levels of the substrate for PDE rises (Rosenweig *et al.*, 1999). Indeed, it can be predicted that the taste of any compound that modulates the level of activation of a protein involved in the taste transduction cascade could be altered if the compound comes into contact with the protein.

Sweet and umami tastes are transduced through GPCR heterodimers consisting of two receptor subunits. T1R1 and T1R3 receptors act in concert to detect umami compounds, whilst T1R2 and T1R3 form the functional sweet taste receptor (Kitagawa *et al.*, 2001; Max *et al.*, 2001; Montmayeur *et al.*, 2001; Sainz *et al.*, 2001; Nelson *et al.*, 2001; Bachmanov *et al.*, 2001; Li *et al.*, 2002). In studies on mice lacking expression of one of the subunits of the receptor, it has been shown that the mice are responsive to certain sweeteners and umami compounds, hence raising the possibility that other receptors or mechanisms are involved in sweet and umami taste transduction in mice (Damak *et al.*, 2003; Zhao *et al.*, 2003). A GPCR called mGluR4 has been hypothesized as an umami receptor (Chaudhari *et al.*, 2000).

In contrast to bitter, sweet, and umami taste, salt and sour taste appear to be transduced by pathways beginning with interaction of tastant with ion channels. For salty taste, evidence suggests that the epithelial sodium channel (ENaC) functions as a salt receptor, as amiloride hydrochloride, a diuretic drug that blocks ENaC, also reduces sensitivity to salt (Doolin and Gilbertson, 1996; Lindeman *et al.*, 1998; Kretz *et al.*, 1999; Lin *et al.*, 1999). The amiloride-insensitive salt taste transduction mechanism appears to be mediated at least in part by a variant of the transient receptor potential V1 (TRPV1) channel, as modulators of TRPV1 have activating and blocking effects on the nerve responses to salts, and mice with TRPV1 genetically ablated lack an amiloride-insensitive nerve response (Lyall *et al.*, 2004, 2005). Both mechanisms appear to be involved in human taste perception (Feldman *et al.*, 2003). Whatever the relative contributions of these pathways to salty taste transduction, the effect of sodium ions on taste cells is depolarization and neurotransmitter release. Sour taste detection appears to be even more complex with a number of pathways possibly involved. Hydrogen ions in acids enter taste cells through proton channels and also interact with a variety of channels including potassium, sodium, calcium and chloride channels (DeSimone *et al.*, 2001 for review). The overall effect is again a depolarization of the taste cell and the generation of a nerve impulse.

11.4   Identifying compounds that decrease the perception of bitterness

The food industry has always used taste modifying compounds to ameliorate the taste of food products. Some examples include the use of salt and sugar to decrease the perception of bitterness and the use of the nucleotides inosine monophosphate and guanosine monophosphate to potentiate the umami taste of monosodium glutamate. Compounds with a particular taste are also used to improve the acceptability of products with an otherwise unacceptable taste, for example the use of menthol in mouthwash, where consumers attribute the astringency of the product to the menthol, not the active ingredients.

With the characterization of the taste system at the molecular level there are a number of available approaches involving taste modification to reduce the levels of sodium in processed foods. The three most likely approaches, in ascending order of technical difficulty, are substitution of potassium salts in place of sodium salts, potentiation of sodium taste, and replacement of sodium with a novel salty tasting molecule. While all three of these approaches can enable reduction in sodium chloride added to foods for taste, the sodium in foods fulfils a number of roles beyond supplying salty taste. Table 11.1 summarizes some of the sodium-containing compounds used in foods along with their purpose. Due to the multiple functionalities of the various sodium-containing ingredients used in modern food processing, it appears that replacement of sodium with potassium-containing salts will enable food manufacturers to continue to add these compounds, while at the same time reducing the level of sodium, as the counterion in many of these ingredients can be either sodium or potassium. In contrast, compounds that potentiate the taste of sodium chloride, or activate the salt taste pathway independently of sodium chloride face the problem that they will almost certainly not replicate the functionality of the other sodium-containing compounds in food systems. The latest dietary guidelines in the United States, released in January 2005 by the Departments of Health and Human Services and Agriculture, recommend potassium intake of 4700 mg per day. However, the average American consumes only 2500 mg of potassium per day. Therefore, in addition to allowing for a reduction in sodium content in foods, the use of potassium chloride as a replacement would increase the amount of potassium in processed foods, hence countering the deficit in potassium intake in the typical diet.

Table 11.1   Some functions of sodium-containing compounds in food systems

Compound	Functions
Sodium chloride	Taste, texture, preservative
Sodium nitrite	Color, taste, protective against *Clostridium botulinum*
Sodium erythorbate	Curing accelerator
Sodium lactate/diacetate	*Listeria monocytogenes* suppression and control of *Clostridium botulinum*
Sodium phosphates	Water retension, texture, antioxidant, control of *Clostridium botulinum*
Sodium bicarbonate	Color in vegetables
Sodium ascorbate	Antioxidant

In addition to potassium chloride, ammonium salts and certain dipeptides and cationic amino acids, such as ornithyl-beta-alanine, lysine-ornithine hydrochloride, arginine and lysine have a salty taste. However, like potassium chloride, the application of these compounds as salt substitutes is limited by bitter taste. In fact, although a number of approaches to reducing salt content are available, they face challenges due to taste, safety or cost of production that means relatively few salt products are currently on the market, particularly when compared to the plethora of low-sugar and low-fat products that are available, where good tasting alternatives have been identified. In 2003, Linguagen Corp. received patent protection on adenosine 5′-monophosphate, which can be used to decrease the bitterness of potassium chloride (more on this shortly). Recently, Prime Favorites has introduced a potassium chloride-containing salt substitute with improved taste, and more recently Givaudan have introduced their own competitor product in this rapidly developing field.

Traditionally, herbs and spices have been used as salt substitutes, including black pepper, curry powder, garlic, onion, tarragon, basil, ginger, cumin, dill seed and coriander. Lemon and vinegar can also be used as salt substitutes.

Enhancers of salty taste include peptides from a variety of hydrolyzed sources including collagen, soybean, wheat, egg white and milk, and the sweeteners thaumatin and trehalose. International Flavors and Fragrances holds patents on alkyldienamides which have been shown to potentiate salty taste and their salt substitute product is the current market leader in reducing sodium content in food ingredients.

Certain food companies are taking a different approach to reducing sodium in their products while maintaining their market. It has been shown that reducing the amount of salt in food in the diet leads to a reduction in the level of salt determined by consumers to be 'just right' (Bertino et al., 1982). In response to regulatory pressures, these companies are gradually reducing the sodium content in the hopes that consumer tastes will adjust. Interestingly, a follow-up study showed that when you increase the level of salt, preference for increased salt content adjusts even more readily (Bertino et al., 1986).

The biopharmaceutical approach to improving taste has already borne fruit in the effort to reduce the salt content of food products. Using an assay that monitors the activation of the taste G protein by bitter tastants, adenosine 5′-monophosphate (AMP) was identified as a compound that reduces taste cell activation by bitter compounds (Ming et al., 1999). Mouse preference studies confirmed that AMP was improving the palatability of bitter solutions and electrophysiological recordings showed a decreased activation of nerve responses by bitter compounds in the presence of AMP.

Human sensory studies have subsequently demonstrated that one of AMP's most robust activities is reducing the bitter taste of potassium. Figure 11.1 shows the effect of AMP on the taste of a low sodium chicken broth in which KCl was added to replace the absent sodium. It can be seen that the broth containing KCl/AMP is perceived as significantly less bitter than the soup with KCl alone. Larger scale consumer testing has confirmed these findings. An interesting

Store brand chicken broth, reduced-sodium + 1.5% KCl

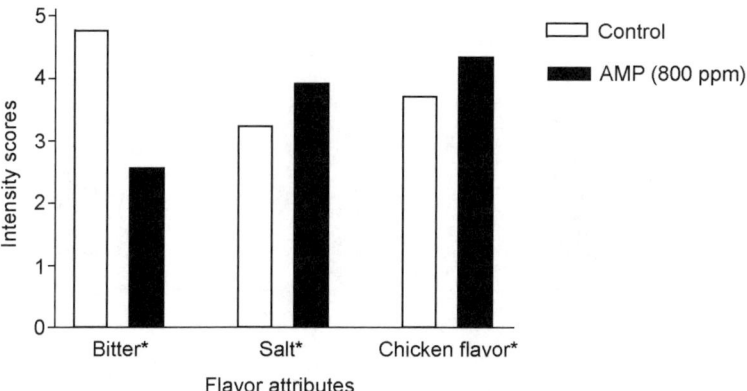

* Statistically significant, 95% CI
19 trained panelists
0–8 intensity scale, 8 = highest intensity

Fig. 11.1   AMP reduces the bitterness of potassium chloride supplemented low sodium
chicken broth, and increases the saltiness and overall flavour.

observation coming out of these studies is that the addition of AMP leads to
increased saltiness and umami in the soup. This is likely due to two factors, the
inherent umami taste of AMP at higher concentrations and the reduction of
bitterness leading to the saltiness and umami taste of the soup being more
pronounced.

Further research at Linguagen has identified a formulation containing KCl,
AMP and other regulatory approved ingredients that works better than KCL and
AMP alone at improving the taste of low sodium, KCl-containing products. This
bitter blocker formulation, known as BetraSalt™, has been shown to effectively
function as a NaCl substitute in a number of applications. Figure 11.2 compares
the effects of Betra™ (the bitter blocker formulation) on the taste of a low
sodium soup containing KCl. Replacing NaCl with KCl in the soup introduces a
significant bitterness. The addition of Betra™ to a 50% reduced sodium, KCl-
containing formulation significantly reduces this bitterness and improves the
saltiness and overall flavor of the soup to the point where it is equivalent to the
full sodium soup.

11.5   Future trends

Salt taste, like other tastes, is perceived through receptor proteins on taste cells,
the receptor proteins for salty taste being ion channels. In the coming years, the
mechanism of salt taste in humans will be fully elucidated and at that point the
same biochemical approaches which have already been used to identify taste
modifiers that affect bitter, sweet and umami taste can be used to identify novel

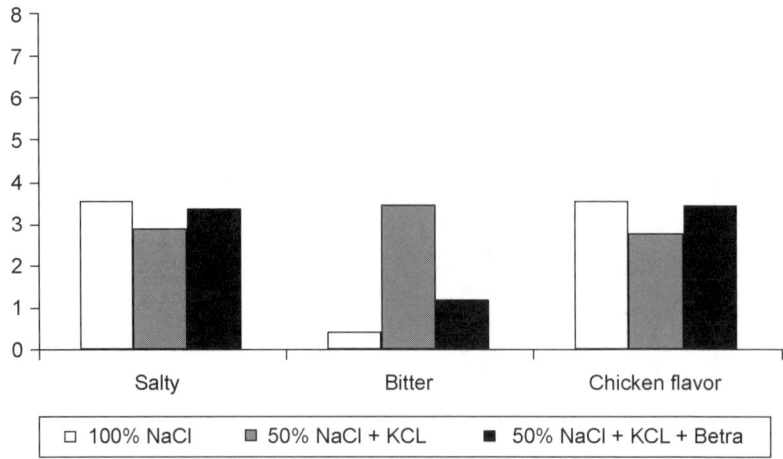

Fig. 11.2   Betra supplemented low sodium chicken soup with potassium chloride gives a
product with taste attributes similar to the full sodium product.

salt substitutes and salt enhancers. While it appears that replacing sodium with a
bitter-blocked potassium formulation would be the preferred method of sodium
reduction, due to the multitude of functionalities of sodium-containing
ingredients in food products, and the interchangeable nature of sodium and
potassium in these ingredients, the availability of alternative methods of sodium
reduction can only improve the choices available to formulators as they produce
healthier, better tasting foods in the twenty-first century.

11.6   Sources of further information and advice

For an introduction to taste, Smith and Margolskee (2001) is a useful primer. A
more detailed review of the genetics of taste perception is Kim *et al.* (2004).
Two recent reviews of the link between dietary sodium and blood pressure are
Weinberger (2004) and Jones (2004).

11.7   References

ADLER, E., HOON, M.A., MUELLER, K.L., CHANDRASHEKAR, J., RYBA, N.J.P. and ZUKER, C.S.
(2000) A novel family of mammalian taste receptors. *Cell* 100: 693–702.
BACHMANOV, A.A., LI, X., REED, D.R., OHMEN, J.D., LI, S., CHEN, Z., TORDOFF, M.G., DE JONG, P.J.,
WU, C., WEST, D.B., CHATTERJEE, A., ROSS, D.A. and BEAUCHAMP, G.K. (2001) Positional
cloning of the mouse saccharin preference (*Sac*) locus. *Chem. Senses* 26: 925–933.
BARRON, L.A., GIARDINA, J.B., GRANGER, J.P. and KHALIL, R.A. (2001) High-salt diet enhances
vascular reactivity in pregnant rats with normal and reduced uterine perfusion
pressure. *Hypertension* 38: 730–735.

BERTINO, M., BEAUCHAMP, G. K. and ENGELMAN, K. (1982) Long-term reduction in dietary sodium alters the taste of salt. *Am. J. Clin. Nutr.* 36: 1134–1144.

BERTINO, M., BEAUCHAMP, G. K. and ENGELMAN, K. (1986) Increasing dietary salt alters salt taste preference. *Physiol. Behav.* 38: 203–213.

BURDE, R., DIPPEL, E. and SEIFERT, R. (1996) Receptor-independent G protein activation may account for the stimulatory effects of first generation H1-receptor antagonists in HL-60 cells, basophils and mast cells. *Biochem. Pharmacology* 51: 125–131.

CHAUDHARI, N., LANDIN, S.D. and ROPER, S.D. (2000) A metabotropic glutamate receptor variant functions as a taste receptor. *Nature Neurosci.* 3: 113–119.

DAMAK, S., RONG, M., YASUMATSU, K, KOKRASHVILI, Z., VARADARAJAN, V., ZOU, S., JIANG, P., NINOMIYA, Y. and MARGOLSKEE, R.F. (2003) Detection of sweet and umami taste in the absence of taste receptor T1r3. *Science* 301: 850–853.

DeSIMONE, J.A., LYALL, V., HECK, G.L. and FELDMAN, G.M. (2001) Acid detection by taste receptor cells. *Respir. Physiol.* 129(1–2): 231–245.

DIBONA, G.F. and SAWIN, L.L. (2002) Effect of arterial baroreceptor denervation on sodium balance. *Hypertension* 40: 547–551.

DOOLIN, R.E. and GILERTSON, T.A. (1996) Distribution and characterization of functional amiloride-sensitive sodium channels in rat tongue. *J. Gen. Physiol.* 107: 545–554.

EGAN, B.M. (2003) Reproducibility of BP responses to changes in dietary salt: compelling evidence for universal sodium restriction? *Hypertension* 42: 457–458.

FELDMAN, G. M., MOGYOROSI, A., HECK, G. L., DeSIMONE, J. A., SANTOS, C. R., CLARY, R. A. and LYALL, V. (2003) Salt-evoked lingual surface potential in humans. *J. Neurophysiol.* 90: 2060–2064.

HARRISON-BERNARD, L.M., SCULMAN, I.H. and RAIJ, L. (2003) Postovariectomy hypertension is linked to increased renal AT_1 receptor and salt sensitivity. *Hypertension* 42: 1157–1163.

HE, F.J., MARKANDU, N.D. and MacGREGOR, G.A. (2001) Importance of the rennin system for determining blood pressure fall with acute salt restriction in hypertensive and normotensive whites. *Hypertension* 38: 321–325.

JONES, D.W. (2004) Dietary sodium and blood pressure. *Hypertension* 43: 932–935.

KIM, U.-K., BRESLIN, P. A. S., REED, D. and DRAYNA, D. (2004) *J. Dent. Res.* 83(6): 448–453.

KITAGAWA, M., KUSAKABE, Y., MIURA, H., NINOMIYA, Y. and HINO, A. (2001) Molecular genetic identification of a candidate receptor gene for sweet taste. *Biochem. Biophys. Res. Commun.* 283: 236–242.

KOPP, U.C., CICHA, M.Z. and SMITH, L.A. (2003) Dietary sodium loading increases arterial pressure in afferent renal-denervated rats. *Hypertension* 42: 968–973.

KRETZ, O., BARBRY, P., BOCK, R. and LINDEMANN, B. (1999) Differential expression of RNA and protein of the three pore-forming subunits of the amiloride-sensitive epithelial sodium channel in taste buds. *J. Histochem. Cytochem.* 47: 51–64.

LI, X., STASZEWSKI, L., XU, H., DURICK, K., ZOLLER, M. and ADLER, E. (2002) Human receptors for sweet and umami taste. *Proc. Natl. Acad. Sci. USA* 99(7): 4692–4696.

LIN, W., FINGER, T.E., ROSSIER, B.C. and KINNAMON, S.C. (1999) Epithelial Na+ channel subunits in rat taste cells: localization and regulation by aldosterone. *J. Comp. Neurol.* 405: 406–420.

LINDEMANN, B., BARBRY, P., KRETZ, O. and BOCK, R. (1998) Occurance of ENaC subunit mRNA and immunocytochemistry of the channel subunits in taste buds of the rat vallate papilla. *Ann. N.Y. Acad. Sci.* 855: 116–127.

LYALL, V., HECK, G.L., VINNIKOVE, A.K., GHOSH, S., PHAN, T.-H.T., ALAM, R.I., RUSSELL, O.F., MALIK, S.A., BIGBEE, J.W. and DeSIMONE, J.A. (2004) The mammalian amiloride-insensitive non-specific salt taste receptor is a vanilloid receptor-1 variant. *J. Physiol.* 558: 147–159.

LYALL, V., HECK, G.L., VINNIKOVA, A.K., GHOSH, S., PHAN, T.-H.T. and DESIMONE, J.A. (2005) A novel vanilloid receptor (VR-1) variant mammalian salt taste receptor. *Chem. Senses* 30: i42–i43.

MATSUNAMI, H., MONTMAYEUR, J-P. and BUCK, L.B. (2000) A family of candidate taste receptors in human and mouse. *Nature* 404: 601–604.

MAX, M., SHANKER, Y.G., HUANG, L., RONG, M., LIU, Z., CAMPAGNE, F., WEINSTEIN, H., DAMAK, S. and MARGOLSKEE, R.F. (2001) Tas1r3, encoding a new candidate taste receptor, is allelic to the sweet responsiveness locus *Sac. Nat. Genet.* 28: 58–63.

McLAUGHLIN, S.K., MCKINNON, P.J. and MARGOLSKEE, R.F. (1992) Gustducin is a taste-cell-specific G protein closely related to the transducins. *Nature* 357: 563–569.

McLAUGHLIN, S.K., MCKINNON, P.J., SPICKOFSKY, N., DANHO, W. and MARGOLSKEE, R.F. (1994) Molecular cloning of G proteins and phosphodiesterases from rat taste cells. *Physiol. Behav.* 56(6): 1157–1164.

MING, D., NINOMIYA, Y. and MARGOLSKEE, R.F. (1999) Blocking taste receptor activation of gustducin inhibits gustatory responses to bitter compounds. *Proc. Natl. Acad. Sci. USA* 96: 9903–9908.

MONTMAYEUR, J.P., LIBERLES, S.D., MATSUNAMI, H. and BUCK, L.B. (2001) A candidate taste receptor gene near a sweet taste locus. *Nat. Neurosci.* 4: 492–498.

NAIM, M, SEIFERT, R., NURNBERG, B., GRUNBAUM, L. and SCHULTZ, G. (1994) Some taste sunstances are direct activators of G proteins, *Biochem. J.* 297: 451–454.

NELSON, G., HOON, M.A., CHANDRASHEKAR, J., ZHANG, Y., RYBA, N.J.P. and ZUKER, C.S. (2001) Mammalian sweet taste receptors. *Cell* 106: 381–390.

OBARZANEK, E., PROSCHAN, M.A., VOLLMER, W.M., MOORE, T.J., SACKS, F.M., APPEL, L.J., SVETKEY, L.P., MOST-WINDHAUSER, M.M. and CUTLER, J.A. (2003) Individual blood pressure responses to changes in salt intake: results from the DASH-sodium trial. *Hypertension* 42: 459–467.

ROSENWEIG, S., YAN, W., DASSO, M. and SPIELMAN, A.I. (1999) Possible novel mechanism for bitter taste mediated through cGMP. *J. Neurophysiol.* 81(4): 1661–1665.

SAINZ, E., KORLEY, J.N., BATTEY, J.F. and SULLIVAN, S.L. (2001) Identification of a novel member of the T1R family of putative taste receptors. *J. Neurochem.* 77: 896–903.

SCLAFANI, A. (2004) The sixth taste? *Appetite* 43(1): 1–3.

SMITH, D.V. and MARGOLSKEE, R.F. (2001) Making sense of taste. *Scientific American* 284(3): 32–39.

TANOUE, A, KOBA, M, MIYAWAKI, S., KOSHIMIZU, T., HOSODA, C., OSHIKAWA, S. and TSUJIMOTO, G. (2002) Role of the alpha1D-adrenergic receptor in the development of salt-induced hypertension. *Hypertension* 40: 101–106.

WEINBERGER, M.H. (2004) More on the sodium saga. *Hypertension* 44: 609–611.

YING, W.Z. and SANDERS, P.W. (2002) Increased dietary salt activates rat aortic endothelium. *Hypertension* 39: 239–244.

ZHAO, G.Q., ZHANG, Y., HOON, M.A., CHANDRASHEKAR, J., ERLENBACH, I. and RYBA, N.J. (2003) The receptors for mamalian sweet and umami taste. *Cell* 115: 255–266.

Part III

Reducing salt in particular foods

12

Reducing salt in meat and poultry products

E. Desmond, AllinAll Ingredients, Ireland

12.1 Introduction

The association between sodium and the development of hypertension has prompted public health and regulatory authorities to recommend reducing dietary intake of salt (NaCl). Hypertension is a major risk factor in the development of cardiovascular disease. Observational epidemiological studies and meta-analyses based on trials of varying methodological rigour in different populations and patient groups have provided grounds for controversy on the link between salt and blood pressure (FSAI, 2005a). The results of the DASH sodium study (Dietary Approaches to Stop Hypertension) showed a graded linear relation between salt intake and blood pressure (Appel *et al.*, 1997). Recent reports from the Food Safety Authority of Ireland (FSAI, 2005a) and the Food Standards Agency in the UK (SACN, 2003) have shown that the average daily sodium (salt) intake from foods in Irish adults has been estimated as 3.25 g (8.3 g salt). This does not allow for all discretionary additions such as during cooking or at the table. It is estimated that about 15–20% of total dietary sodium intake is from discretionary sources and that this would give a total daily sodium intake of about 4 g (10 g salt). In the UK, the estimate of total daily intake for adults is 3.8 g (9.5 g salt) while the average daily intake of sodium in 4 to 18 year olds can be broken down to: 4–6 yr, 1.97 g (5.0 g salt); 7–10 yr, 2.28 g (5.8 g salt); 11–14 yr, 2.49 g (6.13 g salt) and 15–18 yr, 2.79 g (7.1 g salt) (SACN, 2003).

These reports state that meat and fish is one of the main sources of sodium for adults in Ireland contributing 29.8% of the mean daily sodium intake. Of this, cured and processed meats contribute 20.5% to the sodium intake. Similarly, in the UK meat and meat products contribute 20.8% to the sodium intake. The

Table 12.1   FSA (UK) proposed targets of sodium and equivalent salt levels in meat products

Product	Original FSA salt model		FSA proposed targets to be achieved by 2010
	mg Na per 100g	g salt per 100g	g per 100g
Bacon	750	1.9	3.5 g salt/1.4 g Na
Ham/cured meats	750	1.9	2.5 g salt/1.0 g Na
Sausages	550	1.4	1.4 g salt/550 mg Na
Meat pies	300	0.8	
Sausage rolls, pork pies, etc.			1.5 g salt/500 mg Na
Cornish pasties, etc.			1.3 g salt/450 mg Na
Cooked uncured meats	450	1.1	1.5 g salt/600 mg Na
Burgers/grillsteaks (beef, pork & chicken)	300	0.8	1.0 g salt/400 mg Na
Coated poultry products	450	1.1	1.0 g salt/400 mg Na
Canned frankfurters, hotdogs and burgers	550 (sausages) 300mg (burgers)	1.4 (sausages) 0.8 (burgers)	1.4 g salt/550 mg Na

Source: Food Standards Agency UK. Available: http://www.food.gov.uk/multimedia/pdfs/salttargetsapril06.pdf

Food Standards Agency (FSA) and the Institute of Medicine (IOM), National Academy of Science in the USA recommend an upper level of no higher than 2.4 g/100 mmol sodium (6 g) per day by 2010. In October 2003, the FSA produced a salt model in order for the UK food industry to reduce the amount of salt in food products. The FSA model targeted a 50% reduction in bacon and ham, 40% reduction in burgers or patties and a 43% reduction in sausages. However, these reductions where seen by the industry as unachievable. In April 2006, the FSA published revised targets (Table 12.1) to be achieved by 2010 (g salt/sodium per 100 g food, as sold) are a maximum of 3.5 g salt/1.4 g sodium for bacon; 2.5 g salt/1 g sodium for ham/other cured meats; 1.4 g salt/550 mg sodium for sausages; 1.5 g salt/600 mg for cooked uncured meats; 1.0 g salt/400 mg sodium for burgers/patties/grill steaks and 1 g salt/400 mg for coated poultry products (FSA, 2006).

As a result of the ongoing campaign by public health authorities, the following chapter examines the functions of salt in meat and poultry products and the effects a reduction in salt has on these products. This will include the effects salt has on flavour, texture and processing characteristics. The current levels of salt in meat and poultry products will also be examined and details of low- or reduced-salt products that are on the market will be discussed. An outline of the ingredients and processes available to manufactures to produce low- and/or reduced-salt products will be addressed and finally future trends in reducing salt in meat and poultry products will be examined.

12.2   Functions of salt in meat and poultry products

Salt is one of the most commonly used ingredients in processed meat products and has being used since ancient times for the preservation of meat products due to its capacity to reduce water activity and therefore retard microbiological growth. In the modern meat industry salt is used as a flavouring or flavour enhancer and is also responsible for the desired textural properties of processed meats. Salt imparts a number of functional properties in meat products: it activates proteins to increase hydration and water-binding capacity; it decreases fluid loss in vacuum-packaged product that has been thermally processed; it increases the binding properties of proteins to improve texture; it increases the viscosity of meat batters, facilitating the incorporation of fat to form stable batters; it is essential for flavour and is a bacteriostatic at relatively high levels (Terrell, 1983). This author concluded by stating that studies concerning the reduction of salt should address the scientific efforts that it may have on such technological functions as water-holding capacity, fat-binding, texture, sensory, safety, stability and shelf life.

12.2.1   Effect on flavour

Salt has a flavour enhancing effect in meat products, with the perceived saltiness mainly due to the Na^+ with the Cl^- anion modifying the perception (Ruusunen and Puolanne, 2005; Miller and Barthoshuk, 1991). Crehan *et al.* (2000) found that salt reduction decreased the perceived saltiness and overall flavour intensity of frankfurters. Macfarlane *et al.* (1984) noted that salt has the undesirable effect of producing a 'salty taste' at a concentration below that desired for functional effects. Fat and salt jointly contribute to many of the sensory properties in processed meats. Matulis *et al.* (1995) has shown that as the salt levels rise, the increase in saltiness is more noticeable in more fatty products that lean ones. Ruusunen *et al.* (2001a) have shown that the fat content of cooked sausages affects the perceived saltiness, depending on the formulation. Their research has shown that an increase in meat protein content reduced perceived saltiness. More recently, Ruusunen *et al.* (2005) found that the effect of meat content on perceived saltiness was stronger that the effect of fat content. This study showed that more salt was needed in ground beef patties of high lean content to achieve the same perceived saltiness as in products of low meat content. In frankfurters, no difference was found when salt levels where reduced from 3.0% to 2.5% but at 2% or lower the reduction caused a distinctly lower salt flavour intensity score. The frankfurter with the lowest salt content was found to have the highest off flavour intensity, however this was found to be statistically insignificant (Barbut *et al.*, 1988).

   Gillette (1985) found that the addition of sodium chloride to food products does more than simply add a 'salty' flavour. It increases the perception of 'fullness' and 'thickness' giving the impression of a less watery or thin product. A metallic/chemical off-note was decreased or masked by the addition of sodium chloride. It significantly improved the overall flavour balance of products, which was 'smoothed', 'rounded out' and 'fuller' and the overall intensity was greater. This factor was partially a result of decreased off-notes, enhanced mouthfeel and increased flavour strength. The level of salt required for an acceptable product depends on the saltiness expected in the product. Our flavour recognition and flavour tolerance for salt is relatively small. For example, in frankfurters, 2.4% salt is regarded as 'very salty' while 1.6% salt is regarded as 'mildly salty' (Wirth, 1991). This author also states that the high tolerance to common salt often noted in the adult is acquired and therefore one can wean oneself away from this high degree of acceptance of salt without having the flavour of one's food spoilt. Some multinationals, such as Heinz, have adopted this concept. For the past few years Heinz has steadily reduced the salt content of its major branded products such as beans and soup in a step-wise fashion and this has resulted in new recipes for some of their products (Robinson, 2005). Many studies have shown that sodium levels can be reduced by 30 to 50% without affecting taste and consumer acceptability. However, the reduction must happen slowly over several months (Bertino *et al.*, 1982).

In a study on cooked hams, hams with 1.7% NaCl were rated as salty as hams with 2.0 and 2.3% salt, but saltier than those with 1.1 and 1.4% (Ruusunen *et al.*, 2001b). The authors concluded that it would be possible to reduce the salt content of cooked ham to 1.7% NaCl, while still maintaining the normal sensory saltiness of cooked ham. For bologna-type sausages, the salt content impacted on the 'pleasantness' of the sausage (Ruusunen *et al.*, 1991). The lowest salt content in relation to pleasantness was 1.20–1.35% added salt when a sausage with 1.5% added salt is used as a reference. Sausages that scored the highest pleasantness contained 1.65% and 1.80% added salt. Assessors were not willing to consume sausages with only 1.2% salt.

12.2.2   Effect on texture

As can be seen, salt has an impact on the flavour of meat products, however one of its main functions in processed meats is the solubilisation of the functional myofibrillar proteins in meat. This activates the proteins to increase hydration and water-binding capacity, ultimately increasing the binding properties of proteins to improve texture. In a comprehensive review on reducing sodium intake from meat products, Ruusunen and Puolanne (2005) put forward two hypotheses to explain the role of NaCl in water binding in meat. This includes Hamm (1972) who proposed that the Cl^- ions tend to penetrate into the myofilaments causing them to swell, and Offer and Knight (1988) who claim that the Na^+ ions form an ion 'cloud' around the filaments.

The effect of NaCl on meat proteins is most likely caused by the fact that the Cl^- ion is more strongly bound to the proteins that the Na^+ ion. This causes an increase in negative charges of proteins. Hamm (1972) concludes that this causes repulsion between the myofibrillar proteins, which results in a swelling of myofibrils due to the repulsions of individual molecules. The adsorption of Cl^- ions with positively charged groups of myosin results in a shift of the isoelectric point to lower pH, causing a weakening of the interaction between oppositely charged groups at a pH greater than the isoelectric point and therefore an increase in swelling and water-holding capacity (Hamm, 1986). Offer and Knight (1988) base their hypothesis on the selective binding of the Cl^- ion to the myofibrillar proteins. According to these authors, this does not cause a marked repulsion between the filaments but between the molecules of myosin filaments breaking down the shaft of the filament. This causes loosening of the myofibrillar lattice. They propose that the swelling induced by NaCl is entropically rather than electrostatically driven. The process begins with the depolymerisation of myosin in the thick filaments in the presence of moderately high NaCl concentrations. In the H-zone, the myosin tails are flexibly attached to the heads, but in unswollen myofibrils, movement would be severely restricted by the neighbouring thin filaments and the bound myosin heads. Systems, however, tend to move to a state of highest entropy, which is obtained by the myofibrillar lattice swelling, giving the myosin tails the greatest possible freedom of movement (Varnam and Sutherland, 1995).

Offer and Trinick (1983) found that myofibrils are able to swell to at least twice their original volume in salt conditions which are widely used in the processed meat industry and that swelling occurred when a substantial part of the A-band had been extracted. These authors also postulate that NaCl will directly affect the strength of binding of myosin heads to actin. Raising the NaCl concentration weakens the binding of actin and myosin. The addition of NaCl may also lead to the swelling of the myofibrillar by liberation of Mg^{2+} and Ca^{2+} ions from muscle proteins and thus loosening of the microstructure of the tissue may take place (Varnam and Sutherland, 1995) and also to the removal of the structural constraints such as the Z and M lines (Offer and Trinick, 1983). Voyle *et al.* (1984) also showed swelling of peripheral myofibrils and disintegration of Z-lines in muscle treated with 3.5% NaCl.

The extraction of myosin from myofibrils when they swell is of great relevance to meat processors. In processed meats the salt-solubilised myofibrillar proteins form a sticky exudate on the surface of the meat pieces which binds the meat pieces together after cooking. This layer forms a matrix of heat-coagulated protein which entraps free water and binds the meat pieces together. In finely chopped or emulsified products such as frankfurters, bologna, etc. the solubilised protein in the continuous phase form a protein film around fat globules, thereby retaining the fat during cooking (Monahan and Troy, 1997). As can be seen, through its effects on adhesion of meat pieces, on fat retention and on water-binding capacity salt is critical to the texture of processed meat products. These properties are used in products such as emulsion-type sausages where the fibrous nature of the meat is decreased and the product becomes characteristically more gelatinous or rubbery. In this type of product the binding is so strong that the sausage should snap when bent. In coarser products, such as burgers or patties, the effects are reduced and the result is a product that retains its shape during cooking, but retains a crumbly fibrous texture on eating (Hutton, 2002).

Whiting (1984) reported that raw emulsions made with reduced salt levels (<2.0%) had reduced water-holding capacity. Gel strength gradually decreased with salt reductions from 3.5% to 1.5% and the fat was released in emulsions containing <1.0% salt. This is a result of less protein extraction and therefore the system is unable to form an effective water-binding matrix. A minimal amount of salt ($\approx$1.5%) is required to extract protein from a stable matrix (Matulis et al., 1995). Sofos (1983) concluded that a salt concentration in the range of 2.5–2.0% appeared necessary for the manufacture of commercial frankfurters. A reduction in salt content by more than 20% (<2.0%) resulted in frankfurters with reduced emulsion stability and increased cook losses. These products were unacceptable as they had a soft texture. Barbut et al. (1988) reported similar results for turkey frankfurters. In terms of rheology and gelation properties, Barbut and Mittal (1989) found that for beef, pork and poultry batters, the shear stress decreased with a decrease in salt levels from 2.50% to 1.25%. The increase in stiffness with increase in salt indicated that more salt-soluble proteins were extracted at the higher salt levels. Similar results were obtained for the modulus of rigidity (G), with the exception of the poultry batter. The peak G value for the poultry batter containing the 1.25% salt was the highest, followed by batters with 2.5% salt and then the salt-free treatment. These authors suggest that isolated myosin from dark poultry meat produced a gel with much greater strength at low ionic mediums (0.1–0.2 M NaCl) than at a higher molarity (0.6 M). The same observation was reported for isolated myosin from red and white bovine muscle (Fretheim et al., 1986). As can be seen, salt is essential for the creation of comminuted or emulsified meat products. Maximum water-holding capacity is with 0.8–1.0 M NaCl (4.6–5.8% salt) (Offer and Trinick, 1983), but 0.4–0.6M (2.3–3.5%) is generally sufficient for good functionality (Trout and Schmidt, 1983).

Both Hamm (1986) and Offer and Knight (1988) highlight how whole muscle products behave in a different manner than comminuted meat. After comminution, the cell wall of the muscle fibres is, to a large extent, destroyed. Therefore differences between extracellular and intracellular water are eliminated and the water binding power of the thick filaments or myosin primarily determines the WHC of the meat. After comminution the swelling capacity of the myofibrillar system is much less limited (Hamm, 1986). Offer and Knight (1988) stated that if myofibrils are exposed to large pools of salt solutions, especially in the presence of phosphates, myosin molecules formed by depolymerisation of the thick filament will tend to be extracted and this will not result in swelling. However, in finely comminuted products salt, water and phosphates are able to directly attach to the filaments in every part of the batter (Puolanne et al., 2001).

In whole muscle product, such as cooked hams, salt and phosphate contribute to the disruption of muscle fibres caused by massaging, solubilisation of the myofibrillar proteins and the production of an exudate rich in solubilised proteins. This protein-rich exudate binds meat chunks together. Theno et al. (1978) showed that salt and phosphate dramatically increased binding and product quality in sectioned and formed ham rolls using photomicrographs. At a level of 1% salt alone the rolls did not exhibit acceptable characteristics. The junctions between the meat pieces showed no evidence of protein alignment and consisted of mainly fat and cellular fragments. The addition of 0.5% phosphate at the 1% salt level improved binding to acceptable levels after 4hrs of continuous massaging. At the 2% level, several hours of massaging must be done before the product exhibited acceptable binding. Large amounts of proteins were solubilised during processing when levels of 2% salt and 0.5% phosphate were added. Increasing the salt concentration to 3% further facilitated protein extraction while the addition of phosphate at this level resulted in products with acceptable binding characteristics and highly structured aligned junctions.

More recently, Ruusunen et al. (2001b) reported that cooked ham with added salt levels below 1.4% had higher cook losses. As a consequence the lower salt hams had lower moisture levels and higher protein levels. The authors point out than in products with low salt content with a high amount of added water it is necessary to add extra protein or other functional ingredients to increase yield. Reducing the salt content of cooked hams by reducing brine strength, while leaving the manufacturing technology unchanged, leads to a clear increase in cooking losses which results in a drier uneconomical product (Muller, 1991). Ruusunen et al. (2002) found salt levels between 1.3 and 1.7% had no effect on cooking losses. However, the higher salt level increased the water-binding capacity in particular in combination with a potassium phosphate mixture. No significant differences in appearance or texture were observed between the hams.

The beneficial effect of sodium chloride on the binding of poultry meat pieces has been demonstrated as far back as the 1970s. Vadehra et al. (1973) investigated various salt concentrations on the WHC of various chicken parts. As the salt concentration increased in poultry loaves cook losses were reduced and 2% salt was the optimum level. The increase in salt concentration also increased the binding scores and the shear force values of the poultry loaves (Schnell et al., 1970). These authors explained that salt may have had an osmotic effect causing cellular disruption thereby liberating intercellular material for binding. Colmenero-Jiménez et al. (1998) found that salt concentration had a less pronounced effect than cooking temperature in chicken batters. However, at a cooking temperature of 70°C, a reduction in salt concentration by 1% resulted in a significantly higher cooking loss of 1.5%. A reduction in salt concentration also resulted in batters that were less chewy and hard as determined by Texture Profile Analysis.

12.2.3   Effect on microbial stability

The preservation and shelf life of processed meats is of vital importance when reducing the salt levels. Reducing NaCl levels below those typically used without any other preservative measure has shown that product shelf life is shortened (Sofos, 1983, 1985; Madril and Sofos, 1985). Whiting et al. (1984) found that reducing the level of salt by 60% to 1.5% resulted in a more rapid growth in natural flora of frankfurters. Reducing the salt level by 50% to 1.25% in ground pork resulted in slight increases in the growth of *Lactobacillus* spp. (Terrell, 1983). The use of various chloride salts (NaCl, KCl, MgCl$_2$) did not affect the microbial growth of refrigerated pork sausage (Terrell *et al.*, 1982). According to Maurer (1983) higher levels of NaCl provided better control of *Clostridium botulinum* i.e. less nitrite was needed to supplement the NaCl. There are significant interactions between NaCl and nitrite in cured meats (Collins, 1997) while hams (without sodium nitrite) are unsuitable for any range of salt reduction (Wirth, 1991). In terms of nonmicrobial aspects of shelf life, studies have shown that when NaCl was reduced by 50% to 1.25% or replaced by KCl or MgCl$_2$ in ground pork, no significant differences were found for TBA values. In fact, data suggest that NaCl may accelerate the development of rancidity in comparison to KCl or MgCl$_2$ (Rhee *et al.*, 1983). In summary, research has shown that it is important to examine the microbial shelf life and safety of processed meat products before NaCl levels are reduced or replaced by other ingredients.

12.3   Salt content in meat and poultry products

For the most part, the Food Standards Agency in the UK has been leading the way in trying to get both consumers and the food industry to reduce their sodium intake and usage in foods. Table 12.2 shows the nutritional composition of typical meat products in both Ireland/UK and the USA. Sodium occurs naturally in beef, pork and poultry meats ranging from 50 to 70 mg per 100 g (Table 12.2). Most processed meat products contain variable amounts of salt and are associated with high salt contents. In addition to the usual source of sodium in processed meats that everyone readily identifies (salt), other ingredients also affect the sodium content; sodium tripolyphosphate (31.24% Na), sodium nitrate (27.05% Na), sodium ascorbate or erythorbate (11.61% Na), sodium nitrite (33.21% Na), MSG (13.6% Na) and HVP (18.0% Na) (Maurer, 1983). However, according to Breidenstein (1982) in a typical meat product containing 2% salt, the salt contributes approximately 79% of the sodium in the final meat product.

The sodium content of processed meats in Ireland and the UK is quite variable with cured meats having higher sodium/salt contents. Burgers or patties contain 290–590 mg sodium per 100 g or 0.7–1.5 g salt per 100 g. Sausages contain 600–1180 mg sodium or 1.5–3.0 g salt per 100 g. Cured meats, such as hams and frankfurters, contain 720–1200 mg sodium per 100 g or 1.8–3.0 g salt per 100 g while bacon rashers contain 1000–1540 mg sodium per 100 g or 2.5–3.9 g salt per 100 g. Breaded chicken products contain 200–600 mg sodium per 100 g or 0.5–1.5 g salt per 100 g. In a study carried out by Desmond (1992) the composition of a number of processed pork products in Ireland was determined. The sample number ranged from 150–250 depending on the product. Salt levels ranged form 2.8% for cooked ham; 3.2% for back rashers; 3.5% for back bacon; 4.1% for streaky bacon; 3.2% for a gammon joint

Table 12.2 Nutritional composition (per 100 g) of typical meats and meat products

Product	Moisture (g)	Protein (g)	Fat (g)	Sodium (mg)	Salt (g)
Irish/UK products					
Beef	71.9	22.5	4.3	**63**	**0.16**
Pork	74.0	21.8	4.0	**70**	**0.18**
Chicken	74.2	24.0	1.1	**60**	**0.15**
Turkey	74.9	24.4	0.8	**50**	
Beef burgers	56.1	15–17	21–25	**290–400**	**0.7–1.0**
Economy beef burgers	57.1	13.7	21.2	**590**	**1.5**
Grillsteaks	50.1	22.1	2.39	**710**	**1.8**
Sausages	49.4	11–12	25–36	**600–1080**	**1.5–2.7**
Rashers	–				
Frankfurters	54.2	13–15	15–25	**720–920**	**1.8–2.3**
Cooked ham	73.2	18–22	3–4	**900–1200**	**2.3–3.0**
Bacon/rashers	63.9	16–17	14–16	**1000–1540**	**2.5–3.9**
Salami	33.7	20.9	39.2	**1800**	**4.6**
Reduced fat sausages	50.1	12–16	14–17	**1000–1180**	**2.5–3.0**
Chicken goujons		18.2	8.6	**600**	**1.52**
Breaded chicken	53.2	18.0	9–12	**200–420**	**0.5–1.1**
Chicken nuggets	–	16.0	5.5	**600**	**1.5**
Crispy chicken	–	17.4	14.5	**300**	**0.8**
US products					
Beef patties	58.7	17.1	23.2	**68**	**0.17**
Pork sausage	56.2	15.1	26.5	**636**	**1.6**
Frankfurters	56.0	11.5	27.6	**1120**	**2.8**
Oscar Myer Weiners	53.2	11.4	30.3	**1025**	**2.6**
Cured ham	64.5	22.6	9.0	**1500**	**3.8**
Corned beef	66.6	14.7	14.9	**1217**	**3.1**
Hormel Canadian bacon	73.0	16.9	4.9	**1016**	**2.6**
Beef bologna	54.3	10.3	28.2	**1080**	**2.7**
Salami	34.6	21.7	37.0	**1890**	**4.8**

Sources: FSA (2002). The Royal Society of Chemistry, Food Standards Agency and Institute of Food Research.
USDA Food Nutrient Database Ver. SR18. Available from http://www.nal.usda.gov/fnic/foodcomp/Data/SR18/sr18.html
Data also sourced from products retailing in Irish supermarkets

The proposed targets set out by the FSA (UK) are shown in Table 12.1. Cured meat products and sausages have the highest levels of salt/sodium above the FSA proposed targets while burgers and some coated poultry products are quite close to the proposed targets. Therefore meat processors will have to reduce their salt levels and offer alternative products to their current ranges in order to meet the FSA proposals. Currently there are only a limited number of products available to the consumer at retail level (Table 12.3). The US has a larger range, but for a market of this size, the number of products is quite small. In some cases the products that are currently available do meet the FSA target;

Table 12.3 Nutritional composition (per 100 g) of typical 'reduced-salt or sodium' meat products

Product	Moisture (g)	Protein (g)	Fat (g)	Sodium (mg)	Salt (g)
Irish/UK products					
Reduced salt sausages	–	10.5	24.5	**750**	**1.9**
Reduced fat/salt sausages	–	11.5	12.0	**550**	**1.4**
Low fat/salt rashers	–	18.0	8.0	**900**	**2.3**
US products					
Low sodium frankfurter	56.7	12.0	28.5	**311**	**0.8**
Low sodium ham	67.7	20.9	5.5	**969**	**2.5**
Salami 50% less sodium	40.5	21.8	26.4	**936**	**2.4**
Reduced sodium beef bologna	54.8	11.7	28.4	**682**	**1.7**
Reduced sodium luncheon meat	55.5	12.5	25.1	**946**	**2.4**
Reduced sodium bacon (cooked)	12.3	37.0	41.8	**1030**	**2.6**

Sources: USDA Food Nutrient Database Ver. SR18. Available from http://www.nal.usda.gov/fnic/foodcomp/Data/SR18/sr18.html
Data also sourced from products retailing in Irish supermarkets

however, for some sausage and cured meat products further reductions are needed in order to meet the FSA proposals. These reductions will have processing implications in terms of texture and flavour of products and shelf life of these products. The Food Safety Authority of Ireland (FSAI, 2005b) has published a list of salt reduction undertakings by the food industry in Ireland. The Irish Food and Drink Industry have given an undertaking for a number of products to have the following sodium levels: bacon at or below 1.3 g sodium/ 100 g; sausages at or below 0.88 g/100 g; cooked ham at or below 0.99 g sodium/100 g and burgers at or below 0.5 g sodium/100 g. In some cases companies point out that some of their products within their product portfolio have 20–50% less salt then their standard products and that further reductions in products are underway. All the major retail multiples have committed themselves to reducing salt levels in their products and are working with the FSA (UK) and through the British Retail Consortium (BRC). One retailer in Ireland, Superquinn, has had a salt reduction policy since March 2004 and has targeted a reduction in salt in bacon and ham from 5% to 2% in 2007; sausages 5% to 3% in 2007. In the UK, the British Meat Processors Association (BMPA) and the Food and Drink Federation Meat Group have made a commitment to have 1.1 g salt/100g in burgers; 2 g salt/100 g in uncured cooked meats; 1.1 g salt/100 g in coated poultry; however, no target has been set for bacon or cured meats. As a consequence of these reductions processors will have to develop strategies to reduce the sodium levels in their products. Some of these are outlined in the next section.

12.4   Strategies for salt reduction in meat and poultry products

Apart from lowering the level of salt added to products, there are currently three major approaches to reduce the salt content in processed foods. First, and probably the most widely used, is the use of salt substitutes, in particular, potassium chloride (KCl). Masking agents are commonly used in these products. Second, the use of flavour enhancers which do not have a salty taste, but enhance the saltiness of products when used in combination with salt. This allows less salt to be added to the products. Third, optimising the physical form of salt so that it becomes more taste bioavailable and therefore less salt is needed (Hanley, 2005).

12.4.1   Use of salt substitutes

Potassium chloride is probably the most common salt substitute used in low- or reduced-salt/sodium foods. However, at blends over 50:50 sodium chloride/ potassium chloride in solution, a significant increase in bitterness and loss of saltiness is observed. In Ireland, the FSAI scientific committee (FSAI, 2005a) was of the opinion that the use of low-sodium salts incorporating potassium salts could not be endorsed at this time. Concerns were raised about the possible vulnerability of certain population sub-groups (including those with Type I diabetes, chronic renal insufficiency, end stage renal disease, severe heart failure and adrenal insufficiency) to high potassium load from these salt substitutes. It was also noted that the use of salt substitutes does not address the need to reduce salt taste thresholds in the population.

Notwithstanding the FSAI's concerns most research has focused on reducing the sodium intake with the partial replacement of salt with KCl. According to Ruusunen et al. (2005) the use of mineral salt mixtures is a good way to reduce the sodium content in meat products. The same perceived saltiness can be achieved with salt mixtures at lower sodium content. Some of these mixtures such as Pansalt® have been commercialised. Pansalt® is a patented salt replacer that can be used as a food ingredient or as a tabletop salt, which delivers the real taste of salt with only about half the levels of sodium compared to common salt. It has been on the market since 1987 in Finland where it was invented by a medical doctor. Almost half of the sodium is removed and replaced with potassium chloride, magnesium sulphate and the essential amino acid L-lysine hydrochloride. According to the manufacturer, the patented usage of the amino acid enhances the saltiness of the salt replacer and masks the taste of potassium and magnesium, while increasing the excretion of sodium from the human body. However, Ruusunen et al. (2005) state that the Mg^{2+} ions in this salt form an insoluble salt with added phosphates; therefore Pansalt is more suitable for meat products without added phosphates.

Other commercially available mixtures of NaCl and KCl include Lo® salt, Saxa So-low salt and Morton Lite salt® amongst others. Studies by Morton Salt found that ham, bacon and turkey ham products manufactured with Morton Lite

Salt® with a 60:40 mixture of NaCl:KCl had similar flavour scores to the control salt products. Further studies have found that the Lite Salt® maintained protein hydration in meat products (Morton Salt, 1994). Bonorden *et al.* (2003) filed a patent comprising a combination of NaCl, KCl and potassium or calcium sulphate. They also have other combinations with magnesium chloride or calcium sulphate. The patent claims that 67% of panellists preferred the flavour-enhancing blends compared to a NaCl/KCl mixture in meat broths. In clam chowder soups the salt flavour-enhancing mixture was described as saltier with a better flavour balance than the control. Vasquez (2004) developed low-sodium and sodium-free salt substitutes that contain calcium chloride, potassium citrate, citric acid, rice flour, ginger oil and garlic powder.

It has previously been stated that it is the chloride ion that has been primarily responsible for the functional efficiency of NaCl in meat products (Hamm, 1972); therefore several studies (Hand *et al.,* 1982; Terrell, 1983; Maurer, 1983) have concentrated on investigating alternative chloride salts such as calcium, potassium, lithium and magnesium. Lithium chloride has been investigated as a possible salt replacer, however, LiCl is not food approved and is considered to be toxic and therefore has not been commercialised. Research indicates that 25–40% replacement appears to be the range at which the flavour impact is not as noticeable. As the flavour intensity of some flavour increases such as salty, acidic or spice, a higher proportion of KCl may be acceptable (Price, 1997). In cooked hams it was found that a 50% replacement with KCl gave superior bind and acceptable sensory scores (Frye *et al.*, 1986). In sectioned and formed hams, Collins (1997) stated that using a 70% NaCl/30% KCl or 70% NaCl/30% $MgCl_2$ mixtures were not different in terms of flavour, tenderness and overall acceptability compared to hams made with 100% salt. However, formulations using 30% $CaCl_2$ had lower pH, moisture content, yield and residual nitrite. Monahan and Troy (1997) reported that meat batters containing divalent chloride ions had lower brittleness and formed less stable batters than those containing monovalent salts. Terrell *et al.* (1982) investigated replacing NaCl with a number of other chloride salts and found that replacement of NaCl with any chloride salt, except $CaCl_2$, significantly decreased moisture loss in raw and cooked beef clod muscles.

In fermented sausages, Gou *et al.* (1996) found no significant alteration in texture of the products; a bitter taste was detected at 30% level of substitution although panellists did not consider its intensity important until the 40% level was reached. These authors also found that in dry-cured loins, KCl and potassium lactate could substitute 40% of NaCl without any significant detrimental effect to flavour. These results agreed with Keeton (1984) who concluded that a one-third substitution of NaCl by KCl is possible without significantly altering the products characteristics in country-style hams. More recently the concentration level of NaCl (1.5 and 3.0%) was found to have a considerable effect on the amount of volatile compounds in fermented sausages, but the effect was highly related to the ripening stage (Olesen *at al.,* 2004). According to Ruusunen and Puolanne (2005), simple salt reduction in fermented products cannot be made due to the low A_w that has to be reached in order to control the microbial flora. As a result the techno-logical and microbial safety as well as the sensory properties of the substitutes compared to NaCl will determine the extent NaCl can be reduced. This will limit the lowering of NaCl to >2.0%. A process has been developed to produce a low sodium cured meat product by injecting the meat with a brine containing KCl in combination with calcium citrate, calcium lactate, lactose, dextrose, potassium phosphate, ascorbic acid and sodium nitrite (Riera et al., 1996). The patent claims that the process produces a low-sodium cooked ham with zero weight-loss and a flavour identical to cooked ham with a normal sodium content.

Research has also demonstrated that phosphates can be very useful in lowering the NaCl content in meat products (Sofos, 1985; Puolanne and Terrell, 1983; Trout and Schmidt, 1984; Barbut at al., 1988). More recently Ruusunen et al. (2002) and Ruusunsen et al. (2005) have investigated the use of phosphates in reduced-sodium cooked meat products. Phosphates are generally used in meat products to enhance water-holding capacity and improve cook yield. They increase water-holding capacity in fresh and cured meat products by increasing the ionic strength, which frees negatively charged sites on meat proteins so the proteins can bind more water. The functionality of phosphates is greatly affected by the addition of salt and both of these ingredients act synergistically. It is thought that phosphates exert more effect on pH and protein solubility and salt exerts more effect on ionic strength and water-holding capacity.

In terms of sodium reduction, some phosphates are sodium salts; however, the usage rate is substantially lower than NaCl. Sodium polyphosphate contains 31.24% Na compared to 39.34% in NaCl and is typically used at 0.5% compared to 2–4% usage rate for salt. The potassium salts of phosphate are also commercially available and are equally effective in terms of water binding, gelation or ionic strength as the sodium salts (Price, 1997). Barbut et al. (1988) found that the incorporation of sodium tripolyphosphate (STPP), sodium hexametaphosphate (SHMP) or sodium acid pyrophosphate (SAPP) at levels of 0.4% were effective with 20 and 40% reduced salt formulations and improved emulsion stability and sensory properties. SAPP appeared to provide greater benefits than either STPP or SHMP and enhanced the salt flavour intensity of the turkey frankfurters. According to Trout and Schmidt (1984), tetrasodium pyro-phosphate (TSPP) is generally considered superior to sodium tripolyphosphate, tetrapolyphosphate (TTPP) and hexametaphosphate in enhancing meat func-tional characteristics. However, Madril and Sofos (1985) rated TSPP and STPP similarly in comparison to SHMP. These authors found that in reduced salt frankfurters (1.25% vs 2.5%) the addition of phosphate improved yields by 10% in comparison to low-salt frankfurters with no phosphate. The addition of TSPP, STTP or TTPP had higher yields than the control (2.5% salt) frankfurter and had similar if not better flavour and texture. More recently, Baublits et al. (2005) found that STPP and TSPP performed better than SHMP in terms of yield and binding in beef enhanced roasts containing 2% salt.

Ruusunen et al. (2002) found it is possible to produce reduced-salt (1.0–1.4%) bolognas and cooked hams provided that phosphates are added.

Further reduction of sodium content in reduced-salt meat products is possible by replacing sodium phosphate with potassium phosphate. The extent of sodium reduction depends on the phosphates used and their sodium content, being equivalent to a sodium content of 0.2% NaCl or more. These authors also recommend alkaline phosphates in very low-salt products. In low-sodium ground beef patties, it was found that the use of tetrapotassium phosphate had no marked effects on perceived saltiness, but it effectively reduced cooking losses. At a sodium content of 400 mg/100 g (1% salt), the cooking losses in patties made without phosphate increased by about 8% points compared to 5% points with patties made with phosphate (Ruusunen et al., 2005). These authors conclude that it is possible to prepare meat patties with lower sodium contents and higher yield when phosphate is used. They also state that the flavour intensity of patties, with the lowest sodium contents containing phosphate, was slightly reduced when the fat content of the patties increased.

In addition to phosphate, other ingredients have been investigated in low-salt meat products. These are mainly binding agents that, in the absence of or at reduced salt levels, replace salt soluble proteins. These ingredients enhance the binding of meat pieces in restructured or reformed meat products and/or increase the water-binding capacity of the finished product. There is a wide variety of ingredients that can be used for this purpose and include non-meat proteins such as soya, milk and deheated mustard; blood proteins; hydrocolloids such as carrageenan, gums and alginates; starches such as potato, corn, wheat, waxy maize, tapioca or rice. The gel matrix formed with these alternative ingredients provides binding through a combination of protein coagulation and gel formation rather that direct interaction with muscle proteins (Collins, 1997). Studies have also been carried out investigating the effects of combinations of microbial transglutaminase, with caseinate, KCl and wheat fibre as salt replacers in low-sodium frankfurters with added walnuts (Colmenero-Jiménez and Carballo, 2005).

12.4.2   Use of flavour enhancers and masking agents

There are a number of flavour-enhancing and masking agents commercially available. These include yeast extracts, lactates, monosodium glutamate and nucleotides amongst others. Ruusunen et al. (2001a) found that flavour intensity of 'bologna-type' sausages was stronger when MSG or Ribotide (5′-ribonucleotides IMP and GMP) was added to the formulations. However, after 17 days' storage no significant difference was found, even though MSG and Ribotide had a higher panel score than the control. The perceived saltiness was greater when sausages contained MSG than Ribotide or without flavour enhancers. Consumers also rated the MSG sausages more palatable. Pasin et al. (1989) found that it was possible to reduce the NaCl by 75% in pork sausage patties using a modified KCl salt, co-crystallised with Ribotide. However, the addition of MSG in these products also decreased the acceptable level of salt replacement and additional spices used did not have any effect. The nucleotides mentioned above were subject to a patent by Cornelius *et al.* (1978).

Dipeptides also offer the potential for replacing either NaCl or MSG. In the process of studying the structure-taste relationship of a bitter peptide (Arg-Gly-Pro-Pro-Phe-lle-Val) obtained from bacterial proteinase treatment of casein, Tada *et al.* (1984) found that certain analogues of the peptide have an umani or salty taste. Thus, substituting either ornithine or lysine for arginine at the N-terminal position permitted preparation of Orn-Gly and Lys-Gly, both of which exhibit a salty and umani taste (Anon, 1987). However, these claims have been disputed by Tuong and Philippossian (1987) who claim that the saltiness resulted from NaCl present as an artefact of the method of preparation. More recently, Lioe *et al.* (2005) reported that L-phenylalanine and L-tyrosine, both aromatic amino acids, at their sub-threshold concentrations showed significant umani-enhancing effect on the umani taste of MSG/NaCl mixtures. The authors conclude that this is a novel phenomenon for the so-called bitter amino acids. Linguagen, a US company, received patent protection and regulatory approval for a bitter blocker, adenosine 5′-monophosphate (AMP). AMP works by blocking the activation of the gustducin in taste receptor cells and thereby preventing taste nerve simulation (McGregor, 2004). This bitter blocker, marketed under the name Betra^{TM}, can be used to improve the taste of NaCl/KCl mixtures. According to McGregor (2004) this bitter blocker is only the first of what will become a stream of products that are produced due to the convergence of food technology and biotechnology.

Encapsulated ammonium salts such as chlorides, phosphates, citrates and lactate amongst others when added to foods can reduce the amount of NaCl required and will enhance or potentiate the salty taste of the food (Lee and Tandy, 1994). Other combinations such as lysine and succinic acid have been used as salt substitutes (Turk *et al.*, 1993). This compound has a salty flavour and also some antimicrobial and antioxidative properties and may be used to replace up to 75% of the NaCl from a flavour perspective. However, other water binders such as phosphates, starches or gum may have to be used to maintain the water-binding function lost due to the salt reduction. The use of sodium or potassium lactate with a corresponding reduction in NaCl tends to maintain certain saltiness while reducing the sodium content in products to some degree (Price, 1997). Gou *et al.* (1996) investigated the potential of glycine and potassium lactate as potential salt substitutes. In fermented sausages it was possible to substitute 40% of NaCl with either potassium lactate or glycine, above this level a slight potassium lactate or an unacceptable sweet taste was detected. In dry cured loins results showed that KCl and potassium lactate could substitute 40% of NaCl without any significant detrimental effect to flavour. While 30% was the maximum substitution level for glycine.

More recently, a US company Prime Favourites has launched NeutralFres$^{®}$, a natural ingredient formula, which removes the metallic, bitter taste of KCl while maintaining a similar taste to sodium salt. The company claims that NeutralFres® naturally neutralises the characteristic taste or off-flavour of KCl. The product significantly eliminates the alkaline, bitter off-flavour (PrimeFavourites, 2005). Quest are also looking at technologies to cut salt levels and claim that processors using its new ImpaQ taste technology could cut salt levels by 50%. This has led to Quest submitting patents for completely new flavour molecules (FoodNavigator.com, 2005). Givaudan's new, customised Natural Flavour System modifies off-notes exhibited by KCl and enhances the saltiness overall. Other products such as Magnifique Salt Away or Mimic, produced by Wixon Fontrome, claim to mask the bitterness and metallic character of KCl. This product is a water-soluble, natural flavouring and can be used in processed meats at a level of 0.1–0.3% to reduce the metallic character of potassium based salt substitutes. Wild Flavours Inc. has introduced SaltTrim™. Bases on proprietary technology, the company claims that this product simultaneously blocks the negative tastes of KCl while keeping the true taste and mouthfeel of salt. SaltTrim™ creates a complete eating experience by adding much of the taste and texture unique to salt and allows for a 50% substitution of salt with KCl without impacting taste.

Products derived from mycoprotein, such as Mycoscent; claim to have the ability to impart a salty taste without the addition of salt. According to the manufacturer, it has a synergistic effect with dairy and savoury notes. It is possible to have a 50% sodium reduction in biscuits and snack foods and 25% sodium reduction in savoury dishes. Mycoscent 400 has a darker, richer brothy taste and can be used to deliver a succulent, cooked taste in meat applications (Mycoscent website, 2005).

Yeast autolysates are also commonly used in low salt preparations. In particular they mask the metallic flavour of KCl. Synergy Flavours manufacture specific yeast extracts for low-salt applications, e.g. Saltmate™ and Saporese YE. They claim to have shown that a 20%+ salt reduction is possible using their products. In some cases, companies blend the yeast extracts with KCl to offer a complete solution. Provesta® Flavour Ingredients have a number of low-sodium yeast autolysates. The patented co-processed combination with potassium or aluminium chloride and the yeast autolysates is significantly less bitter than KCl alone. Typical application rates are 0.2 to 0.6% of the final product.

Another yeast extract that is available is Aromild produced by Kohjin Co. Ltd in Japan. According to the company, Aromild contains an abundance of natural 5′-inosinic acid (5′-IMP) and 5′-guanylic acid (5′-GMP) and is capable of enhancing flavour of foods with the effect of reducing the salt content. It can be used as an alternative to MSG and HVP and as it is in such a concentrated form, only around a tenth of the amount is needed. Recommended usage rates in meat are 0.01–0.1%. Research from Kohjin has shown that Aromild is ten times stronger than MSG and has a longer lasting taste. At AllinAll Ingredients we have produced reduced salt cures using a range of these ingredients and have been able to produce reduced salt cooked hams similar to the flavour and texture of normal salt hams, without any increase in cooking losses.

12.4.3 Optimising the physical form of salt

The perception of salt in the solid form is affected by crystal size and shape. Research has been carried out using various forms (flaked versus granular) as a method of reducing salt content in meat products. Flake-type salt has been shown to be more functional, in terms of binding, increasing pH, increasing protein solubilisation and improving cooking yield, in model emulsion systems (Campbell, 1979). However, Sofos (1983) found no difference between flake and regular granulated salt. The author does state that the difference between the two studies may be due to the different meat systems used in each study. Flake salt has better and more rapid solubility than granular salt, and this may be critical where no water is added to formulations and therefore flake salt may be beneficial in products where no water is added such as dry-cured products.

More recently, Leatherhead Food International have been investigating optimising the physical form of salt and looking at changing the physical form of salt so that it becomes more taste bioavailable and therefore less can be added to the products. This involves increasing the efficiency of the salt, changing the structure and modifying the perception of the salt. Results of this collaborative project were published in 2006 (Angus *et al.*, 2006). A number of companies, such as Morton Salt and Cargill Salt, manufacture various forms of salt and claim that they can be used at reduced levels and therefore there is potential to reduce the sodium content in meat products. Morton Salt describes its Star Flake® dendritic salt as a 'hybrid' combing the most useful features of vacuum granulated salt and grainer flake salt. The dendritic crystals are branched or star-like in shape and exhibit the low-density, high specific surface area and rapid dissolution properties of fine grainer salt. A unique feature is the cavitation or macroporosity of the crystal (Morton Salt, 1997). With an extremely high exposed surface area per unit weight, dendritic salt exhibits a rapid dissolution rate and theoretically should dissolve twice as fast as granulated salt, but it depends on the dissolving parameters. According to Cargill Salt its Alberger® range has a larger surface area and imparts a more rapid flavour release and therefore for topical applications less salt may be required due to increased salt perception. One of Cargill's products has an unusual pyramidal shaped flake, with numerous facets which give food producers a salt with lower bulk density. According to Cargill, pound for pound the Alberger brand salt gives you more salt flavour (Cargill Salt, 2001).

Another product from the Alberger® process, which has a cube agglomerate structure, Alberger® Fine Flake Salt, is recommended for use in meat products. Lutz (2005) has shown that flake salt (such as the Alberger Fine Flake Improved Salt) can produce red meat batters with superior fat- and water-binding properties using Dendritic or regular vacuum evaporated salts. This study found that a blending time of six to eight minutes coupled with a post-blending storage time of one day resulted in a product with improved yield, increased protein functionality and on less detrimental effects in sensory quality of bologna-type meat products. The pH of the batters made with the Alberger® salt was significantly higher than the pHs of batters made with other salts which, in theory, should lead to a greater degree of solubilisation of the myofibrillar proteins. The results also showed that the moisture of the batters with the Alberger® salt was more tightly bound and exhibited superior binding properties than the other salts examined. Cook losses were also reduced in comparison to the other salts. This increase in functionality using the Alberger® Salt may lead to the possibility of using less salt and producing products of similar quality but with less sodium content.

12.5   Conclusions and future trends

The ultimate goal of ingredient suppliers and meat processors is to produce great-tasting reduced-sodium meat products that consumers can enjoy as part of an ongoing healthier diet and lifestyle. The strategies for and consequences of salt reduction are discussed in this chapter. This was by no means an all-encompassing discussion, but an attempt to review some of the technological aspects of reduced-salt meat products.

Because there is no panacea in terms of a single ingredient that can be used to replace salt in meat products, a range of functional ingredient combinations must be developed and/or optimised. Products need to be recreated that will continue to appeal to consumers. Leatherhead Food International believes that a wide-spread reduction of salt in processed foods by >10% requires new technological approaches (Hanley, 2005). With the food industry working together with the regulatory authorities and consumer groups, the ultimate aim of reduced salt in the diet can be achieved, but only, in my opinion, if a cooperative approach is established and a full understanding of the technological problems associated with salt reduction is realised. The food industry needs to produce reduced salt products that are similar, in terms of texture and flavour, to regular products that the consumer is familiar with. Government agencies need to continue educating consumers in terms of salt and health, as 15–20% of salt intake is coming from discretionary sources (SACN, 2003).

The FSAI believes that food makers should not be looking at simply replacing salt with KCl and this method should only be used as a short-term measure. Progress is being made and meat processors have committed to reducing salt levels in their products. Ingredient firms, such as AllinAll Ingredients, are researching new methods, such as combining various ingredients, to help the industry fulfil its commitments. As an ingredient company supplying to meat processors AllinAll has been actively working with our customers in order that they can supply reduced-salt alternatives that are cost effective and that don't affect the quality of the products. Innovation demands increased technological awareness, through continued research in the field of food ingredients and additive technology companies like AllinAll offer their customers the opportunity to lower the technical barriers to manufacture reduced- or low-salt products. This includes producing various combinations of ingredients, depending on customer and product requirements, to be used in low- or reduced-salt applications.

Any alteration in salt content of meat products requires ingredient reformulation or manipulation. Some companies have produced products that are successful in replacing or substituting sodium in processed products, others have been less so. Monahan and Troy (1997) conclude that research should also be directed towards the meat system itself and methods of enhancing the functionality of the meat system to low-salt formulations merit investigation, such as use of pre-rigour meat and high pressure technologies. Ruusunen *et al.* (2005) conclude that any increase in meat protein content would reduce the perceived saltiness.

Research is continuing to look at various flavours, in particular more savoury/ umani taste to enhance the flavour of reduced salt products. Both through the use of these flavours and by adding more aromatics such as herbs and spices a reduction in salt can be achieved; however, salt cannot be totally eliminated due to its functionality regarding binding and texture of products, unless other ingredients such as phosphates, hydrocolloids, etc. are added. It should also be recognised that a number of companies and retailers have produced successful reduced-salt meat products and that processed meats provide variety, taste and nutritional value to consumers.

12.6 Sources of further information and advice

There is an extensive range of literature in the area of salt reduction, in particular during the 1980s and there are particularly good reviews published in *Food Technology*. More recently reviews have been published by Ruusunen and Puolanne (2005). Both of these authors from the University of Helsinki have carried out a great deal of research in this area and have a number of published papers, some of which have been include in this text. The Ashtown Food Research Centre and University College Cork, in Ireland, have also worked on salt reduction, in particular looking at minimal processing techniques such as high pressure technology. As mentioned in the text Leatherhead Food International has also carried out a comprehensive study on salt reduction and results of this collaborative project were published in 2006. *Production and Processing of Healthy Meat, Poultry and Fish Products, Advances in Meat Research* Vol. 11 examines the nutritional principles behind the drive for reductions in fat, salt and cholesterol in our diet, and illustrates formulations and procedures utilised to produce such products.

12.7 References

ANGUS, F., PHELPS, T., CLEGG, S., NARAIN, C., DEN RIDDER, C. and KILCAST, D. (2006). *Salt in Processed Foods: Collaborative Research Project.* Leatherhead Food International.

ANON. (1987). Salt-free salt. *Nutrition Reviews* 43 (11) 337–338.

APPEL, L.J., MOORE, T.J., OBARZANEK, E., VOLLMER, W.M., SVETKEY, L.P., SACKS, F.M., BRAY, G.A., VOGT, T.M., CUTLER, J.A., WINDHAUSER, M.M., LIN, P.H. and KARANJA, N. (1997). A clinical trial of the effects of dietary patterns on blood pressure. DASH Collaborative Research Group 1997. *New England Journal of Medicine* 336 1117–1124.

BARBUT, S., MAURER, A.J. and LINDSAY, R.C. (1988). Effects of reduced sodium chloride and added phosphates on physical and sensory properties of turkey frankfurters. *Journal of Food Science* 53 (1) 62–66.

BARBUT, S. and MITTAL, G.S. (1989). Effects of salt reduction on the rheological and gelation properties of beef, pork and poultry meat batters. *Meat Science* 26 177–191.

BAUBLITS, R.T., POHLMAN, F.W., BROWN, A.H. and JOHNSON, Z.B. (2005). Effects of sodium chloride, phosphate type and concentration, and pump rate on beef *biceps femoris* quality and sensory characteristics. *Meat Science* 70 205–214.

BONORDEN, W.R., GIORDANO, D.A. and LEE, B.L. (2003). Salt flavour enhancing compositions, food products including such compositions, and methods for preparing such products. US Patent 6541050.

BREIDENSTEIN, B.C. (1982). Understanding and calculating the sodium content of your products. *Meat Processing* 21 (5) 62.

BERTINO, M., BEAUCHAMP, G.K. and ENGELMAN, K. (1982). Long-term reduction in dietary sodium alters the taste of salt. *American Journal of Clinical Nutrition* 36 1134–1144.

CAMPBELL, J.F. (1979). Binding properties of meat blends, effects of salt type, blending time and post-blending storage. PhD Thesis, Michigan State University.

CARGILL SALT (2001). Wow-Alberger Brand Salt. Product Brochure.

COLLINS, J.E. (1997). Reducing salt (sodium) levels in process meat poultry and fish products. In: A.M. Pearson and T.R. Dutson (eds), *Advances in Meat Research Volume 11: Production and Processing of Healthy Meat, Poultry and Fish Products*. London: Blackie Academic & Professional, pp. 283–297.

COLMENERO-JIMÉNEZ, F. and CARBALLO, J. (2005). Physicochemical properties of low sodium frankfurter with added walnut: effect of transglutaminase combined with caseinate, KCl and dietary fibre as salt replacers. *Meat Science* 69 781–788.

COLMENERO-JIMÉNEZ, F., FERNANDEZ, P., CARBALLO, J. and FERNANDEZ-MARTIN, F. (1998). High pressure cooked low-fat pork and chicken batters as affected by salt levels and cooking temperature. *Journal of Food Science* 63 (4) 656–659.

CORNELIUS, D.A., EBERTS, N.J. and STERNBERG, M.M. (1978). Sodium chloride flavour substitute composition and use thereof. US Patent 4066799.

CREHAN, C.M., TROY, D.J. and BUCKLEY, D.J. (2000). Effects of salt level and high hydrostatic pressure processing on frankfurters formulated with 1.5 and 2.5% salt. *Meat Science* 55 123–130.

DESMOND, E.M. (1992). *Compositional analysis of cured pig meat products in Ireland*. Teagasc, The National Food Centre. Internal report.

FOODNAVIGATOR.COM (2005). Quest flavour technology targets salt reduction. Available: http://www.foodproductiondaily.com/news/printNewsBis.asp?id=60683

FRETHEIM, K., SAMEJIMA, K. and EGELANDSDAL, B. (1986). Myosins from red and white bovine muscles: Part 1 – Gel strength (elasticity) and water-holding capacity of heat-induced gels. *Food Chemistry* 22 (2) 107–121.

FRYE, C.B., HAND, L.W., CALKINS, C.R. and MANDIGO, R.W. (1986). Reduction or replacement of sodium chloride in a tumbled ham product. *Journal of Food Science* 51 836–837.

FSA (2002). *McCance & Widdowson's The Composition of Foods, Sixth Summary Edition*. Cambridge: The Royal Society of Chemistry.

FSA (2006). Salt reduction targets. Available: http://www.food.gov.uk/multimedia/pdfs/salttargetsapril06.pdf.

FSAI (2005a). *Salt and Health: Review of the Scientific Evidence and Recommendations for Public Policy in Ireland*. Food Safety Authority of Ireland.

FSAI (2005b). *Salt Reduction Undertakings by the Food Industry – Update 31 August 2005*. Food Safety Authority of Ireland. http://www.fsai.ie/industry /salt/salt_undertakings.pdf

GILLETTE, M. (1985). Flavour effects of sodium chloride. *Food Technology* 39 (6) 47–52, 56.

GOU, P., GUERRERO, L., GELABERT, J. and ARNAU, J. (1996). Potassium chloride, potassium lactate and glycine as sodium chloride substitutes in fermented sasuages and in dry-cured pork loin. *Meat Science* 42 (1) 37–48.

HAMM, R. (1972). Importance of meat water binding capacity for specific meat products. In: *Kolloidchemie des Fleisches* 215–222.

HAMM, R. (1986). Functional properties of the myofibrillar system. In: P.J. Bechtel (ed.), *Muscle as Food*, New York: Academic Press, pp. 135–200.

HAND, L.W., TERRELL, R.N. and SMITH, G.C. (1982). Effects of chloride salts on the physical, chemical and sensory properties of frankfurters. *Journal of Food Science* 47 (6) 1800–1802.

HANLEY, B. (2005). Salt, sugar and fat reduction in foods – Technological challenges. Presentation given at Fat, Sugar and Salt: A question of balance event, Dublin April 2005. Leatherhead Food International.

HUTTON, T. (2002). Sodium: Technological functions of salt in the manufacturing of food and drink products. *British Food Journal* 104 (2) 126–152.

KEETON, J.T. (1984). Effects of potassium chloride on properties of country-style hams. *Journal of Food Science* 49 (1) 146–148.

LEE, E.C. and TANDY, J.S. (1994). Taste enhancement of sodium chloride reduced compositions. US Patent 5370882.

LIOE, H.N., APRIYAMOTONO, A., TAKARA, K., WADA, K. and YASUDA, M. (2005). Umami taste enhancement of MSG/NaCl mixtures by subthreshold L-α-aromatic amino acids. *Journal of Food Science* 70 (7) 5401–5405.

LUTZ, G.D. (2005). Personal Communication: Alberger Salt improves protein functionality in meat blends: Technical Bulletin.

MACFARLANE, J.J., MCKENZIE, I.J. and TURNER, R.H. (1984). Binding of comminuted meat: Effect of high pressure. *Meat Science* 10 307–320.

MADRIL, M.T. and SOFOS, J.N. (1985). Antimicrobial and functional effects of six polyphosphates in reduced NaCl comminuted meat products. *Lebensmittel Wissenschaft und Technologie* 18 (5) 316–322.

MATULIS, R.J., McKEITH, F.K, SUTHERLAND, J.W. and BREWER, M.S. (1995). Sensory characteristics of frankfurters as affected by fat, salt and pH. *Journal of Food Science* 60 (1) 42–47.

MAURER, A.J. (1983). Reduced sodium usage in poultry muscle foods. *Food Technology* 37 (7) 60–65.

McGREGOR, R. (2004). Taste modification in the biotech era. *Food Technology* 58 (5) 24–30.

MILLER, I.J. and BARTHOSHUK, L.M. (1991). Taste perception, taste bud distribution and spatial relationship. In: T.V. Geychell, R.L. Doty, L.M. Barthoshuk and J.B. Snow (eds), *Smell and Taste in Health Disease*, New York: Raven Press, pp. 205–233.

MONAHAN, F.J. and TROY, D.J. (1997). Overcoming sensory problems in low fat and low salt products. In: A.M. Pearson and T.R. Dutson (eds), *Advances in Meat Research Volume 11: Production and Processing of Healthy Meat, Poultry and Fish Products*. London: Blackie Academic & Professional, pp. 257–281.

MORTON SALT (1994). Morton Lite Salt® Mixture. The best alternative to salt. Product brochure.

MORTON SALT (1997). The unique physical properties of Morton® Star Flake® dendritic salt. Product brochure.

MULLER, W.D. (1991). Cooked cured products: Influence of manufacturing technology. *Fleischwirtschaft* 71 544–550.

MYCOSCENT (2005). Mycoscent product details. Available http://www. Mycoscent.co.uk

OFFER, G. and KNIGHT (1988). The structural basis of water-holding in meat. In: R.A. Lawrie (ed.), *Developments in Meat Science 4*, London: Elsevier Applied Science, pp. 173–243.

OFFER, G. and TRINICK, J. (1983). On the mechanism of water-holding in meat: the swelling and shrinking of myofibrils. *Meat Science* 8 245–281.

OLESEN, P.T., MEYER, A.S. and STAHNKE, L.H. (2004). Generation of flavour compounds in fermented sausages – the influence of curing ingredients, *Staphylococcus* starter culture and ripening time. *Meat Science* 66 675–687.

PASIN, G., O'MAHONY, G., YORK, B., WEITZEL, B., GABRIEL, L. and ZEIDLER, G. (1989). Replacement of sodium chloride by modified potassium chloride (co-crystallised disodium-5′-inosinate and disodium-5′-guanylate with potassium chloride) in fresh pork sausages. *Journal of Food Science* 54 (3) 553–555.

PUOLANNE, E.J. and TERRELL, R.N. (1983). Effects of rigor-state, levels of salt and sodium tripolyphosphate on physical, chemical and sensory properties of frankfurter-type sausages. *Journal of Food Science* 48 (4) 1036–1038, 1047.

PUOLANNE, E.J., RUUSUNEN, M. and VAINIONPÄÄ, J.I. (2001). Combined effects of NaCl and raw meat pH on water-holding in cooked sausage with and without phosphate. *Meat Science* 58 1–7.

PRICE, J.F. (1997). Low-fat/salt cured meat products. In: A.M. Pearson and T.R. Dutson (eds), *Advances in Meat Research Volume 11: Production and Processing of Healthy Meat, Poultry and Fish Products*. London: Blackie Academic & Professional, pp. 242–256.

PRIMEFAVOURITES (2005). Available http://www.primefavourites.com

RHEE, K.S., TERRELL, R.N., QUINTANILLA, M. and VANDERZANT, C. (1983). Effect of addition of chloride salts on rancidity of ground pork inoculated with a Moraxella or a Lactobacillus species. *Journal of Food Science* 48 (1) 302–303.

RIERA, J.B., MARTINEZ, M.R., SALCEDO, R.C., JUNCOSA, G.M. and SELLART, J.C. (1996). Process for producing a low sodium meat product. US Patent 5534279.

ROBINSON, T. (2005). Heinz: Salt reduction. Presentation at Fat, Sugar and Salt: A question of balance. Dublin, April 2005. Leatherhead Food International.

RUUSUNEN, M. and PUOLANNE, E. (2005). Reducing sodium intake from meat products. *Meat Science* 70 531–541.

RUUSUNEN, M., NIEMISTO, M. and PUOLANNE, E. (2002). Sodium reduction in cooked meat products by using commercial potassium phosphate mixtures. *Agricultural and Food Science in Finland* 11 199–207.

RUUSUNEN, M., SIMOLIN, M. and PUOLANNE, E. (2001a). The effect of fat content and flavour enhancers on the perceived saltiness of cooked bologna-type sausages. *Journal of Muscle Foods* 12 107–120.

RUUSUNEN, M., TIRKKONEN, M.S. and PUOLANNE, E. (2001b). Saltiness of coarsely ground cooked ham with reduced salt content. *Agricultural and Food Science in Finland* 10 27–32.

RUUSUNEN, M., TIRKKONEN, M.S. and PUOLANNE, E. (1991). The effect of salt reduction on taste pleasantness in cooked 'bologna-type' sausages. *Journal of Sensory Studies* 14 263–270.

RUUSUNEN, M., VAINIONPÄÄ, J., LYLY, M., LäHTEENMÄKI, L., NIEMISTÖ, M., AHVENAINEN, R. and PUOLANNE, E. (2005). Reducing the sodium content in meat products: The effect of the formulation in low-sodium ground meat patties. *Meat Science* 69 53–60.

SACN (2003). *Salt and Health*. Scientific Advisory Committee on Nutrition. The Stationery Office, Norwich, UK.

SCHELL, P.G., VADEHRA, D.V. and BAKER, R.C. (1970). Mechanism of binding chunks of meat. Effect of physical and chemical treatments. *Journal of Canadian Institute of Food Technology* 3 (2) 44–48.

SOFOS, J.N. (1983). Effects of reduced salt levels on sensory and instrumental evaluation of frankfurters. *Journal of Food Science* 48 1683–1691.

SOFOS, J.N. (1985). Influences of sodium tripolyphosphate on the binding and antimicrobial properties of reduced NaCl comminuted meat products. *Journal of Food Science* 50 1379.

TADA, M. SHINODA, I. and OKAI, H. (1984) L-Ornithyltaurine, a new salty dipeptide. *Journal of Agricultural and Food Chemistry* 32 (5) 992–996.

TERRELL, R.N., CHILDERS, A.B. and KAYFUS, T.J. (1982). Effect of chloride salts and nitrite on survival of trichina larvae and other properties of pork sausages. *Journal of Food Protection* 45 281.

TERRELL, R.N. (1983). Reducing the sodium content of processed meats. *Food Technology* 37 (7) 66–71.

THENO, D.M, SIEGEL, D.G. and SCHMIDT, G.R. (1978). Meat massaging: Effects of salt and phosphate on the microstructure of binding junctions in sectioned and formed hams. *Journal of Food Science* 43 493–498.

TROUT, G.R. and SCHMIDT, G.R. (1983). *Utilisation of phosphates in meat products*. 36th Annual Reciprocal Meat Conference 36 24–27.

TROUT, G.R. and SCHMIDT, G.R. (1984). Effect of phosphate type and concentration, salt level and method of preparation on binding in restructured beef rounds. *Journal of Food Science* 49 (3) 687–694.

TUONG, H. and PHILIPPOSSIAN, G. (1987). Alleged salty taste of L-ornithyltaurine monohydrochloride. *Journal of Agricultural and Food Chemistry* 35 (1) 165–168.

TURK, R. (1993). Metal free and low metal salt substitutes containing lysine. US Patent 5229161.

VADEHRA, D.V., NEWBOLD, M.W., SCHNELL, P.G. and BAKER, R.C. (1973). Effect of salts on the water holding capacity of poultry meat. *Poultry Science* 52 (6) 2359–2361.

VARNAM, A.H. and SUTHERLAND, J.P. (1995). *Meat and Meat Products: Technology, Chemistry and Microbiology*. London: Chapman and Hall, pp. 167–210.

VASQUEZ, R.E. (2004). Salt substitute compositions. US Patent 6743461.

VOYLE, C.A., JOLLEY, P.D. and OFFER, G.W. (1984). The effect of salt and pyrophosphate on the structure of meat. *Food Microstructure* 3 (2) 113–126.

WHITING, R.C. (1984). Stability and gel strength of frankfurter batters made with reduced NaCl. *Journal of Food Science* 49 1350–1354.

WIRTH, F. (1991). Reducing the fat and sodium content of meat products: What possibilities are there? *Fleischwirtschaft* 71 (3) 294–297.

13

Reducing salt in seafood products

S. Pedro and M. L. Nunes, Instituto Nacional de Recursos Biológicos (INRB/IPIMAR), Portugal

13.1 Introduction

Fish and shellfish have been acknowledged for being high protein, low calorie foods rich in essential polyunsaturated fatty acids. They are also considered a valuable source of minerals and vitamins. However, they are quite perishable, mainly due to their intrinsic composition and habitat. This fact has contributed to the development and improvement of seafood preserving methods since ancient times.

In fact, the early human communities soon learned that fish spoiled much faster than meat from land animals and consequently the search for processes to avoid putrefaction was an early practice. Drying in the sun and with wood smoke were possibly the first methods for preserving fish products. The use of salt is another of the oldest methods for fish preservation since human beings living in the coastal areas soon recognised salt helped to delay the deterioration.

According to several authors (Bligh, 1980; Sainclivier, 1985), salting was developed over the Bronze Age, since the utilization of both salt and salted products had to be based on an advanced economic organization. Some of the first mentions about salted fish are linked to Egyptian civilization where 'ukas', the name of salted fish, was an important complement to the staple diet of bread. Throughout the Iron Age a significant development in the trading of salted fish, especially the 'tarichos', was registered in the east Mediterranean area from Sicily up to the strait of Bosphorus. During the time of the Greeks and Romans, fish salting was quite relevant and several recipes such as 'tursio', a cake prepared with sturgeon; 'insicia', a kind of sausage; 'garum' and 'alec', fermented fish products; were intensively produced and marketed inside clay pots jars.

In the Middle Ages the centre of fisheries was moved to the north of Europe where the herring fishery was dominant and most of the catches were salted. As a result, several progressive improvements were introduced and some of them have been kept almost unchanged up to the present day.

In the 15th century, with the discovery of Newfoundland, the industry of salted and dried cod was initiated, and different processes followed by French, British and Portuguese fishermen led to the preparation of several salted products which were sold under various popular names ('bacalao', 'bacalhau', 'baccala' or 'morue'). At the turn of the 17th to the 18th century, the lean fish salting industry was well established (Fig. 13.1) and some innovations, regarding smoking and the use of sugar and spices, were introduced in the salting of fatty herring (Voskresensky, 1965).

By the 19th century, other preservation methods emerged, including canning, refrigeration and freezing, and the interest for technological development of the fish industry was greatly stimulated at the end of the century (Valdimarson and James, 2001). Throughout the 20th century the rapid distribution of fresh fish was extensively enlarged as well as the use of other techniques on a larger scale, such as quick freezing and frozen storage. On the other hand, several innovations were introduced in most of traditional processes and a strong transference of technology within fisheries and from other food sectors took place.

Nowadays, salt is still used for seafood preservation, mainly in developing countries, although the majority of salted fish and shellfish products are lightly salted and for preservation other hurdles are used as well. Moreover, based on

THE GEORGE'S BANK COD FISHERY.
Dressing cod on deck of fishing schooner. (Sect. v, vol. 1, pp. 156, 180, 195.)
Drawing by H. W. Elliott and Capt. J. W. Collins.

PLATE 33.

Fig. 13.1   Dressing cod on deck of fishing schooner.

the scientific information, the seafood industry and consumers have become more aware of the relationship between sodium and hypertension and, therefore, in many countries, the demand for a variety of low-salt seafood products has increased. Nevertheless, salt continues to have a definitive role in fish processing, storage stability, and flavour acceptability in addition to the function in the preparation of ready-to-cook or ready-to-eat seafood products, being necessary to define strategies to allow the reduction of salt intake.

Thus, in this chapter the functions of salt in seafood products and some strategies to reduce the present levels are discussed; the role of all stakeholders to reduce salt content, and some recommendations to achieve such goals are presented as well. For this purpose 'seafood' or 'fishery products' comprise all wild or farmed seawater or freshwater animals or parts thereof, including fresh, frozen and cooked fish, crustaceans and molluscs; breaded and prepared products; salted, dried, smoked, fermented and marinated products; sushi and minced fish flesh-based products, surimi and seafood analogues (excluding aquatic mammals, frogs and particular aquatic animals).

13.2   Consumption of seafood products

The consumption of fish products has been associated with human development. Thus, according to the speculations of Crawford *et al.* (1993) the differentiation of man from apes was possibly due to a number of essential fatty acids, preferentially accessible from aquatic organisms. Since that time, when fish exploitation was done in the waters near the land, wider fish resources have been progressively exploited. So, after the Second World War an extraordinary growth in fish production was observed and the catches increased from fewer than 20 to more than 60 million metric tonnes between 1950 and 1970.

In recent years the marine production has levelled at around 100 million metric tonnes and the contribution of aquaculture has significantly increased, exceeding 20 million metric tonnes in 2002 (FAO, 2004). In spite of both wild and farmed seafood products having a significant capacity for processing, the fish processing industry has been very conservative, and, as a consequence, about one-third of the world fishery production is marketed in fresh form; the remaining two-thirds experience some form of processing.

Since 1994 there has been a tendency to increase the proportion of fisheries production used for direct human consumption rather than for other purposes. In 2002, about 76% of estimated world fish production (about 101 million tonnes) was used for direct human consumption, and the remainder was used mainly in the manufacturing of fish meal and oil.

The world average use of fish products reached 16.2 kg *per capita* in 2002 (FAO, 2004), however, consumption can vary greatly between different regions of the world, from 1 kg to more than 100 kg per person per year. Seafood commodities, namely canned tuna, frozen fish fillets and fish fingers are consumed worldwide. Some specific seafood products, such as fish roe, caviar and anchovies are considered as delicacies. Regional products, like canned mackerel in tomato sauce, molluscs, smoked saithe, dried salted cod and fermented fish sauces and pastes, are much appreciated in certain areas. The latter products are widely used in Africa and Asia in the preparation of traditional staple foods as a source of animal protein and also as an essential ingredient, which enhances the flavour of the meal. In Europe, dry salted lean species and smoked products are much appreciated.

As regards fish for direct human consumption, fresh fish has been the most important product, with a share of about 45%, followed by frozen fish (29%), canned fish (14%) and cured fish (12%). The latter products comprise those prepared by drying, dry salting/brining or smoking, which may be used alone or in various combinations. Fresh fish has increased in volume from 25 million tonnes in 1988 to more than 47 million tonnes (live-weight equivalent) in 2002. Processed fish (frozen, canned and cured) increased from 46 million tonnes in 1988 to more than 52 million tonnes live-weight equivalent in 2002. Frozen fishery products (including whole fish, fillets, shellfish, minces and surimi), increased from 24 million tonnes in 1988 to 27 million tonnes in 2002. Canning also reported limited progress, expanding from 12 million tonnes to 13 million tonnes and cured fish production increased from 10 million in 1988 to about 11 million tonnes in 2002.

13.3   Technological functions of salt in seafood

Salt (sodium chloride) is widely used in various amounts during seafood preparation and processing and its role can be divided into three broad objectives. First, it acts as a preserving agent either by inhibiting microbial growth or by promoting indirect changes. Second, it operates as a flavour modifier by giving a salty taste or improving the sensory attributes of other constituents. Third, it plays specific and essential roles in the production and processing of several fish products. However, it should be emphasized that in many situations it fulfils all the three functions and in some cases the separation between them is not clear.

13.3.1   Improvement of sensory aspects

Saltiness is an important sensory attribute of many foods, including seafood, and salt itself contributes to enhancing the characteristic flavour of many food types. Salt is added to fish products to develop the desirable characteristic palatability and also to enhance or modify the flavour perception of other ingredients (Gillette, 1985). However, it is essential to take into account that the sensitivity of consumers to salt is influenced by several endogenous and exogenous variables, the salivary sodium concentration being a key element (Ruunsunen and Puolanne, 2005). Besides, the way in which sodium chloride is added affects the perception of saltiness. For instance, if a given amount of salt is used to sprinkle fish after cooking instead of added to the cooking water, the salty taste is more intense in the former case. It is also known that, usually, fat presence increases salt perception and a higher awareness of salt is obtained when water is weakly bound, especially in the case of fish sausages and gels.

In addition, it has been shown that the presence of sodium chloride alone or combined with other sodium-containing compounds can reduce the sensation of bitterness, the perception of dryness and help to balance any metallic or chemical aftertaste (Hutton, 2002). Regarding the flavour enhancing role, it is probably related to the effect of salt on water activity, since it binds water molecules, increasing the effective content of flavour molecules in solution and enhancing their volatility. Concurrently, salt also increases the ionic strength of the water phase and influences the binding of non-solute molecules within a particular system (Delahunty and Piggott, 1995), thereby improving tenderness.

In most cases the precise influence of salt in taste depends not only on the levels used but also on the freshness, condition and chemical composition of seafood products. These factors can interfere with the enzymatic breakdown of proteins and fat, producing low molecular weight compounds, e.g. peptides, amino acids and fatty acids, which are relevant to the development and perception of flavour.

13.3.2   Processing, preservation and safety

Salting has been commonly used separately or in combination with other processes (such as air-drying, smoking, marinating and fermentation) to preserve the quality and assure the safety of seafood by inhibiting the growth of both spoilage and pathogenic micro-organisms. The aim is to obtain stable final products through water activity (a_w) and pH reductions, although it has been reported that salt content in fish muscle enhances oxidation of the highly unsaturated lipids.

Fish can be salted in different ways – dry, wet and mixed salting. The fish being salted with dry crystalline salt characterises dry salting. Wet or salt brine type of salting is a process by which the fish is immersed in a previously prepared salt solution. Mixed salting is a hybrid method by which the fish is placed in dry salt and becomes immersed in the solution of the salt and the liquid leached out from the fish, additional brine may be added or not. The action of salt in seafood is a thorough combination of direct microbial inhibition, enzyme inhibition and a significant dehydration effect of the fish tissue (Connell, 1980).

During salting, two main fluxes take place: the uptake of salt and other possible curing compounds, and the loss of water together with a number of associated solute molecules. Salt uptake by seafood is influenced by two main factors: the nature of the raw material (particularly size, shape and composition) and the kind of salting technique used. Usually salt uptake is slightly lower for fatty species than for lean ones, as high fat content causes greater resistance to the transfer of an aqueous solute such as sodium chloride. Additionally, smaller size pieces usually present slightly higher salt levels, due to increased salt diffusion. Moreover, for the same seafood product type, the salt content is usually slightly higher for those prepared from frozen raw material than from fresh material, especially for leaner fish. This is a result of the freezing step, which slightly modifies cell structure, increasing salt diffusivity. Products salted in saturated brines generally have higher salt content than dry-salted samples because of better contact between surfaces and the salting medium.

For preserving purposes, salting fish with saturated brine is specially applied to fatty fish, as immersion in this medium keeps the air out and, as such, prevents lipid oxidative rancidity (Aubourg and Ugliano, 2002). The preparation of ripened herring, anchovies and other small fatty fish species is achieved by mixing fish with salt inside plastic or wooden barrels followed by the addition of salt brine prior to the closing of barrels. Ripening takes from a few weeks up to several months and the salt concentration varies usually between 7 and 18% according to the procedure used and the product foreseen. Dry salting is mainly applied to cod and other lean species. Fish is packed in layers with solid salt to form stacks, and the saturated brine leached out from the fish is allowed to drain away. The salt content in the final product usually ranges from 16–25% and depends on the type and grain size of salt, and on freshness, condition and thickness of fish. Most of these products are intended for further drying. In several Latin countries, dried salted cod products, have great popularity and long tradition, in spite of needing rehydration before consumption – desalting (Fig. 13.2); the latter process is traditionally performed by the consumers at room temperature, taking more than 24 h and allowing the product to reach final salt contents of approximately 2% (Pedro et al., 2002, 2004).

Fig. 13.2 Dried salted and desalted cod.

In the preparation of smoked fish products, salt also plays an important role, since it decreases a_w. There are two methods of making smoked fish, by cold and hot smoking. Cold smoking means curing fish by using air temperature not higher than 33°C to avoid cooking the flesh or coagulating the protein. In hot smoking the temperature at some stage in the process has to be around 80°C in order to cook the fish meat. Hot smoked fish products do not require any further cooking before consumption while some cold smoked products are usually cooked, smoked salmon being an exception. Smoked fish products usually present salt levels between 2 and 5.5%; however, the salt content can be up to 8% in the water phase in order to prevent the risk associated to *Clostridium botulinum* (Gram, 2001a). This latter level is very often unacceptable to the consumer. In mild hot smoked fish the NaCl content in the flesh is usually low, about 2.5%, which corresponds to about 3.3% in the water phase (Sikorski and Kolodziejska, 2002). The combination of salt and temperature used in smoked fish products is not sufficient to guarantee the absence of growth for *Listeria monocytogenes*, the use of additional hurdles being necessary (Gram, 2001b). In all smoked products the salt content is critical in terms of safety, given that the concentration in the water phase has to be high enough to inhibit the growth of pathogenic micro-organisms.

Salt is also used in fish marinades, where it is used together with a variety of compounds such as spices, herbs, vegetables and organic acids, namely acetic acid. Herring is the principal marinated fish, rollmop herring being one example. In these products the salt content is usually around 5–8%. Other marinated fish products, like fish in 'escabeche', 'seviche' and marinated mussels present variable amounts of salt.

In Japan and other countries in South-East Asia many kinds of fermented fish products are traditionally made by acid production from added cereals, being used mostly as sauces in rice dishes, fish pastes and a few consumed as snacks (Saisithi, 1994). Lactic acid bacteria (LAB) are found as the dominant micro-organisms in these products, which primary role is to ferment the available carbohydrates causing a decrease in pH. In addition, salt and spices (such as garlic, pepper or ginger) may be supplementary safety factors. The amount of added carbohydrate and the salt concentration mainly control the extent of acid fermentation and quality of fermented fish products (Lee, 1997). The preparation of most products is based on the use of strong amounts of salt (20–25%). This is the case of 'Jeotkal' products, a traditional Korean fish product, which preparation requires the addition of high levels of salt (20–30%) before fermentation, depending on the raw materials. The latter can include marine or freshwater fish, shellfish, and crustaceans (Jo et al., 2004b).

In canned seafood products, a liquid-based medium is usually added to the product for efficient transfer of heat in the processing. That cover medium can be a vegetable oil, other sauces or even light brine, containing 1–2% of salt. Additionally, fish can be immersed in 2–6% salt brine, as a pre-treatment, in order to wash out slime, blood residues and small pieces of guts. In heat-treated fish products the previous immersion in brine is quite relevant since it confers the characteristic flavour, improves the texture and contributes to decrease the fluid lost during cooking and also the leaching out of aqueous exudates during storage.

Amounts of salt between 2 and 3% are usually used in the preparation of restructured fish products. The techniques most used for restructuring are based on solubilizing and extracting myofibrillar proteins with salt by cutting, tumbling or blending to obtain sticky exudates. These are then used to bind fish pieces or to obtain gels. The level of salt employed during massaging is important because it influences the amount of exuded protein obtained, which acts in binding and affects the protein aggregation mechanism. Usually a decrease in salt level to below 2% has a negative effect on the functional and mechanical properties of fish products; 1.5% salt has been considered as the minimum level required for obtaining fish products without excessively decreasing functional properties.

Salt can also contribute to increase shelf life of chilled seafood products and reduce the ability of some strains of *L. monocytogenes* to grow at chill temperatures. Thus, under optimal conditions, a level of about 10% salt in the water phase is required to prevent such growth. In vacuum-packed or modified atmosphere packed (MAP) fish products, if the temperature is not kept below 3.3°C, the risk of growth of *Cl. botulinum* exists. A salt level of at least 3.5% in the aqueous phase can provide an adequate preservative action in combination with the chill temperature. In the application of MAP to ready-to-eat fish meals and the levels of salt used, several aspects have to be taken into account, namely the risk of botulism, as described by Shaw (1997) and by the Advisory Committee on the Microbiological Safety of Foods (ACMSF, 1995). Moreover the preservative effects of MAP can be enhanced by dipping fish portions in sodium chloride solutions prior to MAP storage. Experiments run by Pastoriza *et al.* (1998) have shown that such dips (5 min in 5% NaCl solutions) delayed the microbiological and sensory deterioration of slices of hake.

Fish roe products are very often treated with salt to make a wide range of products, some of them extremely valuable, among which caviars are the best known. Caviars are made from fish roe after the eggs have been graded, sorted, singled-out, salted or brined, and cured. Most caviar is marketed as a refrigerated or frozen food. Several types of caviar from different fish species are marketed as shelf-stable products. Salmon eggs are usually brine salted, giving caviar 3–4% sodium chloride, although some producers and consumers claim that higher salt contents, in the range of 4–6%, are required to obtain high-grade 'ikura' (Bledsoe *et al.*, 2003). Sturgeon eggs are usually dry salted and levels of 3–3.5% are achieved in the final product after the excess of natural brine has been drained off the product (Bledsoe *et al.*, 2003). The salt content in lumpfish caviar ranges from 3 to 5%, however, when the lumpfish roes are intended to be delivered and further reprocessed, desalted and repacked in retail packages, the concentration of salt may be as high as 10–14% (Bledsoe *et al.*, 2003).

Fish sauces are a clear brown liquid hydrolysate from salted fish and are an important source of condiment for people living in Asia and Africa. These sauces are produced by adding significant amounts of salt (fish/salt ratios, 2:1 or 3:1) to fish, generally, low-priced small fish such as anchovies, round scad, or sardines. High salt levels inhibit the growth of pathogenic bacteria in the sauce and also contribute to its characteristic flavour.

Furthermore, salting is also used in the killing of fish parasites. Thus, for instance the immersion of herring in 21% salt brines for ten days destroys some of the common parasites, in particular the herring worm (*Anisakis simplex*), but 12 weeks are required if the salt concentration is 5–6% (Doyle, 2005).

13.4   Current salt intake from seafood

The most relevant constituents of salt are sodium and chloride, which are dietary essentials. In the case of sodium, it is widely acknowledged that current intake exceeds the nutritional recommendations in several industrialized countries. It has been estimated that 15% of dietary sodium is naturally present in unprocessed foods; that 15–20% is added during cooking or at the table; and therefore 65–70% is incorporated during manufacture and processing (SACN, 2003). Most sodium in the diet comes from sodium chloride, which contains about 40% sodium by weight. In spite of some food additives being sodium compounds and contributing sodium to some manufactured seafood products, the contribution of sodium from other sources may be considered negligible taking the diet as a whole.

In general, a diet consisting of finfish and molluscs represents a low sodium chloride intake, as these raw materials are naturally low in sodium, and even

Table 13.1   Sodium content in selected raw seafood per common serving portion

Species common name	Serving portion (g)	Sodium content (mg)	Daily intake* (%)
Crustaceans			
Crab, Alaska king	172	1438	60
Crab, blue	85	249	10
Shrimp, mixed species	85	128	5
Fish			
Anchovy, European	85	88	4
Cod, Atlantic	231	125	5
Haddock	193	131	5
Herring, Atlantic	28	25	1
Salmon, chinook	198	93	4
Salmon, chum	198	99	4
Tuna, bluefin	85	33	1
Molluscs			
Clam, mixed species	227	127	5
Scallop, mixed species	30	48	2

Current daily recommended maximum sodium intake is 2400 mg (IOM, 2004).
Source: data collected from http://www.nutritiondata.com.

those species with the highest sodium levels contain less than 110 mg per 85 g portion, which is less than 5% of the current daily recommended maximum sodium intake of 2.4 g (IOM, 2004). On the other hand, most crustaceans generally have more sodium, ranging from less than 140 mg to more than 500 mg per 85 g serving (Table 13.1).

As a rule, frozen fish, crustaceans and molluscs have a sodium level comparable to fresh seafood, however, some frozen prepared seafood products (i.e., battered, breaded, restructured) have added sodium and a few processed seafood products can even present high to very high salt contents, such as canned anchovies (Table 13.2). Surimi or imitation shellfish products (approximately 600–715 mg per 85 g portions), smoked fish (between 300 and 870 mg per 85 g portions), frozen fish sticks (163 mg per stick), frozen fish portions

Table 13.2 Sodium content in selected processed and cooked seafood per common serving portion

Species common name	Product	Serving portion (g)	Sodium content (mg)	Daily intake* (%)
Crustaceans				
Crab, Alaska king	Cooked, moist heat	134	1436	60
Crab, Alaska king	Imitation, made from surimi	85	715	30
Crab, blue	Canned	135	450	19
Crab, blue	Cooked, moist heat	118	329	14
Crab, blue	Crab cakes	60	198	8
Shrimp, mixed species	Canned	128	216	9
Shrimp, mixed species	Cooked, breaded and fried	85	293	12
Shrimp, mixed species	Cooked, moist heat	85	190	8
Shrimp, mixed species	Imitation, made from surimi	85	599	25
Fish				
Anchovy, European	Canned in oil, drained solids	28	1036	43
Cod, Atlantic	Canned, solids and liquid	312	680	28
Cod, Atlantic	Cooked, dry heat	180	140	6
Cod, Atlantic	Dried and salted	28	1985	83
Fish portions and sticks	Frozen, preheated	57	332	14
Haddock	Cooked, dry heat	150	130	5
Haddock	Smoked	28	216	9
Herring, Atlantic	Cooked, dry heat	143	164	7
Herring, Atlantic	Kippered	28	259	11
Herring, Atlantic	Pickled	140	1218	51
Salmon, chinook	Cooked, dry heat	154	92	4
Salmon, chinook	Smoked, (lox), regular	28	565	24
Salmon, chinook	Smoked	136	1066	44
Salmon, chum	Canned, without salt, drained solids with bone	369	277	12
Salmon, chum	Cooked, dry heat	154	99	4
Salmon, chum	Canned, drained solids with bone	369	1797	75
Sardine, Atlantic	Canned in oil, drained solids with bone	149	752	31

Table 13.2   Continued

Species common name	Product	Serving portion (g)	Sodium content (mg)	Daily intake* (%)
Sardine, Pacific	Canned in tomato sauce, drained solids with bone	89	368	15
Tuna	Salad	205	824	34
Tuna	Fresh, bluefin, cooked, dry heat	85	42	2
Tuna, white	Canned in oil, drained solids	178	705	29
Tuna, white	Canned in oil, without salt, drained solids	178	89	4
Tuna, white	Canned in water, drained solids	172	648	27
Tuna, white	Canned in water, without salt, drained solids	172	86	4
Molluscs				
Clam, mixed species	Canned, drained solids	160	179	7
Clam, mixed species	Canned, liquid	240	516	22
Clam, mixed species	Cooked, breaded and fried	85	309	13
Clam, mixed species	Cooked, moist heat	85	95	4
Scallop (bay and sea)	Cooked, steamed	100	265	11
Scallop, mixed species	Cooked, breaded and fried	31	144	6
Scallop, mixed species	Imitation, made from surimi	85	676	28

* Current daily recommended maximum sodium intake is 2400 mg (IOM, 2004).
Source: data collected from http://www.nutritiondata.com.

Fig. 13.3   Seafood products.

(332 mg per 57 g portions), caviar (240 mg per tablespoon), and canned products that have salt added during processing (approximately 300–350 mg per 85 g edible portions) contain higher levels than fresh seafood (Fig. 13.3). Products that are frozen in sodium brines, such as crab legs, may contain as much as 800–1000 mg of sodium per serving.

It is also interesting to notice that the culinary preparation can influence the sodium content of seafood; for instance, the same portion size shrimps breaded and fried present higher salt levels than those cooked by moist heat.

13.5   Salt reduction in seafood

Salt intake is determined by a large number of factors, including taste preferences, attitudes, personality and the intake of water and other nutrients (Shepherd and Farleigh, 1986). Changing the salt content of the diet of a consumer population, that most often has been adapted to a high-salt diet, is not easy and will involve a number of strategies. Potential sodium chloride reduction depends on the type of product, its composition, the type of processing required and the preparation conditions. These factors determine the type of product that can be modified and the technological limitations of salt reduction. As most sodium chloride in the diet comes from processed foods, the greatest scope for significant reduction of salt intake is focused in these products. There is evidence that significant gradual reductions in the salt concentration of processed food (up to 10% per year) can be achieved without adverse effects on taste (MacGregor and Sever, 1996). However, reductions in the sodium content of certain products can be limited by consumer taste preferences and, ultimately, lack of consumer demand. So, such reductions of the salt content in processed foods also have to be combined with consumer information and education, promoting changes in diet and food preparation.

Nevertheless, the food manufacturing industry has been responsive to consumer requirements and many manufacturers have developed and marketed lower sodium versions of standard products, or have reduced sodium content over a period to meet the demands of consumers. Many manufacturers have adopted a strategy of gradual reductions in the salt content of foods in which the main function of salt is to deliver flavour. In particular, some UK companies have committed to achieve salt levels in several fish products, namely breaded fish, fish-based ready meals, fish fingers and canned fish, which are close to the targets established by the Food Standards Agency (FSA, 2005). A long-run strategy entails combining a gradual salt reduction approach with a more technological one, towards delivering appropriate salt perception at a lower salt content.

The reduction of the salt content of seafood by exclusively lowering the amount of salt has not been straightforward because the properties of low-salt seafood products are often different from the normal-salt, in particular in what concerns flavour and shelf life. Nevertheless, the formulations and processing of several seafood products have been reviewed, using alternative ingredients and new technologies, in order to decrease the salt content in the final product. Besides, with the advent of modern food preservation methods, there is less need for the preservative effect of salt; while cured products cannot be produced without salt, there are many products on the market prepared with less salt.

13.6   Strategies for lowering salt content in seafood

In general, the main approaches adopted for reducing the sodium content in processed seafood products have been based on: (1) lowering the level of added sodium chloride (NaCl); (2) replacing all or part of the NaCl by other chloride salts (KCl, $CaCl_2$ and $MgCl_2$), by flavouring agents, by adding binding agents, preservatives, etc; (3) developing new processing techniques or process modifications; and (4) combinations of any of the above approaches.

13.6.1   Reduction

The formulation of several seafood products has been reviewed in order to decrease the salt content in the final product. In this context the canning industry has attempted to present seafood products without added salt, such as tuna and salmon, either immersed in vegetable oil or water. Even the traditional canned fish product presentations nowadays have lower salt levels, and salt levels lower than 1% have been determined in canned sardine in vegetable oil (Bandarra *et al.*, 2004). In fish roe, lightly salted salmon caviar (2.8–3.5% salt) is becoming more popular as consumer preferences change for lower salt content (Bledsoe *et al.*, 2003). For some cured seafood products, such as dried salted cod, there are in the market alternative ready-to-use presentations, containing low sodium levels. Moreover, some brands commercialize nowadays frozen desalted cod products with 1.5% salt (Vieira, 2005, personal communication).

13.6.2   Replacement and addition of other ingredients

Much research has concentrated on attempting to find other materials delivering saltiness, but the main barrier associated with such substitutes has been the undesirable tastes accompanying saltiness. Nevertheless, a flavour improvement can be obtained by mixing substitutes as suppression of each taste tends to be the rule in mixtures of substances with different tastes (Gou *et al.*, 1996). Most attention has been focused on the use of potassium chloride, and many products on the market utilize this salt as a partial sodium replacer. Substitution of NaCl by KCl can be undertaken without functional loss, but the latter salt imparts some taste defects, a distinct bitter flavour being the most difficult one to overcome. Although a bitter taste is detected at substitutions levels of 30%, in some products it would appear acceptable to substitute 40% of the NaCl content by KCl (Gou *et al.*, 1996). In smoked fish products the substitution of NaCl by some KCl was found feasible for the inhibition of outgrowth and toxin production by *C. botulinum* type E (Pelroy *et al.*, 1982).

Martínez-Alvarez *et al.* (2005) examined the relationship between ionic composition and functional quality of lightly salted cod, using brines for salting in which NaCl was partially replaced by KCl, $MgCl_2$ and/or $CaCl_2$ at two different pHs (6.5 and 8.5). The replacement of around 50% NaCl by KCl in brines for salting considerably reduced the sodium content of the product. In pH 6.5 brines, the presence of KCl did not alter major functional properties to any great extent. However, when brine pH was 8.5, KCl negatively affected protein water-extractability, increasing hardness. Addition of small amounts of divalent salts ($CaCl_2$, $MgCl_2$) to the brines could help to significantly reduce the penetration of sodium ions into the muscle, provided that salting was carried out with pH 6.5 brines. Moreover, according to Rodrigues *et al.* (2005), these latter salts did not adversely affect the microbiological and sensory quality of salted cod.

Some salty synthetic peptides can be used to replace sodium chloride. Nakamura *et al.* (1996) have shown that ornithyltaurine hydrochloride is as salty as salt on a molar basis, and could replace salt in some foods; substitutions of up to 50% of salt in some sauces were suggested. The replacement of salt by herbs and spices is rarely satisfactory but slight increases in acidity can improve the perceived saltiness. For instance, the addition of very high salt concentrations to extend the shelf life of 'Jeotkal' products has limited its use and consequently several attempts have been made to reduce its level by seasoning. Seasoned 'Jeotkal' has advantages to general 'Jeotkal', including a relatively shorter fermentation period (30 days *vs* 3–5 months), lower salt content (commercially 8%), contributing added value and modernized taste for the younger generation. However, the salt content is still high for direct consumption as a side dish (Jo *et al.*, 2004a).

On the other hand, a number of flavour enhancers, some considered to have a umami flavour, can be used in a wide range of seafood applications, in particular in sauces and ready-to-eat meals. Monosodium glutamate (MSG) has the ability to enhance the presence of other taste-active compounds and sensitize taste buds. For an equal sensation of saltiness, the sodium provided by sodium chloride can be reduced and compensated by very low amounts of MSG; thus by adding MSG appropriately the sodium chloride addition can be reduced by 30–40% while maintaining the same perception of saltiness (Yamaguchi and Takashashi, 1984). Also, 5'nucleotides, such as 5'GMP (guanosine monophosphate) and 5'IMP (inosine monophosphate), are very potent flavour enhancers and effective at parts per billion levels. Mixes of IMP and GMP can suppress some bitter and sour notes and enhance sweet and salt perceptions (Woskow, 1969). It is important to note that both glutamates and the nucleotides or their precursors are already present in relatively high levels in biological systems, either in free or bound state. Glutamic acid is a major component of the proteins of most of our foods, and fish is especially rich in IMP while crustaceans and molluscs are rich in adenosine monophosphate (AMP), which eventually serves as the precursor for the formation of IMP (Löliger, 2000). Because of their highly synergistic effect with salt, sodium chloride levels can be reduced, perfectly fitting into the trend of reduced sodium content and healthy seafood. Furthermore, fermented oyster sauce, containing high levels of free amino acids (such as glutamic acid, glycine, lysine and alanine), which are important in the tastes of fish and shellfish sauces, has good taste and nutritional properties, being a potential seasoning agent (Je *et al.*, 2005).

A decrease of salt level in restructured products has a negative effect on the protein extractability and solubility and consequently on the mechanical and functional properties, the addition of binding and texture-modifying agents being necessary in order to develop alternative low-salt products. Partially hydrolysed proteins, such as whey, yeast and plant proteins are valuable as water binders and may contribute to meaty flavours at the same time, being added as ingredient in low-salt restructured seafood products. Moreover, non-muscle proteins (including soy protein, wheat gluten, egg white, casein and beef plasma-thrombin) as well as several food hydrocolloids (such as xanthan, guar, kappa and iota carrageenan and pectins) have been used to improve the mechanical and functional properties of restructured fish products, contributing to the development of similar low-salt products. These binding and texture-modifying agents can be used either alone or in combination with chloride salts and/or with each other. Goméz-Guillén and Montero (1997) improved the thermal gelation of giant squid (*Dosidicus gigas*) mantle proteins at low salt concentrations (1.5% NaCl) by adding non-muscle proteins (2%), starch (5%) and iota-carrageenan (2%). Additionally, the gelation of fresh (unfrozen) raw surimi-type paste from horse mackerel (*T. murphyi*) was studied in the presence of carrageenan (0.5, 1, 2%) and KCl (0.5–2%) as a substitute for NaCl. It was observed that when carrageenan was added as a single ingredient, the gelation properties for the surimi paste containing 2% carrageenan were greatly increased. In spite of KCl enhancing gelation of the 2% carrageenan solution, the addition of this salt to the raw surimi paste containing 2% carrageenan did not improve its gelation (Ortiz and Aguilera, 2004).

Kawano and Nobuhisa (2000) found that addition of salt in the step of grinding fish meat (including that in the form of surimi), which has been considered to be an essential step, can be eliminated by blending a hydrate gel of glucomannan with fish meat ground without added salt. The resulting ground fish meat products or their analogues retained texture resembling or superior to that of the conventional gel products, such as kamaboko.

On the other hand, studies have shown that a non-calcium dependent microbial transglutaminase (MTGase), widely used in the food industry, can also be used to improve the mechanical properties of fish products, inducing covalent crosslinking of proteins. Low-salt (1%) fish restructured products were produced from fish paste of silver carp (*Hypophthalmichthys molitrix*) containing MTGase by massaging and cooking in ham presses. Thus, according to Ramirez *et al.* (2002) low-salt, homogeneous restructured meat blocks, resembling ham, were obtained by adding 1% NaCl and 0.3% MTGase.

Additionally, low-salt fish restructured products (1%) were obtained using mechanically deboned fish meat from filleting wastes of silver carp and 0.3% MTGase (Téllez-Luis *et al.*, 2002). The resulting low-salt restructured silver carp products presented improved mechanical and good functional properties. However, it was not feasible to obtain restructured products by adding just MTGase in the absence of salt. Nevertheless, caseinate, when treated with MTGase, becomes viscous, and the viscous caseinate acts as a glue to hold different foodstuffs together. So a meat-binding system using MTGase and caseinate simultaneously was developed and fish fillets could be prepared from their smaller pieces without salt, using this system (Motoki and Seguro, 1998). Moreover, the mechanical properties of non-salted and low-salt restructured silver carp (*Hypophthalmichthys molitrix*) products have been improved by using dairy proteins combined with MTGase. Restructured fish products from the filleting waste of silver carp were obtained using sodium caseinate (1%), whey protein concentrate (1%) or MTGase (0.3%) at 0% and 1% salt levels (Uresti *et al.*, 2004). Sodium caseinate, that entails a very small quantity of sodium in comparison to common salt, had a greater effect in improving mechanical properties than whey protein concentrate while MTGase increased expressible water.

13.6.3   New processing techniques or process modifications

Taking into account the commercial importance of dried/wet salted cod and the actual consumer preference for ready-to-use food products, several studies have been developed to propose new cod-desalting processes which are able to decrease the salt content to suitable levels without significant changes in the sensory attributes (Andrés *et al.*, 2002, Barat *et al.*, 2004a,b). Recently, Andrés *et al.* (2005) investigated the role of different variables (water hardness and temperature, raw material origin, stirring level, thickness and batch size, additives used in the desalting water, etc.) that interfered in the cod-desalting process namely on net weight changes, water uptake, salt loss and water-holding capacity. They proposed a cod-desalting process at low temperature, without stirring. Moreover, they found that the application of vacuum pulses (15 min vacuum pulse at 50 mbar) could improve the kinetics of the process, as the use of salt an additive in the desalting solution could improve the product's shelf life and process control and the presence of skin did not influenced kinetics. Bjørkevoll *et al.* (2004) developed a three-step rehydration method for salt-cured cod which included injection of tap water followed by tumbling in water and tumbling in a 2% NaCl brine. They found that the method ensured rehydration times of 3 h and weight gains similar to those obtained during the traditional process. Sensory, chemical and microbiological analysis showed that rehydration with the new rapid method gave fillets of almost similar quality as by the use of traditional rehydration. The use of injection and tumbling steps resulted, however, in slightly less cohesive rehydrated products, but improved the products with regard to a more homogeneous NaCl concentration. Increasing the number of injection points/cm^2 from 2 to 5 reduced the salt content after injection from 11% to approximately 5%.

The injection-salting technology is gradually acquiring popularity due to its time saving potential and the higher processing yields obtained when compared to conventional salt-curing techniques (Chiralt *et al.*, 2001). Its application to fish products has been done, although the effects of its use on quality parameters of some fish products are not yet fully documented. However, it has been claimed that such a process is able to cause uniform distribution of salt and to lower salt concentrations as well. Chiralt *et al.* (2001) studied the influence of different process variables (length of vacuum pressure period, temperature, sample structure and dimensions) in terms of kinetic data and process yields, for several products, including salmon and cod. In general, they found that the process implies a notable reduction of salting time, increasing the process yields in line with the greater values of the ratio of salt gain to water loss. Likewise, samples lose natural gas or liquid phases entrapped in their structure and reach a flatter salt concentration profile than that obtained in the conventional salting methods.

Regarding smoking salmon, several studies have been undertaken to compare the effect of dry salting and brining injection on the characteristics, product yield and quality parameters of the final product. Thus, Røra *et al.* (2004) studied the effect of the two salting techniques on the yield, salt uptake and salt distribution of pre-rigor fillets salted and smoked at different stages *post mortem*. They found that early salt and smoke processing of pre-rigor fillets resulted in lower muscle salt concentrations compared to traditional *post rigor* processing and, on the other hand, brine injection of *pre rigor* filleted salmon resulted in a more uniform distribution of salt associated to lower salt levels than dry salting. Other authors, Birkeland *et al.* (2004), investigated the effects of brine injection on salmon muscle gaping, texture and colour and concluded that injected salted fillets had higher gaping score and softer texture when compared to dry salted product.

The use of additional processing, ensuring not only the preservation but also the safety of seafood, can be necessary in lowering salt content in seafood. For instance, the use of pasteurization, when applicable to caviar, has to be conducted at mild temperatures (50 to 70°C) due to the sensitivity of the product to heat and such a process can contribute to a certain extent to reducing the salt level (Bledsoe *et al.*, 2003).

The application of high pressure processing (HPP), as an alternative method for food preservation (Murchie *et al.*, 2005), fulfils consumer requirements regarding the preferences for minimally processed and additive-free products and allows the inactivation of pathogenic and spoilage micro-organisms while maintaining sensory and nutritional properties of foods (Torres and Velásquez, 2005). Currently, HPP has several commercial applications, including inactivation of some bacteria (Smelt, 1998; Berlin *et al.*, 1999; Voisin, 2001, 2002), viruses (Kingsley *et al.*, 2002), and shelf life extension of seafood (Hurtado *et al.*, 2000), therefore contributing to the development of seafood products with lower salt content.

Besides the interest of HPP in fish preservation, other advantages regarding this processing over conventional thermal processing have been demonstrated especially for surimi and kamaboko, traditional Japanese products made from fish mince, due to the superior quality of HPP-induced gels. Such utilization can also help to reduce the salt content commonly used to make the gellification of myofibrillar proteins because HPP-induced gels are described as glossier, more uniform and smoother than gels produced by heat treatment in the presence of 2–3% salt (Ohshima *et al.*, 1993). Additional physical processes, such as sectioning, forming, massaging and tumbling can improve water-added content in restructured products enhancing rheological properties. Kawano and Nobuhisa (2000) claimed that ground fish meat products or their analogues could be obtained from fish meat ground without added salt either by increasing the cutting time when grinding with a food-cutter, or by high-speed rotation of rotating blades when grinding using a ball cutter.

For fermented seafood, significant effects of gamma irradiation on micro-biological inhibition have been reported, allowing the production of products with reduced salt content. Thus, fermented squid (*Todarodes pacificus*) with low salt content was successfully produced by using gamma irradiation as a sub-sequent hurdle. The combination of 10% salt concentration and gamma radiation (10 kGy), without any food additives was effective in processing low salted fermented squid and extending its shelf life compared to control (non-irradiated product with 20% salt) (Byun *et al.*, 2000). The effects of gamma radiation (5–10 kGy) on sensory quality, microbial population, and chemical properties were also investigated in low-salted fermented shrimp (*Acetes chinensis*) 'Jeotkal'. Lee *et al.* (2002) observed in the low-salted (15–20%) irradiated products no adverse sensory quality and improved microbial shelf-stability compared to control (non-irradiated product with 30% salt addition). Positive effects of irradiation were also observed in a fermented and seasoned Alaska pollock (*Theragra chalcogramma*) intestine with a lower salt content (5%). Irradiation at 2.5 kGy lowered the microbial contamination, reduced volatile basic nitrogen and amino nitrogen contents and improved chemical storage stability. A sensory evaluation indicated that 2.5 kGy-irradiated 'Changran Jeotkal' with 5% salt content was more acceptable than the commercial control (8% salt content). So, according to Jo *et al.* (2004a,b) irradiation can be applied for the development of low salted, seasoned 'Changran Jeotkal' without adverse effects on the quality.

13.7   Recommendations

Reducing salt intakes related to seafood consumption is a challenge facing governments and industries, and co-operation appears to be a major contributor to potential success.

Food safety agencies and other relevant agencies should have dual ap-proaches combining a consumer education and information programme with the promotion of new seafood product development and reformulations. The latter can involve meetings with key players and seeking commitments in the form of salt reduction plans from individual companies. In the development of programmes for reducing salt consumption, advice should be targeted for the population and also for individual adults. However, concerning seafood, campaigns should be balanced and also emphasize the health benefits of consuming fresh fish, in particular fatty species, which are low-salt foods and contain omega-3 fatty acids. The particular vulnerability of children and the elderly to the adverse effects of high salt intake needs to be highlighted in order to accelerate the development of new seafood products and the reformulation of existing ones.

The food industry should work in consultation with food safety agencies and other relevant agencies to achieve a gradual, sustained and universal reduction in the salt content of processed foods and foods prepared by the seafood sector. Moreover, high priority should be given to the development of research aimed at addressing technological, shelf-life, preservation and taste issues in relation to the reduction of salt content in seafood processed products. Efforts should be made to deliver a lower sodium alternative in each processed seafood product range. Clear and accurate nutrition labelling information about sodium should be provided, in accordance with legislative requirements, enabling consumers to exercise informed choice and to select products for a balanced diet according to their own specific dietary preferences and requirements.

Retailers should also recognize the relevance of delivering appropriate information to consumers. For instance, the inclusion in shelf cards of information showing whether the salt content of comparable products is high, medium or low, can enable consumers to easily choose between them on the basis of their nutritional value.

Consumers should be enabled to compare between and across seafood products, through more effective use of on-pack labelling and appropriate consistent education. They need to be aware of both natural and added sodium content when choosing seafood to lower the sodium intake and must read the labels when buying prepared and pre-packaged products (Fig. 13.4). It is recommended to consume preferably fresh/frozen fish and reduced sodium or no-salt-added seafood. Traditional canned, smoked, or processed varieties of seafood should be used parsimoniously. Consumption of cured seafood (such as, anchovies), seafood packed in brine, some crustaceans and condiments (such as fish sauce) should be limited. Moreover, for seafood products high in salt, consumers should try to choose brands/recipes that contain less salt. In addition, seafood tinned in brine should be drained and rinsed to reduce their salt content. Among convenience seafood, preference should be given to those lower in sodium, and mixed seafood dishes, which often have a lot of sodium, should be avoided. When cooking at home, consumers should give preference to the use of herbs and spices for enhancing the flavour of seafood, rather than salt, using strong, flavourful spices such as black pepper, coriander, curry powder, cumin, basil, oregano, onion and garlic, according to personal taste. When eating out, preference should be given to low-salt meals.

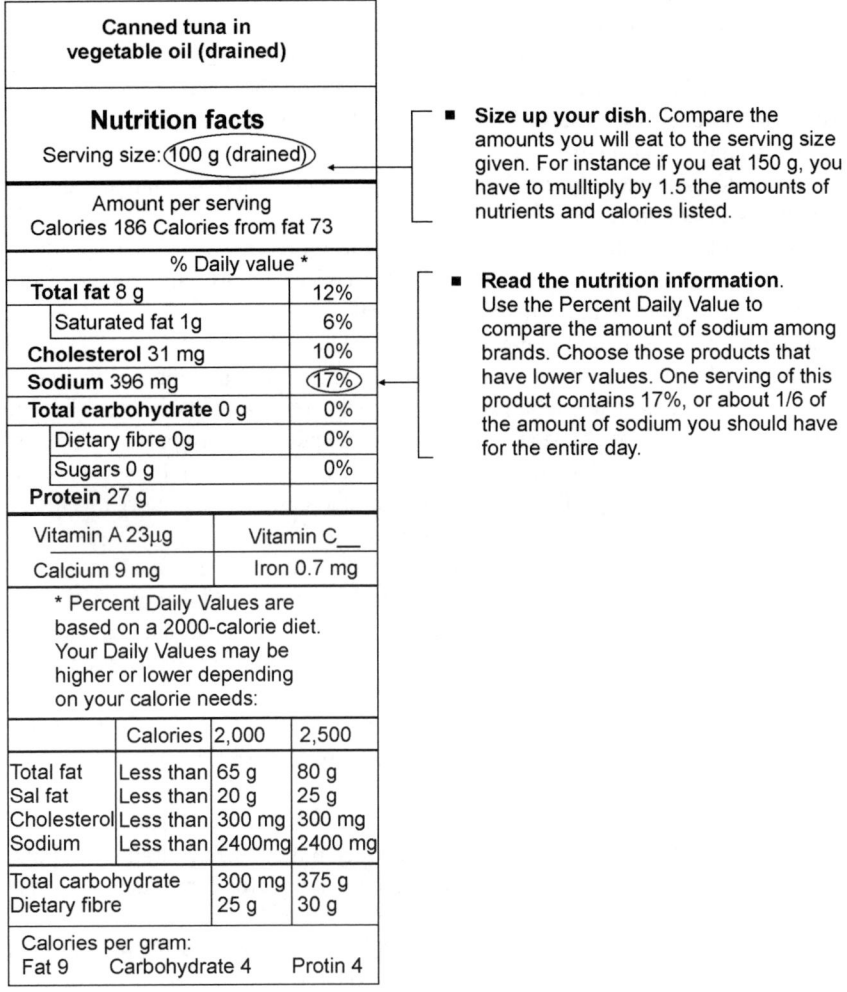

Canned tuna in vegetable oil (drained)		
Nutrition facts		
Serving size: 100 g (drained)		
Amount per serving Calories 186 Calories from fat 73		
% Daily value *		
Total fat 8 g		12%
Saturated fat 1g		6%
Cholesterol 31 mg		10%
Sodium 396 mg		17%
Total carbohydrate 0 g		0%
Dietary fibre 0g		0%
Sugars 0 g		0%
Protein 27 g		
Vitamin A 23µg	Vitamin C__	
Calcium 9 mg	Iron 0.7 mg	

* Percent Daily Values are based on a 2000-calorie diet. Your Daily Values may be higher or lower depending on your calorie needs:

	Calories	2,000	2,500
Total fat	Less than	65 g	80 g
Sal fat	Less than	20 g	25 g
Cholesterol	Less than	300 mg	300 mg
Sodium	Less than	2400mg	2400 mg
Total carbohydrate		300 mg	375 g
Dietary fibre		25 g	30 g

Calories per gram:
Fat 9   Carbohydrate 4   Protin 4

- **Size up your dish**. Compare the amounts you will eat to the serving size given. For instance if you eat 150 g, you have to mulltiply by 1.5 the amounts of nutrients and calories listed.

- **Read the nutrition information**. Use the Percent Daily Value to compare the amount of sodium among brands. Choose those products that have lower values. One serving of this product contains 17%, or about 1/6 of the amount of sodium you should have for the entire day.

Fig. 13.4   How to read a seafood label.

13.8   The way forward

The international trade of seafood has increased spectacularly in recent years and more than 37% of fish products produced for human purposes is traded worldwide. European Union, United States and Japan transact more than 75% of global fish traded (FAO, 2004). This worldwide circulation of fish products has accelerated the transference of technology from other foods and within the fish industry and simultaneously contributed to a greater responsiveness regarding quality and safety. This has led the fish industry to produce more added value or convenient products associated with a healthy image. Several studies have shown a beneficial effect of fish consumption in the prevention of cardiovascular diseases, particularly due to their content in long-chain omega-3 polyunsaturated fatty acids (namely, eicosapentaenoic C 20:5ω3 and docosahexaenoic C 22:6ω3). Besides, most farmed and wild fish species presents quite low salt contents. In preserved and processed fish products the salt content is quite variable, ranging from less than 1% up to 30%. However, these heavy salted products are usually desalted before cooking or eaten in small amounts as appetizers or condiments.

On the other hand, consumers have become more conscious of the importance of a health-oriented diet, in particular concerning sodium intake. Consequently the consumers' demand for low-salt seafood products has increased and the industry is facing some pressure to reduce the actual levels. There is scope for reducing salt content in several fish products through the application of adequate strategies that encompass not only the reduction itself, but also the substitution of sodium chloride by other salts and additives or applying different methods or technologies. However, for scheduling a suitable strategy the involvement and co-operation of international and national health organisations, governments, industry, retailers and consumers is imperative.

The health organisations have strongly recommended a reduction in salt intake from the present levels to values around 6 g per day (Tilston et al., 1993; IFST, 1999; Gibson et al., 2000), based on extensive medical studies showing that the salt reduction in food can lower blood pressure, perhaps even more substantially than with pharmaceutical drugs. Nevertheless, such organisations have to be sufficiently dynamic to propose and co-ordinate programmes related to gradual salt reduction, based on modern research on salt health effects more than salt reduction strategies.

Regarding governments, their awareness on these issues, through the food agencies, supporting and financing programmes aimed at encouraging populations to eat low salt products, namely processed seafood products, is quite important.

Seafood processors have already developed some lower salt products, but pressure for further salt reduction is increasing. However, high salt reductions in processed seafood present limitations associated not only with the loss of flavour but also preservative effects, even though many technological developments might help to reduce or eliminate such dependence. On the other hand, some food producers have been opposed to the utilization of low levels of salt given that it contributes to increase the product yield. Moreover, some industry sectors are demanding stronger evidence relating salt and hypertension, despite all extensive medical research that has been developed, the several papers published and the re-analysis of many results.

Retailers have had different approaches, so, while some of them remain inactive or are responding slowly, a few, such as a number of the major retailers of the United Kingdom (Gibson et al., 2000), have answered positively the concerns related to high-salt content of food products.

Finally, consumers have huge responsibilities and play an important role, since they are the key element for the success of a food product. Therefore, they have to be informed and motivated to reduce the salt intake in their diets, and to persuade producers to reduce the current high salt content in processed foods and to provide more low-salt alternatives. Moreover, their commitment in demanding food salt labelling and their ability to read and understand food labels is of utmost importance.

13.9   Sources of further information and advice

Matters related to preserving, processing, nutritional value, safety, and quality assurance of fishery and farmed products have been published in several articles and publications, the books being edited during the 1960s by Borgstrom a reference in these areas. In parallel, the interest for reducing salt content in foods, including fish and shellfish, has gained special relevance. The following books and articles are recommended for further information.

1961–5. Borgstrom G (ed.), *Fish as Food*. Vols. I–IV. Academic Press New York.

1991.   Regenstein J M and Regenstein C E, *Introduction to fish technology*. Van Nostrand Reinhold, New York, 269 p.

1997.   Luten J B, Børresen T. and Oehlenschläger J. (eds.), *Seafood from producer to consumer, integrated approach to quality*. Elsevier Science B. V., Amsterdam, 712 p.

1997.   Shahidi F, Jones Y and Kitts D D (eds.), *Seafood safety, processing, and biotechnology*. Technomic Publishing Co., Inc., Lancaster, 266 p.

1997.   Pearson, A M, Dutson, T R (eds), *Production and processing of healthy meat, poultry and fish products*. Advances in Meat Research Series Vol. 11, Blackie Academic & Professional, London, 367 p.

2001.   Chair P, Busta F F, Bledsoe G E, Flick Jr G J, Gram L, Herman D, Jahncke M L, Ward D R, Processing parameters needed to control pathogens in cold-smoked fish, *J Food Sci*, 60 (Suppl.), 80 p.

2002.   Bremner H A (ed.), *Safety and quality issues in fish processing*. Woodhead Publishing Limited, Cambridge, England, 520 p.

2004.   Sakaguchi M (ed.), *More efficient utilization of fish and fisheries products*. Developments in Food Science Vol. 42, Elsevier Science & Technology Bookstore, Amsterdam, 478 p.

As the Internet has become a valuable and easy source of information, the following addresses about this matter are suggested in Table 13.3.

13.10   Acknowledgements

The authors wish to thank to Freshwater and Marine Image Bank as the source of the image in Fig. 13.1 and express our appreciation to Irineu Batista and Carlos Cardoso, from Instituto Nacional de Recursos Biológicos (INRB/IPIMAR), for the kind support provided during the preparation of this chapter.

Table 13.3   Internet addresses of interest

Subject	Address
British Nutrition Foundation	http://www.nutrition.org.uk
Consensus Action on Salt and Health (CASH)	http://www.actionsalt.org.uk
European Food Safety Agency	http://www.efsa.eu.int
Fish, Seafood and Production Section of	http://www.inspection.gc.ca/english/
Canadian Food Inspection Agency	anima/fispoi/fispoie.shtml
Fisheries Department of Food and Agriculture	http://www.fao.org/fi/default.asp
Organization of United Nations	
Information on fish species worldwide	http://www.fishbase.com
Nutrition facts, including food sodium content	http://www.nutritiondata.com/
Resource gateway for the fisheries and aquatic	http://www.onefish.org
research and development sector	
Role of omega-3 polyunsaturates from fish and	http://www.fish-foundation.org.uk
fish oils in nutrition and health	
Scientific Advisory Committee on Nutrition	http://www.sacn.gov.uk
US FDA/Center for Food Safety and Applied	http://www.cfsan.fda.gov/list.html
Nutrition	

13.11   References

ACMSF (ADVISORY COMMITTEE ON THE MICROBIOLOGICAL SAFETY OF FOOD) (1995), 'Vacuum Packaging and Associated Processes', Annex III from the Annual Report, London, http://www.foodstandards.gov.uk/dept_health/archive/acmsf.htm#pres, consulted on November 2005.

ANDRÉS A, RODRÍGUEZ-BARONA S, BARAT J M (2005), 'Analysis of some cod-desalting process variables', J Food Eng, 70, 67–72.

ANDRÉS A, RODRÍGUEZ-BARONA S, BARAT J M, FITO P (2002), 'Note: mass transfer kinetics during cod salting operation', Food Sci Technol Int, 8, 309–314.

AUBOURG S P AND UGLIANO M (2002), 'Effect of brine pre-treatment on lipid stability of frozen horse mackerel (Trachurus trachurus)', Eur Food Res Technol, 215, 91–95.

BANDARRA N M, CALHAU M A, OLIVEIRA L, RAMOS M, DIAS M G, BÁRTOLO H, FARIA M R, FONSECA M C, GONÇALVES J, BATISTA I, NUNES M L (2004), Composição e valor nutricional dos produtos da pesca mais consumidos em Portugal, Publicações Avulsas do IPIMAR, 11, 103 p.

BARAT J M, RODRÍGUEZ-BARONA S, ANDRÉS A (2004a), 'Modelling of cod desalting operation', J Food Sci, 67, 1922–1925.

BARAT J M, RODRÍGUEZ-BARONA S, CASTELLÓ M, ANDRÉS A, FITO P (2004b), 'Cod desalting process as affected by water management', J Food Eng, 61, 353–357.

BERLIN D L, HERSON D S, HICKS D T, HOOVER D G (1999), 'Response of pathogenic Vibrio species to high hydrostatic pressure', Appl Environ Microbiol, 65, 2776–2780.

BIRKELAND S, RØRA A M, SKARA T, BJERKENG B (2004), 'Effects of cold smoking procedures and raw material characteristics on product yield and quality parameters of cold smoked Atlantc salmon (Salmo salar L.) fillets', Food Res Int, 37, 273–286.

BJØRKEVOLL I, OLSEN J-V, OLSEN RL (2004), 'Rehydration of salt-cured cod using injection and tumbling technologies', *Food Res Int*, 37, 925–993.

BLEDSOE G E, BLEDSOE C D, RASCO B (2003), 'Caviars and fish roe products', *Crit Rev Food Sci Nutr*, 43 (3), 317–356.

BLIGH E G (1980), 'Methods of marketing, distribution and quality assurance' In Connell J J *Advances in fish science and technology*, Farnham, England, Ed. Fishing News Books Ltd, 48–55.

BYUN M W, LEE K H, KIM D H, KIM J H, YOOK H S, AHN H J (2000), 'Effects of gamma radiation on sensory qualities, microbiological and chemical properties of salted and fermented squid', *J Food Protect*, 63 (7), 934–939.

CHIRALT A, FITO P, BARAT J M, ANDRÉS A, GONZÁLEZ-MARTÍNEZ C, ESCRICHE I, CAMACHO M M (2001), 'Use of vacuum impregnation in food salting process', *J Food Eng*, 49 (1/2), 141–151.

CONNELL J J (1980), *Advances in Fish Science and Technology*, Farnham, England, Fishing News Books Ltd, 138 p.

CRAWFORD M A, CUNNANE S C, HARBIGE L S (1993), 'A new theory of evolution: quantum theory' in Sinclair A J, Gibson R, *Proceedings of the third international congress on essential fatty acids and eicosanoids*, Adelaide, AOCS Press, 87–95.

DELAHUNTY, C M, PIGGOTT J R (1995), 'Current methods to evaluate contribution and interactions of components to flavour of solid foods using hard cheese as an example', *Int J Food Sci Technol*, 30 (5), 557–570.

DOYLE E, 'Foodborne Parasites. A Review of the Scientific Literature', http://www.wisc.edu/fri/briefs/parasites.pdf, consulted on November 2005.

FAO YEARBOOK (2004), 'Fishery statistics', Vol. 95. FAO, Rome, 224 p.

FSA (FOOD STANDARDS AGENCY) (2005), 'Salt in processed food', http://www.food.gov.uk/healthiereating/salt/saltmodel, consulted on November 2005.

GIBSON J, ARMSTRONG G, McILVEEN H (2000), 'A case for reducing salt in processed foods', *Nutr and Food Sci* 30 (4), 164–173.

GILLETTE M (1985), 'Flavor effects of sodium chloride', *Food Technology*, 56, 3947–3952.

GOMÉZ-GUILLÉN M C AND MONTERO P (1997), 'Improvement of giant squid (*Dosidicus gigas*) muscle gelation by using gelling ingredients', *Z Leb Unt For A*, 204, 379–384.

GOU L, GUERRERO L, GELABERT J, ARNAU J (1996), 'Potassium chloride, potassium lactate and glycine as sodium chloride substitutes in fermented sausages and in dry-cured pork loin', *Meat Sci*, 42 (1), 37–48.

GRAM L (2001a), 'Potential Hazards in Cold-Smoked Fish: *Clostridium botulinum* type E', *J Food Sci*, 60 (7), Suppl.: S1082–S1087.

GRAM L (2001b), 'Potential Hazards in Cold-Smoked Fish: *Listeria monocytogenes*', *J Food Sci*, 60 (7), Suppl.: S1072–S1081.

HURTADO J L, MONTERO P, BORDERIAS A J (2000), 'Extension of shelf life of chilled hake (*Merluccius capensis*) by high pressure', *Food Sci Technol Int*, 6 (3), 243–249.

HUTTON T (2002), 'Sodium: Technological functions of salt in the manufacturing of food and drink products', *Br Food J*, 104, 2126–2152.

IFST (INSTITUTE OF FOOD SCIENCE AND TECHNOLOGY) (1999), 'IFST: Current Hot Topics – Salt, Position statement', http://www.ifst.org/hottop17.htm, consulted on November 2005.

IOM (INSTITUTE OF MEDICINE – UNITED STATES) (2004), 'Dietary reference intakes for water, potassium, sodium, chloride and sulphate', *Panel on dietary reference intakes for electrolytes and water, Standing committee on the scientific evaluation of dietary reference intakes*, http://www.nap.edu/execsumm_pdf/10925.pdf, consulted on November 2005.

JE J, PARK P, JUNG W, KIM S (2005), 'Amino acid changes in fermented oyster (*Crassostrea gigas*) sauce with different fermentation periods', *Food Chem*, 91, 15–18.

JO C, KIM D H, KIM H Y, LEE W D, LEE H K, BYUN, M W (2004a), 'Studies on the development of low-salted, fermented, and seasoned *Changran Jeotkal* using the intestines of *Therage chalcogramma*', *Radia Phys Chem*, 71, 121–124.

JO C, LEE W D, KIM D H, AHN H J, BYUN M W (2004b), 'Quality attributes of low salt *Changran Jeotkal* (aged and seasoned intestine of Alaska Pollock, *Theragra chalcogramma*) developed using gamma irradiation', *Food Control*, 15, 435–440.

KAWANO AND NOBUHISA (2000), 'Process for production of ground fish meat products or their analogues', Issued November 14th, 2000, US Patent 6,146,684.

KINGSLEY D H, HOOVER D G, PAPFRAGKOU E, RICHARDS G P (2002), 'Inactivation of hepatitis A virus and a calicivirus by high hydrostatic pressure', *J Food Protect*, 65, 1605–1609.

LEE C-H (1997), 'Lactic acid fermented foods and their benefits in Asia', *Food Control*, 8 (5/6), 259–269.

LEE K H, AHN H J, JO C, YOOK H S, BYUN M W (2002), 'Production of low salted and fermented shrimp by irradiation process', *J Food Sci*, 67, 1772–1777.

LÖLIGER J (2000), 'Function and Importance of Glutamate for Savory Foods', *J Nutr*, 130 (4), 915S–920S.

MacGREGOR G A, SEVER P S (1996), 'Salt-overwhelming evidence but still no action: can a consensus be reached with the food industry?', *BMJ*, 312, 1287–1289.

MARTÍNEZ-ALVAREZ O, BORDERÍAS A J, GÓMEZ-GUILLÉN M C (2005), 'Sodium replacement in the cod (*Gadus morhua*) muscle salting process', *Food Chem*, 93, 125–133.

MOTOKI M AND SEGURO K (1998), 'Transglutaminase and its use for food processing', *Trends Food Sci Technol*, 9, 204–210.

MURCHIE L W, CRUZ-ROMERO M, KERRYB, J P, LINTONC M, PATTERSONA M F, SMIDDYB M, KELLY A L (2005), 'High pressure processing of shellfish: a review of microbiological and other quality aspects', *Innov Food Sci and Emerg Technol*, 6, 257–270.

NAKAMURA K, KURAMITU R, KATAOKA S, SEGAWA D, TAHARA K, TAMURA M, OKAI H (1996), 'Convenient synthesis of L-ornithyltaurine HCl and the effect on saltiness in a food material', *J Agric Food Chem*, 44, 2481–2485.

OHSHIMA T, USHIO H, KOIZUMI C (1993), 'High pressure processing of fish and fish products', *Trends Food Sci Technol*, 4, 370–375.

ORTIZ J AND AGUILERA J M (2004), 'Effect of kappa-carrageenan on the gelation of horse mackerel (*T. Murphyi*) raw paste surimi-type', *Food Sci Technol Int*, 10 (4), 223–232.

PASTORIZA L, SAMPEDRO G, HERRERA J J, CABO M L (1998), 'Influence of sodium chloride and modified atmosphere packaging on microbiological, chemical and sensorial properties in ice storage of slices of hake (*Merluccius merluccius*)', *Food Chem*, 61 (1/2), 23–28.

PELROY G A, EKLUND M W, PARANJPYE R N, SUZUKI, E M, PETERSON M E (1982), 'Inhibition of *Clostridium botulinum* types A and E toxin formation by sodium nitrite and sodium chloride in hot-processed (smoked) salmon', *J Food Protect*, 45 (9), 833–841.

PEDRO S, MAGALHÃES N, ALBUQUERQUE M M, BATISTA I, NUNES M L, BERNARDO F M (2002), 'Preliminary observations on spoilage potential of flora from desalted cod (*Gadus morhua*)', *J Aqua Food Prod Technol*, 11 (3/4), 143–150.

PEDRO S, ALBUQUERQUE M M, NUNES M L, BERNARDO F M (2004), 'Pathogenic bacteria and indicators in salted cod (*Gadus morhua*) and desalted products at low and high temperatures', *J Aqua Food Prod Technol*, 13 (3), 39–48.

RAMÍREZ J, URESTI R, TELLEZ S, VÁZQUEZ M (2002), 'Using salt and microbial transglutaminase as binding agents in restructured fish products resembling hams', *J Food Sci*, 67 (5), 1778–1784.

RODRIGUES M J, HO P, LÓPEZ-CABALLERO M E, BANDARRA N M, NUNES M L (2005), 'Chemical, microbiological, and sensory quality of cod products salted in different brines', *J Food Sci*, 70 (1), M1–6.

RØRA A M, FURUHAUG R, FJÆRA S O, SKJERVOLD P O (2004), 'Salt diffusion in pre-rigor filleted Atlantic salmon', *Aquaculture*, 232, 255–263.

RUUSUNEN M, PUOLANNE E (2005), 'Reducing sodium intake from meat products', *Meat Sci*, 70, 531–541.

SAINCLIVIER M (1985), *L'industrie alimentaire halieutique: Des techniques ancestrales a leurs réalisations contemporaines*, Rennes, ENSA, 366 p.

SAISITHI P (1994), 'Traditional fermented fish: fish sauce production'. In: *Fisheries processing*. Martin AM (ed), Chapman & Hall, London, 111–131.

SACN (SCIENTIFIC ADVISORY COMMITTEE ON NUTRITION) (2003), 'Salt and Health Report', http://www.food.gov.uk/multimedia/pdfs/sacnforumreport.pdf, consulted on November 2005.

SHAW R (1997), 'Cook-chill meals: opportunities for MAP', *Food Review,* 24 (4), 23–31.

SHEPHARD R, FARLEIGH C A (1986), Attitudes and personality related to salt intake', *Appetite*, 7, 343–354.

SIKORSKI Z E, KOLODZIEJSKA I (2002), 'Microbial Risks in Mild Hot Smoking of Fish', *Crit Rev Food Sci Nutr*, 42 (1), 35–51.

SMELT J P P M (1998), 'Recent advances in the microbiology of high pressure processing', *Trends Food Sci Technol*, 9, 152–158.

TéLLEZ-LUIS S J, URESTI R M, RAMÍREZ J A AND VÁZQUEZ M (2002), 'Low-salt restructured fish products using microbial transglutaminase as binding agent', *J Sci Food Agric*, 82 (9), 953–959.

TILSTON C, NEALE R, GREGSON K AND BOURNE S (1993), *Salt: a challenge to food manufacturers*, Food Marketing Research Group, University of Nottingham, Horton Publishing, Bradford, 5 p.

TORRES J A, VELÁSQUEZ G (2005), 'Commercial opportunities and research challenges in the high pressure', *J Food Eng*, 67 (1/2), 95–112.

URESTI R M, TÉLLEZ-LUIS S J, RAMÍREZ J A, VÁZQUEZ M (2004), 'Use of dairy proteins and microbial transglutaminase to obtain low-salt fish products from filleting waste from silver carp (*Hypophthalmichthys molitrix*)', *Food Chem*, 86, 257–262.

VALDIMARSSON, G, JAMES, D (2001), 'World fisheries utilisation of catches', *Ocean & Coastal Management*, 44, 619–633.

VOISIN E (2001), 'Process of elimination of bacteria in shellfish and of shucking shellfish', Issued April 17th, 2001, US Patent 6,217,435.

VOISIN E (2002), 'Process of elimination of bacteria in shellfish and of shucking shellfish', Issued July 30th, 2002, US Patent 6,426,103.

VOSKRESENSKY N A (1965), 'Salting of herring', In Borgstrom G, *Fish as Food* Vol III,

New York, Academic Press, 489 p.

WOSKOW M H (1969), 'Selectivity in flavor modification by 5'nucleotides', *Food Technol*, 23, 1364.

YAMAGUCHI S AND TAKASHASHI C (1984), 'Interaction of MSG and NaCl on saltiness and palatability of clear soups', *J Food Sci*, 49, 82.

14

Reduced salt in bread and other baked products

S. P. Cauvain, BakeTran, UK

14.1 Introduction

The term 'baked products' is applied to a wide range of food products which includes breads, cakes, pastries, cookies and crackers and many other uniquely named products. The most commonly identified link between the different product types is that they all use recipes which are based on wheat flour. This definition may be expanded to include wheatless breads (for example, the gluten-free breads associated with the coeliac digestive disorders) which are still considered to be baked products even though they are based on cereals other than wheat. Baked products are commonly defined by having undergone some form of heat processing – usually baking – which causes changes in both ingredient chemistry and physical form and structure. Within the overall heading of baked products there are a number of 'sub-groups', each of which has distinctly different natures defined by a combination of recipe, process and baked form. The individual sub-groups are broadly spread so that overlapping does occur, nevertheless there are some key differences and these are summarised in Table 14.1.

A key feature which distinguishes baked products from other cooked foods is the formation, to varying degrees, of a cellular-like structure in the baked product which confers a specific eating character to the products. This cellular structure arises from the incorporation of air and the subsequent retention of the air bubbles within the unbaked matrix. Later the gas bubble structure so created will be expanded as supplementary gases, usually carbon dioxide, are introduced during processing. In the early stages of baking the 'foam-like' bubble structure is disrupted and an open, 'sponge-like' structure is formed in the baked products.

Table 14.1 A broad classification of bakery products

	Bread	Cakes and sponges	Biscuits, cookies and crackers	Pastry
Main ingredients	Flour, water, yeast and salt	Water, sugar, flour, fat, egg, baking powder, salt	Flour, sugar, fat, water, baking powder, salt	Flour, fat, water, salt
Unbaked form	Dough – air bubbles trapped in a gluten matrix	Batter – air bubbles trapped by fat in an aqueous medium	Paste – relatively dense mixture with little trapped air	Paste – relatively dense mixture with little trapped air
Special processing	Development of gluten network by mixing (energy) and expansion on fermentation	Mixing to incorporate necessary gas bubble nuclei	Sheeting or moulding of paste to create required shape	Sheeting or blocking of paste to required shape before insertion of filling
Key aeration mechanisms	Incorporation of air, expansion through carbon dioxide gas from yeast fermentation	Incorporation of air, expansion through carbon dioxide gas from baking powder reaction	Generation of water vapour during baking	Generation of water vapour during baking
Key baked characteristics	Hard, crisp crust, soft resilient crumb with cellular structure and chewy eating quality	Crumb with cellular structure and friable, moist eating character	Dense structure with low moisture content and hard, flinty eating character	Dense structure with flinty eating character

14.2   The technological functions of salt in the processing of baked products

Salt has become a common ingredient in baked products and contributes a number of sensory and technological functions. The sensory function is mostly related to flavour but there is more than one technological function which needs to be considered. The main technological functions of salt in baked products may be summarised as follows:

- impacting the development of gluten structures in the mixing of bread and other fermented products,
- inhibition of bakers' yeast in the fermentation of bread doughs,
- control of water activity in the baked product.

In the manufacture of bread and other fermented products it is the development of an extensible gluten network in the dough which is an essential feature of the process. The gluten network contributes to the entrainment of air bubbles in the dough and is an integral factor in retaining them during the fermentation and baking stages (Cauvain, 2000). The gluten network begins to form following the hydration with water of the gluten-forming proteins (the glutenins and the gliadins) present in wheat flour and is further developed by any mixing action which imparts energy to the flour-water mixture. During the whole of the mixing process there is a considerable change in the rheological properties of the flour-water mixture. After first being formed, the mixture is 'tough' and lacks extensibility. As mixing proceeds the resistance of the mixture to being worked increases but eventually reaches a maximum. As this stage approaches the flour-water dough develops a coherent form which is extensible in nature while retaining some elasticity. Further mixing leads to a reduction in dough resistance but an increase in dough extensibility until eventually the mass begins to develop a sticky feel. This is the point at which the dough structure begins to 'break down' and lose its coherence.

These changes in dough properties during mixing are crucial in the understanding of the control of bread quality. Much research work has been devoted to the subject and the reader is referred elsewhere for a more detailed understanding of the dough-making process (e.g., Cauvain and Young, 1998). Overall the effect of increasing mixing time is to increase loaf height (one indication of bread quality) and it is clear that gluten development is a major factor in determining bread quality (Fig. 14.1).

In practice bakers seldom work with a flour-water mixture to make bread. There are some exceptions but commonly bakers add salt, yeast and other functional ingredients to make bread. Thus, hydration and gluten development take place in the presence of a brine solution. The ionic nature of salt means that it readily combines with the water molecule dipole (Cauvain and Young, 2000) and, as such, will restrict the availability of water for the development of a gluten structure. This can be seen as changes in dough rheology with increasing levels of salt addition (Linka et al., 1984). In general the overall effect is small

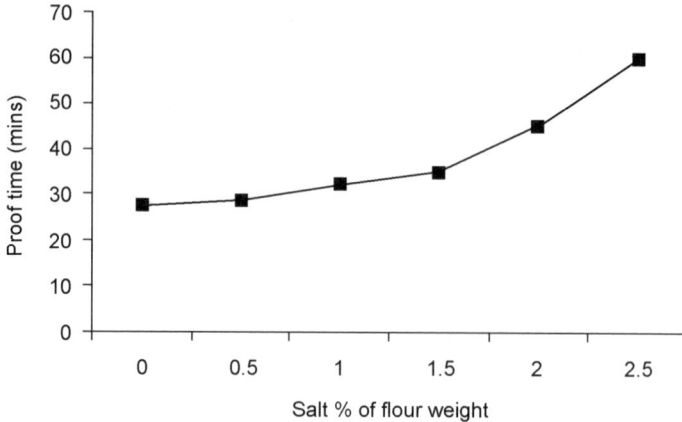

Fig. 14.1   Effect of salt level on yeast activity (proof time to constant dough height in the pan).

compared with the dough development that comes from the energy transferred to the dough during mechanical mixing. The impact of salt on dough rheology is more noticeable with slow mixing and especially if mixing is carried out by hand. In the case of flours which have an inherently poor gluten-forming potential, it is commonplace to delay the addition of salt in order to more readily exploit the potential of the gluten-forming proteins that are present. The readily soluble nature of salt means that it may be quickly incorporated in the last few minutes of mixing when despite its affinity for water, is unable to 'undo' gluten development in the dough to any extent.

Since a technological function of salt is related to its impact on gluten formation it is reasonable to assume that it has little effect in those products which lack significant gluten formation. Thus the impact of salt is most readily observed in bread, fermented goods such as rolls and buns, laminated products such as croissant, Danish and puff pastry. In biscuits, cookies, many pastry products, cakes and sponges gluten formation is limited or absent and there appears to be no significant effect of salt.

The impact of salt on yeast fermentation is well known (Williams and Pullen, 1998) and has been exploited by bakers for many years in the formulation of bread and other fermented products by balancing the levels of these two key functional ingredients. The overall effect of increasing levels of salt is to slow down yeast fermentation. This effect is critical in the closely timed fermentation processes which are used in modern bakers. In particular most bakeries work to a 'fixed' proof time, that is a fixed fermentation period after the dough pieces have been moulded to shape and before they enter the oven. This period is important in bread making because it allows the dough piece to increase in volume which, along with modification of the rheological properties of the developed gluten network present, allows the dough piece to expand uniformly during the baking process in the oven. Once the appropriate proof time has been established it is important that the baker keeps to that time as under- and over-proof leads to loss of product quality. The impact of changing salt levels on dough with a fixed level of yeast is illustrated in Fig. 14.1. The variation in proof time which is illustrated indicates the length of time that the dough piece required in order to achieve a constant volume or height in the pan. The higher the level of salt added to the recipe longer the proof time required for the yeast to produce the required level of gas to achieve a fixed volume or height.

14.3   Control of water activity in baked products

The control of water activity in baked foods is critical to both product quality and safety. Bread and cakes are considered as intermediate moisture foods with moisture contents typically ranging from 18 to 42% (cakes 18 to 25% and bread 35 to 42%). The water activity of the products, expressed as Equilibrium Relative Humidity (ERH), ranges from 70 to 95% (cakes 70 to 88% and bread 88 to 95%). With such moisture contents and ERHs baked goods are susceptible to microbial spoilage with the main effect coming from the growth of various moulds though bacterial growth is possible at the high end of the ERH range.

A key limiting factor in the development of microbial spoilage from all three sources is the ERH of the product. The lower the ERH the longer that it takes for a baked product to exhibit mould or bacterial spoilage (Cauvain and Young, 2000). While moisture is a significant contributor to product ERH, other materials impact on the property of the food. In this context the addition of ionic materials like salt can have a profound effect on the spoilage-free shelf-life of the product with higher levels of added salt lowering the ERH and extending the spoilage-free life of the product.

Salt is particularly effective at extending the spoilage-free shelf-life of food products. It is this concept that is behind the salting of meats to avoid bacterial growth. The effectiveness of salt in the lowering of baked products' ERH can be appreciated by comparing its effects with those of sucrose, another well-known 'preserving' ingredient (e.g., jam making). Using the concept of 'sucrose equivalents' (se) developed by Grover (1947) we would observe that salt has a se of 11 compared with sucrose as 1 (Cauvain and Young, 2000). This means that for equal weights of salt and sugar, salt will be 11 times more effective than sucrose at lowering product ERH. Because of this property salt has become a significant ingredient in the recipes used for extending the spoilage-free life of baked products.

In the past the focus of concern with bakery products has largely been on the growth of moulds. This is because when baked products leave the oven the surface of the product is essentially sterile. It is also the case that the ERH and moisture content of the surface (crust) of baked products is somewhat lower than that at the centre of the product. With time moisture moves from the moist centre to the drier crust and the product equilibrates. During this period the ERH of the product surface is too low to support bacterial growth and so any problems associated with post-baking microbial contamination are largely confined to moulds.

There are some exceptions, such as crumpets, which with their higher ERH can be susceptible to bacterial spoilage very soon after manufacture. The other, less well appreciated area of potential bacterial spoilage, occurs in the centre of bread products. Here ERHs are much higher than at the crust and temperatures during baking stay below 100°C. Typically the core temperature of bread products reaches 92 to 96°C by the end of baking. Under these heating conditions spore-forming bacteria, such as *Bacillus subtilis* or *Bacillus licheniformis*, can survive the baking process and if the subsequent storage conditions are appropriate, can revive and spoilage may occur. The spoilage is manifest in the early stages by a distinctive sweet, pineapple-like odour with the later stages characterised by liquefaction of the crumb and the formation of strands when the bread is cut which give the spoilage its popular name 'rope'. The spoilage bacteria are present in the field and are carried through the flour milling process on the wheat grains. Usually rope is not a problem in modern bakeries because of the good hygiene practices which they exercise but the ERH of the bread crumb is marginal for rope growth and any factor which potentially reduces the ERH 'hurdle' (e.g., lower salt levels) may increase the risk of rope growth.

14.4   Other sodium salts in baked products

Sodium salts other than the chloride form are found in baked products. The main uses for these sodium salts are as microbial spoilage inhibitors or as components of baking powders. Spoilage inhibitors usually take the form of salts of the organic acids such as propionic, sorbic and acetic. While the calcium or potassium salts are commonly used the sodium salts do find some uses, the most common being sodium acetate. The level of these inhibitors is strictly regulated (Williams and Pullen, 1998) and so they make relatively small contributions to the sodium levels in the diet.

The main source of carbon dioxide in baking powder used in the manufacture of baked products comes from the addition of sodium bicarbonate. The bicarbonate is combined with a suitable food grade acid to release the carbon dioxide. Baking powders are mostly used to supplement gas expansion in cakes, biscuits, cookies and pastries, though it may find use in some bread type products. Sodium salts of organic acids have become the most common form of acid used in baking powder. This is because the relatively wide range of forms that may be manufactured can be used to control the rate of reaction with the baking powder in the baked product system. Typical salts which may be encountered are sodium aluminium phosphate (SALP) and sodium acid pyrophosphate (SAPP).

The residual compounds, which are left when the baking powder reaction has been completed, confer characteristic flavours to the final product. The 'phosphate after-taste' does not find favour with many consumers but while alternative acids may be used (e.g., citric, tartaric), control of the rate of reaction with the sodium bicarbonate is more difficult to achieve. Potassium salts are occasionally used in baking powders but usually they contribute only a small part of the necessary carbon dioxide.

14.5   Levels of salt in baked products and targets for reduction

A brief survey of various published baked product recipes reveals that salt levels in UK formulations vary considerably depending on the product and the environment in which it is manufactured. Precise data cannot be given for many products since commercially sensitive recipes are not published. However, a number of examples are given in Table 14.2 to illustrate the range of variation which might be encountered.

For the levels of sodium chloride in baked products as summarised in Table 14.2 two key points emerge. First, it is difficult to make a direct comparison across product types since there appears to be limited commonality in the actual levels of sodium chloride used. In part this arises because other than for UK bread recipes there is too much variation in ingredients and formulations used, e.g. there is no such thing as a 'standard' cake recipe. The wide variation perhaps also reflects the different functions in baked products for sodium chloride discussed above. It is equally difficult to compare UK products with

Table 14.2   Sodium chloride levels in baked products

Product and manufacturing environment	Approximate level of sodium chloride per 100 g baked product*	Typical level of sodium chloride in 100 g dough or batter	Typical level of sodium chloride compared with 100 g flour
UK pan bread	1.15–1.25	1.08–1.19	1.8–2.0
UK crusty breads	1.15–1.34	1.08–1.27	1.8–2.1
UK soft rolls	1.00–1.12	0.94–1.11	1.6–1.9
UK fruit buns	0.72–0.86	0.68–0.80	1.6–1.9
UK plain cake	0.39–0.52	0.37–0.49	1.5–2.0
UK fruited cake	0.32–0.42	0.29–0.39	1.5–2.0
US pan bread[#] (Cauvain, 1998b)	1.04	0.99	2.3
French baguette (Calvel, 2001)	1.23	1.17	2.0

* Absolute levels of salt will vary according the actual moisture content of the product concerned. The values given for a product do not take into account the effects of any fillings or toppings which might come into contact with the products quoted (e.g., icing on a fruit bun) since there will be significant variation in the use of such materials.
[#] US pan bread recipes contain sugar, milk solids and more fat than UK bread and the presence of these ingredients depresses the salt level in the baked product.

those manufactured in other parts of the world though a few examples have been given in Table 14.2.

To some extent the data given in Table 14.2 may not represent the true picture for current sodium chloride levels in UK baked products. As already indicated, the levels of added sodium chloride are taken from published sources and are not derived from commercially sensitive sources. Over the last 15 years or so the level of sodium chloride present in UK baked products has fallen (Federation of Bakers, 2005). Thus, the values given in Table 14.2 may be an over-statement of current levels. This is most likely to be true for UK breads which have been the subject of scrutiny by those sectors of the medical community which consider that a reduction of salt in bread would make a significant contribution to the level of sodium chloride in the average diet.

Bakers have reacted positively to such requests and have made successive reductions in the levels of sodium chloride in their products both across the board and within particular product groups. The current target for salt levels in bread stand at 0.9 g per 100 g product by 2010 (Food Standards Agency, UK, 2005) though levels as low as 0.5 g per 100 g have been proposed (Anon., 2005a) The overall view is that current levels are at around 1.23 g per 100 g product. Thus, there is still a gap between the proposed and the current levels of salt in bread.

The levels of sodium chloride and other sodium salts in flour confectionery and biscuits are even more widely varied than those encountered in bread. Given that such products contribute relatively less to sodium in the diet by virtue of the lower levels of consumption, they have not attracted the same level of interest and pressure for reductions. However, bakers have reacted to the levels of concern expressed by consumers and the medical profession and there has undoubtedly been reduction in the levels of sodium salts where this does not adversely affect important aspects of product quality.

14.6   Methods to reduce salt and sodium levels while retaining quality and safety

14.6.1   Reducing salt levels

In the manufacture of bread the replacement of sodium chloride with potassium chloride does not appear to offer any significant production disadvantages. The impact of potassium chloride on yeast activity and dough rheological properties is similar to that of the sodium salt. Thus, the major disadvantage of using potassium chloride to replace sodium is the adverse impact on flavour once the substitution levels rise beyond 10 to 20% of the sodium chloride. Potassium chloride confers a distinctive 'metallic' taste which cannot be readily disguised from other sources. A number of products are on the market which enable a reduction in sodium chloride. The most well know being Lo-salt[TM] which combines potassium and sodium salts in the ratio of 2:1 (Klinge Chemicals, 2005). The flavour impact of the potassium salt becomes more noticeable as the level of sodium chloride is reduced. This means that when the strategy is to reduce the overall level of sodium chloride used then the opportunities for using potassium chloride are also reduced.

It is clear that sodium chloride is an important part of bread flavour and that reductions in its level of addition will have an adverse effect on consumers' perception on the flavour-fullness of bread. It would be unfortunate if this was to lead to a reduction in the consumption of bread products since they contribute many positive benefits to a healthy diet, for example, fibre – especially wholemeal (Katina, 2003), calcium, iron and vitamins (Rosell, 2003). Bread flavour is highly individualistic and has always been a great subject of debate. In addition to the salt there are two other sources of bread flavour; the baking process through the formation of coloured crust from the Maillard reaction (Wirtz, 2003) and the development of acidity in the crumb through fermentation. One means of increasing the perception of flavour in bread products would be to increase the ratio of crust to crumb in the loaf. In part, this is why French baguette is perceived as having more flavour than the UK sandwich loaf (Cauvain, 1998a). However, while it is perfectly possible to increase the crust formation on a UK sandwich loaf, it will certainly be to the detriment of the other characteristics required by consumers, namely, softness and the retention of crumb softness through spoilage-free life.

14.6.2   Using fermentation*
Given the above comments then one of the options for the baker is to 'replace' the flavour of salt with those acid flavours which are developed during fermentation. This will increase the flavour-fullness of bread but it will mean a change in bread flavour since the source of the flavour will also be different. The type of flavour which would be developed would depend on the control of the fermentation and the types of microorganism used. Both time and temperature play crucial roles in the development of acid flavours in bread. In general, the flavour will become more intense with longer fermentation times. However, since the pH of any natural fermentation of a flour-water mix falls with increasing time, this has the effect of slowing down and even stopping the fermentation process. In practice this means that fermentation times are usually kept below 24 hours. The more common fermentation systems designed to change bread flavour are described in detail by Cauvain (1998b). In summary they are:

- Fermentation of the bulk dough after it leaves the mixer for periods of time between 1 and 8 hours, commonly 2 to 4 hours. This process is commonly described as 'bulk fermentation'.

---

*   The term 'fermentation' is used by bakers to (mainly) refer to the action of bakers' yeast (*Saccharomyces cerevisiae*) which produces carbon dioxide gas through the reaction with sugars present in the mix. Some confusion arises because it has become common practice to refer to fermentation which takes place after the dough units have been moulded as 'proof'. This is still fermentation but commonly takes place at higher temperatures than other periods of fermentation, e.g. after mixing or between moulding steps (Cauvain, 2000).

- Fermentation of part of the dough ingredients (including flour) for up to 16 or 24 hours before they are added to the remainder of the dough ingredients for mixing and processing. This process is commonly described as 'sponge and dough'. The sponge will have a similar consistency to that of the dough. In the USA the sponge only ferments for around 4 hours before the addition of the rest of the ingredients.
- Fermentation of part of the dough ingredients in a matrix which is considerably softer than that of the final dough. Such processes may be referred to as 'brews' or 'ferments' and need not contain any flour during the initial fermentation steps.

As already noted, temperature during fermentation plays a critical role in determining flavour development. In wheat flour large numbers of bacteria are present. These are generally harmless in the breadmaking system. With longer fermentation times they can make a contribution to the flavour of the bread. Groups of lactic acid bacteria are the ones which dominate and therefore contribute most to flavour. Their contribution is significantly affected by the fermentation temperature and is greater when the fermentation temperature is lowered. In general lactic acid bacteria contribute to 'sour' flavour notes. Dried sours may be prepared or purchased from specialist suppliers to influence bread flavour. Such products minimise the problems associated with controlling fermentation in the bakery.

Lowering the pH of bread has an additional benefit in that the lower pH contributes to the inhibition of microbial activity and so makes a contribution to restricting the potential development of the rope-forming bacteria discussed above.

14.6.3   Using flavoured ingredients

There are a number of different materials commonly encountered in breadmaking which can contribute flavour to bread. These include wheat bran, wheatgerm and other cereal flours and brans. The use of such materials has become increasingly common not necessarily as a replacement for salt but in response to the desire to have more flavour-full bread. There have been two main contributors to these developments. One has been the desire to provide increased fibre in breads so that they would appeal more readily to children than appears to be the case with wholemeal bread. This has been achieved through the addition of so-called 'white fibres' from different sources (Katina, 2003).

The second development was perhaps less expected. There has been a dramatic increase in the volume of ready-prepared sandwiches consumed in the UK which has resulted in the manufacture of around 1.8 billion sandwiches (1 sandwich = 2 slices of bread) per year exclusively for that purpose (British Sandwich Association, 2005). It is common practice to keep the sandwiches refrigerated between 4 and 8°C in order to limit risks associated with the growth of spoilage organisms. This temperature range is the one at which bread stales fastest (Pateras, 1998). It is also often the case that the sandwiches may be eaten relatively cold and this reduces the perception of flavour in the bread material. This effect is most noticeable when the bread used is white and less so when other forms of bread are used. Indeed non-white bread now represents the largest proportion of bread used for sandwich making. The same principles of manipulating flavour could be used to compensate for the reduction of salt flavour.

14.6.4 Bread dough formulation and processing

One immediate change coming from a reduction in salt level would be increased activity from the bakers' yeast. This could be compensated for by a reduction in proof time but as the capacity of most provers in bakeries is linked with that of the ovens, the practical solution will be to reduce yeast level and maintain processing times as standard. In the early stages of baking there is a significant increase in gas production by the yeast. This is usually manifest as 'oven spring', that is the lifting of the initial crust which forms on the dough when subjected to the heat of baking. For many technical reasons bakers seek to control the oven spring and sodium chloride plays a part in that controlling mechanism. The lowering of salt levels will lead to increased oven spring and this will need to be controlled by other means. Commonly bakers do this by reducing the contribution of other materials to the gas retention properties of the dough through formulation adjustment because there are no satisfactory processing options.

As noted above, sodium chloride has an impact on the rheological properties of dough and so any reduction may have to be compensated for in the dough-processing plant. Usually the rheological properties of the dough are finely balanced with the mechanical functioning of the plant. All formulation changes have some impact on the rheological properties of the dough and will affect the dough-processing plant interactions. Thus, adjustment of the plant processing parameters may be necessary to satisfactorily process reduced-salt doughs. In some cases equipment re-design may be needed to be able to handle the new dough rheological properties.

14.6.5 Reducing the levels of other sodium salts in non-bread products

The potential for restricting the levels of sodium salts used in non-bread products is limited because in many cases they provide a contribution to product structure formation which cannot be readily matched from other sources, e.g. generation of carbon dioxide from baking powder. Calcium phosphate salts may be used as partial or complete replacement of their sodium equivalent but as noted previously the rates of reaction are different. Most difficult to replace is sodium bicarbonate. The potassium salt may be used but its reaction profile is quite different and the residual flavour unlikely to be acceptable.

Sodium chloride by virtue of its low sucrose equivalent makes a significant contribution to the mould-free shelf-life of cake products (Cauvain and Young, 2000). Its reduction or removal is possible but may require the addition of greater quantities of mould inhibitors to maintain product shelf-lives at existing levels.

14.7   Future trends

There is no doubt that the attention of consumers and governments will remain focused on the contribution of individual foods to a healthy diet and lifestyle for a long, long time. It has always been the case. The contribution of baked foods has become an important part of the 'salt' debate. There has been a considerable focus on bread because of the central role that it plays in the diet of many consumers. The UK baking industry has responded to concerns of consumers and government by making reductions in salt levels used in bread and no doubt will continue to do for as long as positive heath benefits can be demonstrated.

There has been a gradual decline in bread consumption since the 1950s in the UK as a result of changing lifestyles. There has also been a switch, to some degree, from white to non-white breads. Thus, bread has remained a significant contributor to fibre in the diet. A key to continuing to encourage the consumption of bread for desirable fibre in the diet is for the industry to deliver a product that the consumer will enjoy and buy again and again. One of the reasons for buying any baked and all other food products is that they should be enjoyable to eat. A key part of that enjoyment is that the food should be 'tasty'.

There is no value in forcing through lower salt levels in bread and baked goods if the consumer turns away from the product. Not least such a change will lead to a potential reduction in fibre consumption. The issue of taste is a highly individual one but as recent events have shown, making lower salt products in bread does not guarantee sales of the product, even when it is part of a branded product offering (Anon., 2005b).

The challenge for the UK baking industry is to meet the demands of sometimes conflicting interest groups. There is medical opinion that some sectors of the population would benefit by reducing salt levels as part of a controlled diet. The same opinion believes that reduced salt levels in bread can make a significant contribution to lowering the levels of salt in everyone's diet. There is no doubt that the UK baking industry can/has/will find the technological means by which to reduce sodium salts in its products, whether consumers will accept the resulting products only time will tell.

14.8   Sources of further information and advice

For background into the practice of baking:

CAUVAIN, S.P. and YOUNG, L.S. (1998) *Technology of Breadmaking*, Blackie Academic & Professional, London, UK.

CAUVAIN, S.P. (2003) *Breadmaking: Improving Quality*, Woodhead Publishing Ltd., Cambridge, UK,

MANLY, D. (2000) *Technology of Biscuits, Crackers and Cookies*, 3rd edn, Woodhead Publishing Ltd., Cambridge, UK.

STREET, C.A. (1991). *Flour Confectionery Manufacture*, Blackie & Son, Glasgow, UK.

For water control and product shelf-life:

CAUVAIN, S.P. and YOUNG, L.S. (2000) *Bakery Food Manufacture and Quality: Water Control and Effects*, Blackwell Science, Oxford, UK.

For a 'classical' view of breadmaking:
CALVEL, R. (2001) *The Taste of Bread,* Aspen Publishers Inc., Gaitherburg, MD, USA.

For industry views:
Food & Drink Federation, www.fdf.org.uk
Federation of Bakers, www.bakersfederation.org.uk

14.9   References

ANON. (2005a) Plant bakers resistant to FSA's salt proposals. *British Baker,* **203** (39), 3.
ANON. (2005b) Weak response causes RHM to drop Hovis Lower Salt. *British Baker,* **203** (39), 3.
BRITISH SANDWICH ASSOCIATION (2005) www.sandwich.org.uk
CALVEL, R. (2001) *The Taste of Bread,* Aspen Publishers Inc., Gaitherburg, MD, USA.
CAUVAIN, S.P. (1998a) Bread – the product. In S.P. Cauvain and L.S. Young (eds), *Technology of Breadmaking*, Blackie Academic & Professional, London, UK, pp. 1–17.
CAUVAIN, S.P. (1998b) Breadmaking processes. In S.P. Cauvain and L.S. Young (eds), *Technology of Breadmaking*, Blackie Academic & Professional, London, UK, pp. 18–44.
CAUVAIN, S.P. (2000) Breadmaking. In G. Owens (ed.), *Cereals Processing Technology*, Woodhead Publishing Ltd., Cambridge, UK, pp. 204–230.
CAUVAIN, S.P. and YOUNG, L.S. (1998) *Technology of Breadmaking*, Blackie Academic & Professional, London, UK.
CAUVAIN, S.P. and YOUNG, L.S. (2000) *Bakery Food Manufacture and Quality: Water Control and Effects*, Blackwell Science, Oxford, UK.
FEDERATION OF BAKERS (2005) www.bakersfederation.org.uk.
FOOD STANDARDS AGENCY, UK (2005) www.salt.gov.uk.
GROVER, D.W. (1947) The keeping properties of confectionery as influenced by its water vapour pressure. *Journal of the Society of Chemistry and Industry*, **66**, 201–205.
KATINA, K. (2003) High-fibre baking. In S.P. Cauvain (ed.), *Breadmaking: Improving Quality*, Woodhead Publishing Ltd., Cambridge, UK, pp. 487–499.
KLINGE CHEMICALS (2005) www.klinge-chemicals.com
LINKA, Y., HARKONEN, H. and LINKO, P. (1984) Sodium chloride in bread making technology. *International Symposium on Advances in Baking Science and Technology,* 27–28 Sept., Kansas State University, USA.
PATERAS, I. (1998). In S.P. Cauvain and L.S. Young (eds), *Technology of Breadmaking*, Blackie Academic & Professional, London, UK, pp. 240–295.
ROSELL, C.M. (2003) The nutritional enhancement of wheat flour. In S.P.Cauvain (ed.), *Breadmaking: Improving Quality*, Woodhead Publishing Ltd., Cambridge, UK, pp. 253–270.
WILLIAMS, A. and PULLEN, G. (1998) Functional ingredients. In S.P. Cauvain and L.S. Young (eds), *Technology of Breadmaking*, Blackie Academic & Professional, London, UK, pp. 45–80.
WIRTZ, R.L. (2003) Improving the taste of bread. In S.P. Cauvain (ed.), *Breadmaking: Improving Quality*, Woodhead Publishing Ltd., Cambridge, UK, pp. 467–486.

15

Reducing salt in snack products

P. Ainsworth and A. Plunkett, Manchester Metropolitan University, UK

15.1   Introduction

Historically salt has been used in foods for its preservative action as well as its distinctive flavour and ability to enhance the flavour of other ingredients. With a low cost and high availability it is still used extensively within the food industry.

Whilst salt is essential in our diet, its widespread use in processed foods has caused concern. This concern has arisen as a result of studies that have suggested a relationship between high sodium intake and increased blood pressure. A factor that has been linked to the incidence of stroke and coronary heart disease.

In the UK and North America changes in lifestyle and eating habits have seen a rise in the quantity of convenience foods consumed. In the case of salt consumption it has been found that more than 70% of dietary salt intake arises from processed foods. This has prompted groups in the UK and US to put pressure on the food industry to reduce salt levels.

The growth in the consumption of savoury snack foods has increased greatly over recent years. Many of these snacks are perceived as being high in salt and thus possibly contribute to the high salt intake. This chapter covers snack foods, the growth of the snack food market, and salt levels in snacks. It also extends to ways in which different savoury snack foods are manufactured, the function of salt in snacks and possible ways in which salt levels can be reduced.

15.2   Snack foods

15.2.1   Definition of snacks
It has been increasingly apparent over recent years that a growing proportion of the population no longer eats formal meals, but prefers to consume foods 'on the go'. The food industry has catered for this need with a wide variety of products designed to be eaten between meals. These products are referred to as snacks or snack foods. This definition of snack foods covers a wide range of products, but can be subdivided into sweet snacks such as cereal bars, breakfast cereals, chocolate bars, biscuits, etc., and savoury snacks, for example potato crisps. This chapter concentrates on the role of salt in savoury snacks, in particular crisps, pretzels, extruded products and popcorn.

15.2.2   History of snacks
Snack foods have their origins thousands of years ago, but it is only since the commercialisation of snacks and the development of the potato crisp that snack foods, as we know them today, came into being (Snack Food Association, 1987). In Europe the first flavoured crisps appeared 40–45 years ago with cheese and onion, however, potato crisps could be purchased years before this with the familiar blue bag containing salt in every packet. Today consumers have a choice of many flavours, in addition to ready salted, when selecting snack foods, with cheese and onion, salt and vinegar, prawn cocktail, chicken/beef, and smoky bacon being the most popular in the UK (Mintel, 2005).

15.2.3   Types of snacks
The most common savoury snacks consumed in the USA and Europe are potato crisps (potato chips), corn based crisps, tortilla chips, pretzels, popcorn, extruded puffed and dried fried products, snack nuts/seeds and meat snacks for example jerky-type product and pork rinds. However, many countries have their own typical snack products. India and Japan are countries which have a wide variety of traditional snack foods. India has a number of roasted and fried cereal and legume snacks, and fruit and tuber snacks (Mudambi and Rajagopal, 2001). Japanese snacks (Anon, 1997) may be baked, fried, moulded and coated products and a wide variety of noodles are eaten as snacks as well as with the main meal.

15.2.4   The savoury snack market
Consumers in the USA are the greatest purchasers of snack foods followed by the UK (Hilliam, 2002b). It is estimated that in the USA on average each person will consume between three and four kilograms of snack food each year, with the market in 1999 being worth $19.3 billion (Anon, 2000). Overall world sales are nearly twice that of the USA at $30–35 billion annual global sales (McCarthy, 2001). Potato crisps is the leading snack with approximately 24% of the market (Anon, 2000).

In the UK the trends in eating habits have tended to drift towards a snacking culture that has been positive for the market. The market for crisps and snacks in 2005 was estimated to be worth £2.25 billion and is set to grow by 4% to £2.35 billion in 2010 (Mintel, 2005). Children and young adults are key purchasers in this market and with their purchasing power continually rising this means that their potential spending power on crisps and snacks is expected to grow. It is estimated that 84% of children have snacks or crisps in their packed lunches. The consumption of snacks by young adults, with a drink in the pub, is also increasing. There is a view that the adult sector will be the next target for savoury snacks by the snack food manufacturers (Whitworth, 2003).

15.2.5   Salt levels in snacks

The base material used for the production of snack foods is bland and therefore all snack foods need to be flavoured. They are typically seasoned, usually with salt and often with other additional flavourings. Salt is often essential, acting as a carrier for these additional flavourings and ensuring even distribution of flavour compounds that alone are often difficult to handle. In addition to taste, seasonings are often added to impart colour to the snack.

Savoury snacks appear to be salty, primarily due to the presence of the salt and seasoning on the surface, but they are generally not high in salt. The levels of seasonings vary between savoury snacks and may total up to 15% of the weight of the particular snack. The percentage of salt in the seasoning also varies and may be as high as 25%.

Crisps and snacks are eaten in relatively small amounts compared with other sources of salt. Since snacks contribute to less than 2% of overall dietary intake, it has been estimated that a 25 g packet of crisps would contribute to about 5% of the recommended 6 g/day (National Diet and Nutrition Survey, 2003). The salt content for most snacks is between 1 and 2%, with an average 25 g serving of a snack having about 0.25–0.5 g salt.

Salted potato crisps contain approximately 1.4 g salt/100 g but the salt is on the surface and so is detected more easily. Bread contains a similar level and some breakfast cereals 3 g salt/100 g, although these foods are not normally described as tasting particularly salty. Table 15.1 gives some examples of salt and seasoning levels present in snack foods.

15.3   Salt and health

In order to maintain body functions, most people need only 1.25 g salt/day, and it is recommended that no more than 6 g of salt should be consumed by an adult and 4 g by a child (Committee on Medical Aspects of Food Policy, 1994; US Department of Agriculture, 1995; Department of Health, 1996; Department of Health, 2001; Anon, 2001; Food Standards Agency, 2001, 2002), but the

Table 15.1   Average figures for salt and seasoning levels present in snack foods

Snack type	Salt level (%)	Seasoning level (%)
Crisps	1–2	6–8
Extruded snack	1–2	10–15
Tortilla chips	2–2.5	8–10
Crispbreads	1	12
Jerky product	2–4	8–16
Popcorn	1.8–4	12–16
Pretzel	2.2–2.5	2.2–2.5

average person's diet incorporates at least 9 g per day. It is estimated that male intake is between 4 and 18 g salt/day, and female 3 and 14 g/day.

The determinants of why salty foods are consumed and their effect upon the level at which salt is found palatable in them has been studied extensively, with the nature or nurture debate taking prominence. However, it is stated (Yeomans et al., 2004) that most of the sodium chloride intake in humans occurs in a need-free state.

Increasing palatability of foods produced by manufacturers may be a factor in their over-consumption, thus, attempts to produce healthier products may be offset by over-consumption induced by the manufacturers' attempts to make a more attractive product to ensure sales (Yeomans et al., 2004).

It is estimated that 75% of salt consumed is derived from processed foods, 10–15% added by consumers during cooking or at table and 10–15% is naturally present in foods (Gregory et al., 1990; Mattes and Donelley, 1991; Anon, 2004a).

Sodium is essential for health. Sodium and potassium ions are required for the maintenance of extra cellular fluid volume for the generation and transmission of electrical impulses in nerves and muscles and for the uptake of certain nutrients from the small intestine. Chloride ions are also needed for the production of digestive acid in the stomach.

Insufficient salt prevents our bodies from functioning properly, however, a diet high in salt has been linked to the body retaining more fluid, hypertension, a risk factor for strokes, heart attacks and kidney failure, and gastric cancer and osteoporosis, where salt removes calcium from bones which is then excreted in the urine (Gansevoort et al., 1993; Antonios and MacGregor, 1995, 1996; Devine et al., 1995; Joosens et al., 1996; MacGregor, 1998).

Although several factors are known to increase blood pressure, salt has been shown to be one of the most important factors (Dahl, 1960; Gleibermann, 1973; Froment et al., 1979; Tobian, 1991; Committee on Medical Aspects of Food Policy, 1994; Cutler et al., 1995; EU-Salt Position Paper 1, 2004).

The DASH (Dietary Approaches to Stop Hypertension) (Sacks et al., 2001) sodium trial evaluated the relationship between salt intake and blood pressure. Consumers were given either a DASH diet rich in vegetables, fruit, low-fat dairy products, low in total and saturated fat and reduced sodium intake, or a sodium reduced typical US diet. Both diets substantially lowered blood pressure in people with high blood pressure and higher than optimal blood pressure. In both diets the lower the sodium intake, the lower was the blood pressure.

Children's diets can have a direct impact on the development of risk factors such as osteoporosis (Cappuccio *et al.*, 2000) obesity, high blood cholesterol and high blood pressure, which can lead to heart disease and strokes in later adult life (National Diet and Nutrition Survey, 2000; Dyer, 2002).

The United States Department of Agriculture are concerned that snack foods are making up too large a part of children's diets (Roberts, 2004), contributing to 20% of the calories, and may also contribute to additional sodium intake. The Snack Food Association therefore recommend salt levels of no more than 1.5–2.0% salt in snacks.

15.4   Snack food manufacture

15.4.1   Manufacturing methods

Potato crisps

Potatoes selected for crisp manufacture are cleaned, washed, peeled and then sliced ready for frying. Further washing takes place to remove free surface starch that would result in oil breakdown and cause discolouration on the crisp surface (Gould, 1999). Following frying, the crisps are inspected for quality and then seasoned and packaged. The oil content is usually between 25 and 35% and can be controlled by selecting different oils, selecting potatoes with high solids content, thicker slices, and frying at a high temperature for a short time. The oil has an important part to play in the flavour of the crisp. Metallised foil packaging is becoming more common and in combination with low moisture/vapour proof films and an inert gas such as nitrogen or argon, a long shelf life in excess of ten weeks can be achieved.

Fabricated potato snacks

Potato flour (flakes or granules) in combination with water and salt are kneaded to make a dough. Other ingredients with potato flour may include emulsifiers, starches, corn flour and meal, and flavourings (Leipa, 1971, 1976; Weiss *et al.*, 1977). The dough may then be sheeted to form flat pieces, dried and then fried or in many cases it may be extruded to form a pellet. These pellets are then dried to below 14% moisture and then fried in hot oil to give an expanded product. In some cases they may be puffed by heating in hot air.

Extruded snacks

Extrusion cooking offers a wide variety of expanded snacks (Riaz, 1997). They comprise mainly carbohydrates with smaller amounts of protein, lipid and other ingredients such as seasonings.

The raw materials for extruded foods tend to be cereal flours particularly maize, rice and wheat. Other materials may be added to the cereal base to improve the flavour and or texture, for example salt, potato granules, sorghum grits, soya grits, legume flour, vegetable flours and modified starches. After extrusion puffing most snacks require drying to reduce the moisture content to 1.5–3% before the application of seasonings. This drying is necessary to control the texture, appearance and shelf life. In many instances, after drying, extruded snacks are fried in oil or coated in oil to adhere salt, flavourings and seasonings to the product. Crispbread, a popular snack particularly in Scandinavian countries, is mainly produced by extrusion processing (Huber, 2001). The formulation comprises wheat flour 68%, rice flour 20%, sugar 5%, dried milk 4%, salt 1% and oil 2%.

Tortilla chips
Raw materials for tortilla manufacture are maize or dry masa flour, lime, water, oil, salt, and seasonings (Bressani, 1990). The product quality and processing parameters depend on corn characteristics (Quintero-Fuentes *et al.*, 1999), but oil, salt and flavourings have the greatest effect on mouth feel, taste and acceptability. Maize is cooked at 85–100°C with lime (1%) and water (1–3 parts) (Serna-Saldivar *et al.*, 1990). The cooked maize is then drained, ground, sheeted, cut into disks, baked or fried. Salt and seasonings are added immediately after frying usually in rotating drums.

Hard pretzels
Pretzels have a characteristic alkaline exterior and acidic interior. The two pH levels give pretzels their characteristic texture and flavour. The formulation consists of wheat, flour, yeast, salt, and water (Groff, 2001). Non-diastatic malt may be included for flavour and colour. Fat may be added for tenderness. The ingredients are mixed into a dough which is then deposited into an extruder. The dough is cold extruded through a die, shaped and allowed to undergo proofing. The outer surface seals and gives the lustrous appearance of the finished product. The pretzels are cooked by feeding them through a pool of hot water (93°C) and sodium hydroxide (1–1.5%). The surface becomes alkaline and starch gelatinisation and sugar caramelisation during cooking take place. The wet steamy pretzels are then passed through a curtain of salt. They are then baked and dried.

Popcorn
To manufacture popcorn, maize and oil in a ratio of 3:1 are placed in a popping container or kettle. This is heated to approximately 230°C for 2.5–3 minutes allowing the maize to pop. Popcorn can also be made in microwave ovens by allowing maize to pop in special packaging constructed and designed for the popping of the corn. Salt is the most popular flavour for popcorn in the USA whereas in Europe, sugar coated popcorn is in most demand. Salt powder is added to the popping containers with the raw maize and oil. Other flavours are added if required after the corn has popped.

Jerky products
There are several methods for making jerky products (Ocherman, 1989; Davis, 1990), but primarily they consist of taking strips of meat, allowing them to soak in a marinade containing mainly salt and a range of other ingredients which include dextrose, nitrate, soya sauce, garlic, lemon juice and spices. After marinading, the strips are dried and packed.

15.4.2   Application of coatings

A number of systems are available for coating snack foods with the primary objective being to apply seasoning in a uniform and consistent manner. It is often a two-stage process since if not enough liquid is present on the surface of product, a liquid adhesive is first added, before the seasoning is applied.

The inclined drum tumbler is one of the most common pieces of equipment for coating snacks. As the drum rotates, the product is lifted by longitudinal flights. The product reaches a critical height, falls to the bottom of drum and is then lifted again. The product is transported forward since the drum is inclined downwards from entrance to exit. The salt and seasoning are metered into the tumbler and introduced as a curtain of powder. Oil may be sprayed into the tumbler to aid the adhesion of the seasoning to the snack. The tumbler is designed for a uniform flow of product down the length of the unit. The flights turn the snack over while the coating is dispersed inside the tumbler. This is a continuous application in rotating stainless steel drums. The product enters the drum at one end, and is coated on exit at other end. The speed can be changed so that the coating ratio can be varied. Generally crisps are seasoned directly from the fryer to take advantage of the hot oil for binding the seasonings to the crisp surface. With most fried snacks the oil remaining on the surface acts as an adhesive material to bind salt and seasoning and as a carrier for oil-soluble flavours. Non-fried snack products such as extruded snacks are sprayed with oil and then dusted with dry seasonings in the tumbler.

A conveyor-based coating unit is another system for applying coatings to snacks. The product is transported on a conveyor through the unit and the coating applied by a mist system consisting of oil, seasoning and salt at a temperature of 49–55°C. For extruded snacks, a common composition of mist is powdered seasoning and vegetable oil in a ratio of 1:2, and sprayed onto the dried warm snack to give a dose of 12% seasoning and 24% oil.

A more recent method for the coating of seasonings is by electrostatic application. The coating drum has a ground surface and the snack products become ground by contact with the drum. Seasoning is fed into a mixing drum that has an energised electrode, and air is used to blow the seasoning onto the snack. The seasoning receives an electric charge in the mixing area and is carried onto the base product by the air. This gives a more uniform and controlled coating and minimises the possibility of excess salt and seasonings adhering to the surface.

Low and no-fat products may be sprayed with a water based slurry. This consists of salt and water-soluble flavourings dispersed in starches, maltodextrins or gums that act as adhesive agents (Bertram, 1999). Coating and drying in successive steps may take place two or three times, giving an even, consistent, multi-layered coating. Moisture is removed by the drying stages and some flavour may be lost. Dextrin at 30% solids may be used. This has a high degree of stickiness, and is sprayed onto the snack to adhere seasonings and salt. This has a benefit in that it dries quickly. The use of a starch-based hot melt system is also possible. This is a free flowing, pre-blended dry powder, oil and water free. The powder is applied to the hot snack at 120–150°C, which melts the powder, and the snack becomes coated. In the case of salted popcorn, oil, which may be flavoured, is sprayed on at 20–30% finished weight of product, salt is then blown on. Dry roast nuts are wetted by a starch/gum solution, dusted with salt and seasoning, usually containing paprika, and then slowly roasted.

15.5   Function of salt in snacks

The role of salt in products can be complex, serving many functions depending upon the product and level of addition. It not only adds its own distinctive flavour but can also enhance other flavours present (van der Heijden et al., 1983; Breslin and Beauchamp, 1995, 1997). The low cost of salt along with its many functions and distinctive taste have ensured its widespread usage in the snack food industry.

15.5.1   Coatings and flavour

Salt is added to foods as a preservative, to enhance flavour, and to improve texture. In the case of snacks, since one of the most important characteristics of snack foods is taste, salt is primarily present to give the product flavour. In snacks, salt is one of the least costly ingredients, but is usually one of the major components applied as part of the seasoning to enhance and compliment other flavour components. Without salt many favours would be bland and lack intensity.

Salt offers a convenient vehicle for uniformly distributing micro-ingredients such as flavours, colours and possible functional additives, e.g. vitamin and mineral nutrients on the surface of the snack and in some cases throughout the finished product. Salt may be used as a carrier of antioxidants on fried snack products. As well as having tasks of its own, sodium chloride modifies flavours and improves the flavour balance of the product (Gillette, 1985). Balancing the three main flavour components (salt, acidity and sweetness) creates a pleasurable experience.

Salt used in snack foods may differ in the size and shape of the particles. This has been shown to affect consumer acceptance. One of the most common types of salt used is flour salt which is fine and granular and not prone to fall off the surface of snacks, unlike larger granules.

Seasonings are a blend of ingredients to provide flavour sensations, mouth feel, textures and colours to the snack. Seasonings are usually in the dry form but

Table 15.2   Seasoning components commonly used in snack food production

Component	Function	Examples
Colours	Aid colour development/ compensate for processing losses	Paprika, annatto, turmeric, etc.
Fillers	Extend seasoning mix to ensure product coverage	Wheat flour, rusk, corn starch, soya flour, lactose dextrose and maltodextrin, whey powder
Dairy powders, fat powders	Carrier, add mouthfeel	Whey powder
Flavour enhancers	Increase inherent flavour	MSG, ribonucleotides, (disodium guanylate) sodium inosinate, sodium 5-ribonucleotides
Flavourings	Add flavour to bland base materials	Natural and nature identical flavour blends BBQ, cheese, smoky bacon, etc., yeast extract, HVP
Sweeteners	Flavour, can be used as carriers	Dextrose, sucrose, fructose
Vegetable powders	Flavour, colour, fillers	Onion, tomato, garlic
Acids	Flavour, preservative, inhibit colour loss	Citric, lactic, acetic or their salts
Antioxidants	Inhibit colour, flavour and nutritional losses	BHA, TBHQ, tocopherols, rosemary extract, tea extract
Salt	Flavour, carrier, preservative	Fine vacuum dried
Herbs and spices	Flavour	Paprika, mint, dill, oregano, basil
Processing aids	Aid seasoning dispersal, adhesion, anti-caking agent	Vegetable oil, salt, silicon dioxide

can be flavoured oils or two-phase slurries consisting of a dry flavour and liquid carrier. Because not all flavours are soluble in a liquid carrier, a two-phase mixture is formed. The liquid carrier is usually oil. A typical seasoning recipe would consist of salt (15–35%), filler (20–50%), acids, acidity regulator, anti-caking agent and other ingredients.

A more comprehensive list of seasoning components is given in Table 15.2.

15.5.2   Rheological and other properties
The involvement of salt in the rheological and physical properties of snack products varies and is often subject to interactions with the other components

present. Salt added as part of a coating or seasoning has been studied extensively in the context of popcorn production.

The effects of a variety of factors, including the level of oil and salt applied, on the properties of popcorn produced in a microwave oven were studied (Lin and Anantheswaran, 1988). At levels of 2% (of the weight of the kernels), without the presence of oil, salt was found to significantly increase the expansion volume when compared with a control sample. Increasing the salt level to 5% did not result in any further increase in expansion volume. As oil was added, at 30%, during the popping process, the number of unpopped kernels decreased significantly with expansion volume remaining constant. In combination, addition of both oil and salt resulted in a decrease in expansion volume and no change in the number of unpopped kernels indicating some degree of interaction.

In similar studies (Singh and Singh, 1999; Ceylan and Karababa, 2004) the complex interactions of salt during popcorn production were studied in more detail, both studies highlighting the significant effects of relatively small changes in salt addition when combined with other components.

In addition to salt being used as a flavour in the product coating, it may be added to the base recipe of some snacks. Starches are the most important functional ingredients to achieve textural attributes in snack products. Ingredients such as fat, sugars, proteins and salt can affect the properties of starch and thus affect the quality of snack products. During processing, hydroscopic ingredients such as salt compete with starch for moisture, reducing the effective concentration of water available for starch gelatinisation and requiring higher temperatures and longer times to produce the same results.

Several studies have investigated the effects of salt, added to the base material, on the production of extruded snack products. In most extruded snack products texture is of paramount importance and is usually conferred primarily by the presence of starch.

In a study (Chinnaswamy and Hanna, 1990), the effect of salt, sugar and screw speed on the macromolecular and functional properties of corn starch during extrusion were investigated. Addition of salt at 1% on a dry weight basis was found to enhance degradation of the branched component of the starch present giving rise to an increased proportion of linear fractions. This in turn was accompanied by an increase in the level of expansion observed. An optimum mean molecular weight of starch for expansion was proposed, which could be altered to some extent by the addition of salt.

A similar study using corn meal (Hsieh et al., 1990) gave results in agreement, showing an increase in expansion, reduction in bulk density and decrease in break strength as salt was added at up to 3%. Additionally changes in colour were noted with the extrudates becoming lighter, more yellow and less red as salt was added. This was suggested to be a result of interference in browning reactions such as caramelisation and Maillard reaction, probably as a result of lowering the water activity (a_w) away from the intermediate water levels where maximum browning reaction occurs. In a later study (Jin et al.,

1994), similar changes in extrudate colour were noted with the addition of salt at levels of up to 2% when extruding corn meal with added soy fibre. In contrast they found the addition of salt to increase the bulk density and reduce the expansion ratio of the extrudates. This was suggested to be a result of competition for available water and inhibition of gelatinisation. This apparent contradiction in results highlights the complex nature of the interaction between the components of the product with small changes in the ratios of ingredients often leading to marked changes in product characteristics.

15.6   Salt reduction in snacks

Salt is widely regarded as a flavour enhancer for most savoury flavour types. The replacement of salt is a challenge since it is the nerve pathways transmitted by potassium/sodium ion exchange that result in the perception of the salt taste quality. Typically it is the sodium that is being reduced and replacements tend to be potassium, calcium, or magnesium (Denis, 2005).

15.6.1   Salt and the consumer

Processors have been criticised for high levels of salt in processed foods. The food industry has been asked to reduce salt levels in processed foods since the evidence of a link between salt intake and high blood pressure (Kaplan, 2000; Brady, 2002; Anon, 2003a, 2004b, 2005a; Gregory, 2003; Ruusunen and Puolanne, 2005). However, the initial recommendations made by the Food Standards Agency have been made less stringent after industry complaints (Anon, 2005b). About two-thirds of consumers want to eat more healthily and consumer groups have demanded that products should be developed with reduced levels of salt. However, the growth of the snack food market shows that people are still opting for salty snacks (Anon, 2005a).

Lower sodium products have been available to the American market since the early 1990s, however, total sales have been low at around 3–4%. The reasons for this are unclear but it may be that they are unacceptable for taste (Beasley, 1998), are more costly than regular products, have limited availability or salt intake may not be an important concern to American consumers (Evans *et al.*, 1996; Jacobson and Liebman, 1996).

The food industry in the UK is beginning to act on the nutritional concerns of customers, and has reduced salt levels in a number of recipes, however as was experienced in the US, manufacturers have experienced limited demand for foods promoted as being no salt or low salt. It has been possible to reduce salt levels in many products without adversely affecting the quality attributes. However, the level to which salt can be reduced is dependent upon the level of salt present in the diet of the population for which reduction is being attempted. This poses problems for producers marketing to many countries (Ruusunen and Puolanne, 2005).

In the eight years prior to 2002 there was a reduction of approximately 20% in the salt added to potato crisps with reductions of around 5 and 10% respectively for peanuts and tortilla chips (Brady, 2002) bringing the levels down to those seen in the US market (Hazen, 2005). In 2003 the Food Standards Agency in the UK announced guidelines on maximum salt intakes for children. Despite this, a survey of products marketed specifically for children showed almost no change in the levels of salt used (Anon, 2004a). Since the initial decrease in salt levels used in snacks, subsequent reductions have been limited (Anon, 2003a) and may reflect the difficulty in finding a replacement that has the same functionality at a comparable price.

Taste panels show that low salt foods are often unappetising and reducing salt will only work if consumers find the products acceptable. Salt is a very stable material with a long shelf life and is also a low cost item; therefore, any replacement or substitute needs to compete effectively on price.

15.6.2   Formulation changes

Salt reduction has occurred in some snack foods over time by simple incremental changes, allowing the consumer to gradually become accustomed to lower levels. A number of savoury snack products have recently been launched with lower sodium levels, these included unsalted crisps, and snacks with a 50% reduction in salt (Mintel, 2005). Whilst it is possible for consumers to adapt to changes in flavour due to lower salt levels, changes in flowability, shelf life and texture may not be acceptable, requiring significant changes to the formulation of the seasoning.

Base material
The bland flavour of many cereal flours makes it essential that seasonings are added to snacks to give consumer acceptability. Specific combinations of flours can contribute to flavour development. Barley blended with wheat gives a sweet flavour, barley and corn gives a bitter flavour, potato and corn, and potato and rye give meat flavour attributes. If more intense flavours can be developed then this will reduce the level of salt needed.

Flavour enhancers
Using flavour enhancers can increase salt perception. However, flavour enhancers cannot replace salt, only increase the effect of salt present in the product. It is possible that enhancers increase stimulation of the salt receptors due to some interaction.

Many flavour enhancers which may be used are sodium salts such as monosodium glutamate, disodium inosate and disodium guanylate. These enhance the overall flavour impact and give a mouth-watering sensation. The preference for foods containing MSG tends to be higher even when sodium levels are equivalent (Prescott, 2004) and thus MSG addition may enable a reduction in salt level whilst maintaining the desirable characteristics of the product. A typical level of addition of this to food would be around 0.5% (Yamaguchi, 1991). Because of the enhancement of flavour, levels of salt and enhancer would be less than the original salt level and thus sodium levels would be reduced. These enhancers could also be used in their acid form to improve flavour without adding sodium. Autolysed yeast (Anon, 1989) and some organic acids may also be used to enhance flavour, and reduce salt content.

Salt substitutes
Substitution of sodium chloride with salt substitutes has been tried with limited success to produce a 'lower salt' seasoning. A number of salt substitutes are available such as potassium chloride, and some salts of ammonium, calcium and magnesium. An increased intake of foods containing minerals such as potassium, magnesium, calcium can help normalise blood pressure (McCarron, 1997).

Potassium chloride is the most widely used salt substitute, it has a salty taste but also contributes to an off metallic/bitter flavour. It has been found that Italian type spices, will mask the off flavour of potassium chloride (Best, 1989). Combinations of salt and potassium chloride have been used in reduced salt formulations with reasonable results. However, increasing the ratio of potassium to sodium causes the bitter taste to increase. Lo-Salt, a combination of 66.6% potassium chloride and 33.3% sodium chloride is available, and still retains the full salt taste. Mycoscent salt, derived from the production of mycoprotein, has a salty taste, but contains no sodium. It is claimed (Anon, 2003b) to enable a 50% sodium reduction in snack foods without a noticeable aftertaste.

Natural flavour materials
Natural flavour materials such as meat powders, vegetable powders and yoghurt may be added up to 30% without compromising final product textures. Ground spices such as black pepper, dill, cumin, and oregano up to 2% level in the seasonings may be used. These are concentrated in flavour that is released slowly during the eating experience.

Essential oils and oleoresin extracts of spices have more flavour impact than normal ground spices. They are generally spray dried which accentuates flavour release. These may be encapsulated, which increases shelf stability and have a lower microbial risk. They can be used as part of the seasoning or incorporated into the mix prior to extrusion. Encapsulation gives stability to the flavour during the extrusion process. Other materials such as herbs, lime vinegar, lemon juice, and ginger may be used in coatings to flavour foods allowing for salt reduction.

New flavours
New flavoured snacks (Hilliam, 2001, 2002a) rather than the traditional salt coated are having an impact on the snack food market. These include regional and ethnic flavours. Spanish, Indian and Pacific Rim flavours are growing in popularity, as are fresh, vibrant flavours such as fresh fruit flavours. Flavour pairings are appearing including the use of traditional sweet flavours into savoury applications (Ahmed, 2004; Rittman, 2004).

There are now a large number of flavours available by taking familiar flavours from other food sectors such as turkey and stuffing, coronation chicken, sizzling bacon, flaming chicken wings, duck, orange and ginger, chicken and mushroom, port and stilton, roast chicken and thyme (Mintel, 2005; Hilliam, 2001). The formulation of these strong flavours may help in reducing salt levels.

15.6.3   Manufacturing changes

The use of pre-conditioning chambers in extrusion where the mixing, hydration, cooking, pH adjustment, addition of flavours, colours, etc. takes place can improve the development of flavour components in the final product. This may be due to Maillard type reactions. By pre-conditioning the starch and protein matrices, they become more pliable and more easily deformed which makes them less susceptible to damage and melting in the extrusion process. This improves availability of flavour components for quick release during chewing, again allowing for a possible reduction in salt levels.

The use of a more accurate, controlled coating system, such as electrostatic units (Biehl and Barringer, 2003; Ratanatriwong et al., 2003), would reduce excessive depositing of salt on the surface of snacks, giving a lesser, more uniform coating of seasoning on the surface. The use of mock salt, starch particles coated with a thin layer of salt for surface flavouring would also reduce salt levels in coatings.

Supercritical carbon dioxide under high pressure can be used for extrusion puffing at lower temperatures (Huber, 2001), preserving flavours, leading to a snack with a higher retention of flavour. Flavours may also be added with the carbon dioxide and deposited into the extrudate. Different oils (Rowan, 2004) can be used as a frying medium that can impart different flavours to fried snacks possibly allowing for a reduction in salt levels.

15.7   Conclusion

The reductions in salt levels achievable in snack products are considerable; however, these products tend to be consumed as a relatively small, although often frequent, part of the diet and hence reductions may have limited effect on attaining the six grams per day recommended target. Many of the reductions seen have been a result of incremental changes where little influence on the product occurs and few changes to product formulation are required. Future demands for salt reduction may be more difficult to attain and will require a greater understanding of its multiple functions in food products.

It would be valuable to elucidate the detailed mechanism of salt taste reception that could lead to the development of effective salt substitutes and replacers. A future approach might be to identify other natural compounds which are not currently in use as salt replacements, but which provide either organoleptic substitution or textural substitution or both. It has been suggested that salted snacks like crisps should be replaced with fruit and unsalted nuts.

15.8   Sources of further information and advice

British Nutrition Foundation
High Holborn House
52-54 High Holborn
London WC1V 6RQ
Tel: +44 (0) 2074 046504
www.nutrition.org.uk

CASH – Consensus Action on Salt and Health
Blood Pressure Unit
Department of Medicine
St George's Hospital Medical School
Cranmer Terrace
London SW17 0RE
Tel: +44 (0) 2087 252409
www.cash@sghms.ac.uk
www.actiononsalt.org.uk

Cereal Foods World
Published by the American Association of Cereal Chemists
3340 Pilot Knob Road
St Paul
Minnesota 55121
USA
Tel +1 6514 547250
www.aaccnet.org

European Snack Association
6 Catherine Street
London WC2B 5JJ
Tel +44 (0) 2074 207220
www.esa.org.uk

Food Standards Agency
Aviation House
125 Kingsway
London WC2B 6NH
Tel: +44 (0) 2072 768000
www.food.gov.uk

W A Gould 1994
Snack Food Manufacturing & Quality Assurance Manual
The Snack Food Association
1711 King Street
Suite 1
Alexandria
Virginia, 22314-2720
USA
Tel +1 7038 364500
www.sfa.org

Snack Food and Wholesale Baking Magazine
1935 Shermer Road
Suite 100
Northbrook
Illinois 60062-2309
USA
Tel +1 8472 055660
www.packnet.com

15.9   References

AHMED A (2004), 'Meeting munching needs', *World of Food Ingredients*, April/May, 36–38.
ANON (1989), 'Yeast-based enhancers tailored for snack foods', *Snack Food*, 78 (10), 18–19.
ANON (1997), 'Snacks growth in the land of the rising sun', *Food Review (Cape Town)*, 24 (6), 15–16.
ANON (2000), 'State of the industry report 2000', *Snack Food and Wholesale Bakery*, 80 (6), SI-1–SI-74.
ANON (2001), 'Nutrition and diet for healthy lifestyles in Europe: The EURODIET evidence', *Public Health Nutrition*, 4, 2A and 2B.
ANON (2003a), 'Bread, crisps, beans and soup – as salty as ever', *Food Magazine*, 60, 8.
ANON (2003b), 'Cut the salt, keep the taste', *Food Manuf*, April, 61.
ANON (2004a), 'A grain of salt and a grain of sense?', in *EUFIC Food Today* No 25.
ANON (2004b), 'Children's food as salty as ever', *Food Magazine*, 65, 7.
ANON (2005a), 'Better-for-you snacks and drinks', *Grocer*, 228 (7692), 55–60.
ANON (2005b), 'Targets for salt in food relaxed', http://news.bbc.co,uk/1/hi/health/4737707.stm.
ANTONIOS T F T and MacGREGOR G A (1995), 'Deleterious effect of salt intake other than effects on blood pressure', *Clin Exper Pharm Physiol*, 22, 180–184.
ANTONIOS T F T and MacGREGOR G A (1996), 'Salt-more adverse effects', *Lancet*, 348 250–251.
BEASLEY J (1998), 'Salt shake-up', *Snacks Magazine*, Sept, 20, 22–23.
BERTRAM A (1999), 'Snacks and the appliance of starch', *International Food Ingredients*, 4, 30–31.

BEST D (1989), 'Compensating for sodium; the low salt solution', *Prepared Foods*, 158 (2), 97–98.

BIEHL H L and BARRINGER S A (2003), 'Physical properties important to the electrostatic and nonelectrostatic powder transfer efficiency in a tumble drum', *J Food Sci*, 68 (8), 2512–2515.

BRADY M (2002), 'Sodium: Survey of the usage and functionality of salt as an ingredient in UK manufactured food products', *Br Food J*, 104 (2), 84–125.

BRESLIN P A and BEAUCHAMP G K (1995), 'Suppression of bitterness by sodium: variation among taste stimuli', *Chem Senses*, 20, 609–623.

BRESLIN P A and BEAUCHAMP G K (1997), 'Salt enhances flavour by suppressing bitterness', *Nature*, 387, 563.

BRESSANI R (1990), 'Chemistry, technology and nutritive value of maize tortillas', *Food Rev Int*, 62, 225–263.

CAPPUCCIO F P, KALAITZIDS R, DUNECLIFT S and EASTWOOD J B (2000), 'Unravelling the links between calcium excretion, salt intake, hypertension, kidney stones and bone metabolism', *J Nephrol*, 13 (3), 169–177.

CEYLAN M and KARABABA E (2004), 'The effects of ingredients on popcorn popping characteristics', *Int J Food Sci Technol*, 39, 361–370.

CHINNASWAMY R and HANNA M A (1990), 'Macromolecular and functional properties of native and extrusion-cooked corn starch', *Cereal Chem*, 67 (5), 490–499.

COMMITTEE ON MEDICAL ASPECTS OF FOOD POLICY (1994), *Nutritional aspects of cardiovascular disease. Report of the Cardiovascular Review Group*, London, HMSO.

CUTLER J A, PSATY B M, MacMAHON S and FURBURG C D (1995), 'Public health issues in hypertension control: what has been learned from clinical trials', in Laragh J H and Brenner B M, *Hypertension: Pathophysiology, Diagnosis, and Management*, New York, Raven Press, 253–270.

DAHL L (1960), 'Possible role of salt intake in the development of hypertension', in Cottier P and Bock K D, *Essential Hypertension – An International Symposium*, Berlin, Springer-Verlag, 53–59.

DAVIS J M (1990), 'Meat based snack foods', in Booth R G, *Snack Foods*, New York, AVI-Van Nostrand Reinhold, 205–224.

DENIS M L (2005), 'Molecules of taste and perception', in Rowe D E, *Chemistry and Technology of Flavours and Fragrances*, Oxford, Blackwell, 199–243.

DEPARTMENT OF HEALTH (1996), *A progress report from the Nutrition Task Force on the action plan to achieve the Health of the Nation targets on diet and nutrition.* London, HMSO.

DEPARTMENT OF HEALTH (2001), *From vision to reality (action against the big killers)*, London, HMSO.

DEVINE A, CRIDLE A R, DICK I M, KERR D A and PRINCE R L (1995), 'A longitudinal study of the effect of sodium and calcium intakes on regional bone density in postmenopausal women', *Am J Clin Nutr*, 62, 740–745.

DYER O (2002), 'First case of type 2 diabetes found in UK teenagers', *Br Med J*, 324, 306.

EU-SALT POSITION PAPER 1 (2004), '*Salt and Blood Pressure. Controversial and Misunderstood*'. European Salt Producers Association, Avenue de l'Yser 4, Brussels.

EVANS M, COHEN J D, KUMANYIKA S, CUTLER J A and ROCELLA E J (1996), *Implementing recommendations for dietary salt reduction. Where are we? Where are we going? How do we get there? A summary of NHLBI workshop*, Bethesda, National Heart, Lung and Blood Institute.

FOOD STANDARDS AGENCY (2001), 'Should I cut down on salt?', http://www.food.gov.uk/healthiereating/men/whataboutsalt.

FOOD STANDARDS AGENCY (2002), 'Health initiative worth its salt, says agency', http://www.food.gov.uk/news/newsarchive/salt_awareness_day.

FROMENT A, MILON H and GRAVIER C (1979), 'Relationship of sodium intake and arterial hypertension. Contribution of geographical epidemiology', *Rev Epidemiol* Sante Publique, 437–454.

GANSEVOORT R T, DE ZEEUW D and DE JONG P E (1993), 'Long-term benefits of antiproteinuric effect of ACE-inhibition in non-diabetic renal disease', *Am J Kidney Dis*, 2, 202–206.

GILLETTE M (1985), 'Flavour. Effects of sodium chloride', *Food Technol*, 39 (6), 47–52.

GLEIBERMANN L (1973), 'Blood pressure and dietary salt in human populations', *Ecol Food Nutr*, 143–156.

GOULD W A (1999), *Potato processing, production and technology*, Alexandria, CTI Publications.

GREGORY H (2003), 'Hard to swallow', *Grocer*, 227 (7594), 36–37.

GREGORY J, FORSTER K, TYLER H and WISEMAN M (1990), *The dietary and nutritional survey of British adults*, London, HMSO.

GROFF E T (2001), 'Perfect pretzel production', in Lusas W L and Rooney L W, *Snack Foods Processing*, Lancaster, Technomic, 369–384.

HAZEN C (2005), 'New spins on salty snacks', *Food Product Design*, 14 (12), 34–56.

HILLIAM M (2001), 'Have a snack', *World of Food Ingredients*, Sept, 12–14.

HILLIAM M (2002a), 'Snacks and confectionery', *World of Food Ingredients*, Sept, 44, 46, 48.

HILLIAM M (2002b), 'Food on the go', *World of Food Ingredients*, Oct/Nov, 34, 36, 38.

HSIEH F, PENG I C and HUFF H E (1990), 'Effects of salt, sugar and screw speed on processing and product variables of corn meal extruded with a twin-screw extruder', *J Food Sci*, 55 (1), 225–227.

HUBER G R (2001), 'Developments and trends in extruded snacks', *Food Product Design*, 11 (3), 123–142.

JACOBSON M F and LIEBMAN B F (1996), 'Sodium in processed foods', *Am J Clin Nut*, 63, 138.

JIN Z, HSIEH F and HUFF H E (1994), 'Extrusion cooking of corn meal with soy fibre, salt and sugar', *Cereal Chem*, 71 (3), 227–234.

JOOSENS J V, HILL M J, ELLIOT P, STAMLER R, LESFFRE E, DYER A, NICHOLS R and KETELOOT H (1996), 'Dietary salt, nitrate and stomach cancer mortality in 24 countries. European Cancer Prevention (ECP) and the INTERSALT Cooperative Research Group', *Int J Epidemiol*, 25, 494–504.

KAPLAN N M (2000), 'The dietary guideline for sodium: should we take it up?', *Am J Clin Nutr*, 71, 1020–1026.

LEIPA A L (1971), *Preparation of potato chip-type products*, US Patent No. 3,576,647.

LEIPA A L (1976), *Potato chip products and process for making same*, US Patent No. 3,998,975.

LIN Y E and ANANTHESWARAN R C (1988), 'Studies on popping of popcorn in a microwave oven', *J Food Sci*, 53 (6), 1746–1749.

MacGREGOR G A (1998), 'Salt: blood pressure, the kidney, and other harmful effects', *Nephrol Dial Transplant*, 13, 2471–2479.

MATTES R D and DONELLEY D (1991), 'Relative contributions of dietary sodium sources', *J Am Coll Nutr*, 10, 383–393.

McCARRON D A (1997), 'Role of adequate dietary calcium intake in the prevention and management of salt-sensitive hypertension', *Am J Clin Nutr*, 65, suppl 1, 712S–715S.

McCARTHY, J A (2001), 'The snack industry: history, domestic and global status', in Lusas W L and Rooney L W, *Snack Foods Processing*, Lancaster, Technomic, 29–35.

MINTEL (2005), '*Crisps and snacks – UK – May 2005*', London, Mintel International Group Limited.

MUDAMBI S R and RAJAGOPAL M V (2001), 'Snack foods of India', in Lusas W L and Rooney L W, *Snack Foods Processing*, Lancaster, Technomatic, 477–491.

NATIONAL DIET AND NUTRITION SURVEY (2000), *National diet and nutrition survey: young people aged 14–18, vol 1*, London, HMSO.

NATIONAL DIET AND NUTRITION SURVEY (2003), *National diet and nutrition survey: adults aged 19–64 years, 2000–2001, vol 3*, London, HMSO.

OCHERMAN H W (1989), *Sausage and processed meat formulations*, New York, AVI-Van Nostrand Reinhold.

PRESCOTT J (2004), 'Effects of added glutamate on liking for novel food flavours', *Appetite*, 42 (2), 143–150.

QUINTERO-FUENTES X, ALMEIDA-DOMINGUEZ, H D, MCDONOUGH C M and ROONEY L W (1999), 'Ingredient functionality in baked tortilla and corn chips', *Cereal Chem*, 76 (5), 705–710.

RATANATRIWONG P, BARRINGER S A and DELWICHE J (2003), 'Sensory preference, coated evenness, dustiness and transfer efficiency of electrostatically coated potato chips', *J Food Sci*, 68 (4), 1542–1547.

RIAZ M N (1997), 'Technology of producing snack foods by extrusion', *Technical Bulletin, American Institute of Baking Research Department*, 19 (2), 1–8.

RITTMAN A (2004), 'Future snack flavours', *World of Food Ingredients*, Sept, 18, 20–21.

ROBERTS W A (2004), 'Snack times, they are a changin'', *Prepared Foods*, 173 (1), 23–24, 27, 29.

ROWAN C (2004), 'Snack time', *Food and Beverage International*, 3 (3), 20–21.

RUUSUNEN M and PUOLANNE E (2005), 'Reducing sodium intake from meat products', *Meat Science*, 70 (3), 531–541.

SACKS F M, SVETKEY L P and VOLLMER W H (2001), 'Effects on blood pressure of reducing dietary sodium and the dietary approaches to stop hypertension (DASH) diet', *New Eng J Med*, 344, 3–10.

SERNA-SALDIVAR S O, GOMEZ M H and ROONEY L W (1990), 'The technology, chemistry and nutritional value of alkaline cooked products', in Pomeranz Y, *Advances in Cereal Science and Technology, Vol 10*, St Paul, American Association of Cereal Chemist, 243–307.

SINGH J and SINGH N (1999), 'Effects of different ingredients and microwave power on popping characteristics of popcorn', *J Food Eng*, 42, 161-,-165.

SNACK FOOD ASSOCIATION (1987), *50 years: A Foundation for the Future*, Alexandria, Snack Food Association, 10–12.

TOBIAN L (1991), 'Salt and hypertension: lessons from animal models that relate to human hypertension', *Hypertension*, 17 (1 Suppl), 52–58.

US DEPARTMENT OF AGRICULTURE AND US DEPARTMENT OF HEALTH AND HUMAN SERVICES (1995), *Dietary guidelines for Americans*, 4th edn, Washington, US Government Printing Office.

VAN DER HEIJDEN A, BRUSSEL L B P, KOSMEIJER J G and PEER H G (1983), 'Effects of salts on perceived sweetness', *Z Lebensm Unters Forsch*, 176, 371–375.

WEISS V E, CAMPBELL G M and WILSON G L (1977), *Fried formed chip*, US Patent No. 4,032,664.

WHITWORTH M (2003), 'Strictly for the grown-ups', *Food Manuf*, 78 (12), 12–13.

YAMAGUCHI (1991), 'Basic properties of umami and effects on humans', *Physiol and Behav*, 49, 833–841.

YEOMANS M R, BLUNDELL J E and LESHAM M (2004), 'Palatability: response to nutritional need or need-free stimulation of appetite?', *Br J Nutr*, 92, Suppl 1, S3–S14.

16

Reducing salt in cheese and dairy spreads

T. P. Guinee and B.T. O'Kennedy, Moorepark Food Research Centre, Ireland

16.1   Introduction

Owing to its scarcity, salt (NaCl) was a highly prized trade item in early civilizations, being used as a form of currency for goods and labour (salary). Procurement of salt was costly and much of our dietary sodium was acquired from natural, 'unsalted' sources such as fresh meat, fish and milk. With time, salt became plentiful and widely available and was used for preservation and to enhance flavour of various dishes and delicacies. Hence, the evolution of our taste for 'high' salt levels. Today, salt is added to most foods, and in some cases mainly for its flavour-enhancing role. Sodium intake in the modern Western diet is generally excessive and its reduction is a major challenge, as scientific evidence linking excessive dietary sodium with adverse health effects grows.

Sodium is an essential component in the diet, being required for regulation of blood pressure, water transport into and out of cells, tissue osmolality, and transmission of nerve cell impulses (Anonymous, 1980). The recommended daily requirement of sodium for the adult human is ~2.4 g Na, or ~6 g NaCl, per day (Kaplan, 2000), which can generally be met through the indigenous Na content in foods.

Added NaCl is a major source of sodium in modern Western diets, which contain approximately two to three times more Na than is necessary. Excessive intakes of Na have toxic, or at least undesirable, physiological effects, including hypertension and increased calcium excretion, which may lead to osteoporosis (see Abernethy, 1979; Anon., 1980, 1983; Moses, 1980; Schroeder et al., 1988; Denton et al., 1995; Midgley et al., 1996; Beard et al., 1997; McCarron, 1997; Beilin, 1999; Cutler, 1999; Feldman and Schmidt, 1999; Korhonen et al., 1999,

2000; Cappucio *et al.*, 2000; Kaplan, 2000; McCarron and Reusser, 2000; O'Shaughnessy and Karet, 2004). Such concern has led to recommendations for a reduced dietary intake of Na, classification of foods (high, medium, low) according to sodium level, declaration of sodium level on the food labels and an increased demand for reduced-sodium foods, including cheese (Institute of Food Technologists, 1980; Anon, 1993, 1995; Reddy and Marth, 1995; Morris and Dillon, 1992; Narhinen *et al.*, 1998).

Cheese exhibits marked intra- and inter-variety differences in salt content (Table 16.1). Most natural cheese varieties, apart from the low pH (4.5–4.8), fresh, short shelf life, acid-curd types such as Fromage frais, Quark (and related types), contain added salt (NaCl). (In Quark, etc. which is not salted, the presence of salt at a level of ~0.15% (w/w), is due to the presence of the indigenous milk sodium and Cl (~50 and 95 mg/100 mL), in the moisture phase of the cheese.) Salt in cheese results mainly from the direct addition of salt, but it may also be added indirectly by way of dressings or condiments, such as cream dressing in cottage cheese (Guinee and Fox, 2004). The sodium content of pasteurized processed cheeses (including cheese spreads) and substitute/imitation cheese analogues is generally much higher than that of natural cheeses because of the addition of: NaCl at levels of ~0.4 to 1.05%, (w/w), emulsifying salts (sodium phosphates or sodium citrates) at levels up to 3% (wt/wt), and optional condiments and preservatives (e.g., sodium propionate).

The contribution of cheese to dietary sodium depends on types and quantities consumed (Table 16.1), with a high per capita consumption of fresh acid curd cheeses as in Germany (~12 kg per year; Schulz-Collins and Senge, 2004) contributing much less than a high consumption of high salt varieties such as

Table 16.1 Approximate NaCl level in different cheese varieties[a]

	NaCl (%, w/w)	Moisture (%, w/w)	Salt-in-moisture (%, w/w)	Weight of cheese which gives full RDA for Na (g)[b]
Quark	0.15	82	0.19	3158
Emmenthal	0.7	38	1.8	871
Appenzeller	1.3	37	3.6	469
Low-moisture Mozzarella	1.4	46.4	3.1	429
Cheddar	1.7	37	4.6	359
Limburger	2.0	45	4.4	300
Gouda	2.4	39	6.1	254
Danish Blue	3.3	43	7.7	185
Roquefort	4.1	41	10.1	149
Romano-type	4.1	30	13.8	149
Feta	4.5	63	7.1	136
Domiati	6.0	55	10.9	102

[a] Data compiled from various sources
[b] Estimated quantity of cheese which would contribute the full Recommended Daily Allowance of sodium (Na) ~2.4 or 6 g NaCl (Kaplan, 2000).

Feta or Blue cheese (Table 16.1). The quantities of different cheeses that could be consumed in an otherwise salt-free diet before exceeding the recommended daily levels of sodium are shown in Table 16.1. It is clear that the quantity decreases with sodium content of cheese from ~3.2 kg/day for Quark, 0.89 kg/ day for Emmental, to 0.102 g for Domiati. At the highest global per capita cheese consumption rates of ~25.5 kg/year, or ~70 g/day, low (e.g., ~0.7%, w/w, as in Emmental), medium (e.g., ~2.0%, w/w as in Gouda), or high (e.g., ~3.5% in Blue cheese) salt cheeses would contribute ~8, 20 or 40%, respectively of the recommended dietary Na intake.

Salting of cheese and the role of salt on the biochemistry, microbiology and sensory quality of cheese has been extensively researched and reviewed (Guinee and Fox, 2004). In contrast, little information is available on the role of salt, or more specifically, sodium, in dairy spreads including pasteurized processed cheese products (Guinee et al., 2004) or milk fat spreads (Keogh, 1992). This review will briefly discuss the functions of salt in cheese products and dairy spreads, and discuss means by which the sodium levels of these products may be reduced.

16.2   Manufacture and salting of cheese and table spreads

Cheese manufacture is essentially a process of dehydration and acidification whereby the fat and protein (casein) of milk are concentrated between 6- and 12- fold, and the pH is reduced from ~6.6 in milk to between 4.6 and 5.4 in freshly made curd. While the manufacturing protocol differs markedly with variety, the basic manufacturing steps common to all varieties are gelation (coagulation), acidification, dehydration (e.g., by cutting the gel, syneresis/whey expulsion of resultant curd pieces by stirring and cooking in the expressed whey, drainage of whey, pressing of curd mass, continued acidification), moulding/shaping (by placing pieces of curd mass in forms and pressing) and salting. Once the curd has been through the five basic manufacturing steps, it is referred to as cheese. Cheeses may be consumed fresh, as in the case of fresh cheeses (such as Mozzarella, Quark, Cottage cheese), or may be stored (matured, ripened) and ripened for periods ranging from 1 week to 2 years depending on the variety (e.g., Cheddar, Gouda, Parmesan).

All cheeses, apart from Domiati, are salted towards the end of manufacture after the recovered curd is formed and/or moulded and pressed. There are three principal methods of salting cheese curd (Guinee and Fox, 2004):

1. *Brine salting* or *brining* – immersion of moulded curd in brine solution, as for Edam, Gouda, Saint Paulin, Provolone, Swiss-type cheeses, and many other well-known cheeses.
2. *Dry salting* – direct addition and mixing of dry salt crystals to broken or milled curd pieces prior to moulding and/or pressing, as for Stilton, Cheddar and related varieties.

3. *Surface dry salting* – rubbing of dry salt, or salt slurry, to the surface of the moulded curds, e.g. some Blue-type cheeses.

Sometimes, a combination of the above methods is used; e.g. low-moisture part-skim Mozzarella cheese, which was traditionally brine-salted, is now subjected to combinations of dry salting and/or brine salting.

The practice of adding salt to the curd at the end of curd manufacture, rather than to the milk, was undoubtedly deliberate in early cheese manufacture as rennet coagulability and gelation of bovine milk of typical composition (3.3%, w/w, protein) are markedly impaired by the addition of 2% (w/w) salt to milk, and completely inhibited at salt levels of 4% (w/w) (Grufferty and Fox, 1985; Abou-El-Nour, 1998; Guinee and Fox, 2004). Moreover, the addition of salt to milk prior to renneting severely impairs curd syneresis and whey expulsion during curd manufacture (Pearse and Mackinlay, 1989), and leads to excessively high moisture levels in the final cheese. The adverse effects of salt addition are due to solubilization of colloidal calcium phosphate (sodium/calcium ion exchange), and the resultant increase in casein hydration, which impairs casein aggregation. In Domiati cheese, where 5–15%, w/w, NaCl is added to the milk (Abou-El-Nour, 1998), the effects of NaCl in curd formation are off-set by the use of water buffalo milk, which has a higher casein content than bovine milk, or by the fortification of milk with skim milk powder, and/or the addition of $CaCl_2$. The effects of NaCl on casein hydration and the physical properties of cheese are discussed in more detail in Sections 16.4–16.5.

Table spreads, whether of the high- or low-fat type, are usually water-in-fat plastic solids. Low-fat spreads, butter and margarine are water-in-oil emulsions. Butter and margarine must by law contain a minimum of 80% fat, but spreads conventionally contain either 72–80% fat (full fat), 55–60% fat (reduced fat), 39–41% fat (low-fat) or 20–30% fat (very low-fat). The principle ingredients of table spreads are fat, emulsifier, milk protein, stabilizer, sodium chloride and water, and each of these will affect the emulsion and processing behaviour of the final product.

16.3   Functions of salt in cheese and dairy spreads

Salt plays a major role in the regulation of many aspects of cheese and dairy spreads. Its main functions in cheese are preservation and taste (Fig. 16.1). The taste of salt is highly appreciated by many and saltiness is regarded as one of the four basic flavours. Presumably, the characteristic taste of NaCl resides in the Na moiety since KCl has a distinctly different flavour sensation. At least part of the desirability of salt flavour is acquired but while one can easily adjust to the flavour of foods without added salt, the flavour of salt-free cheese is insipid, 'watery' and unnaturally bland even to somebody not 'addicted' to salt; the use of 0.8%, w/w, NaCl is probably sufficient to overcome the insipid taste (Thakur et al., 1975; Lindsay et al., 1982; Schroeder et al., 1988).

The preservation role of salt is due to its effect on water activity, a major

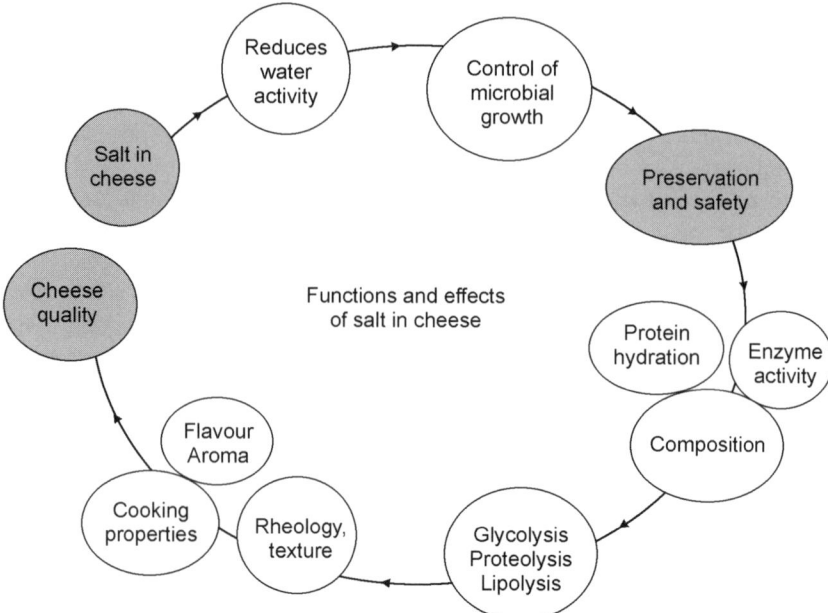

Fig. 16.1   Salt contributes to preservation, safety and overall quality of cheeses, by its effects on water activity, microbial growth, protein hydration/confirmation and enzymatic activities which in turn influence biochemical changes, such as proteolysis.

determinant of microbial growth. Additionally, salt content has major effects on cheese composition, microflora dynamics, protein hydration and enzymatic activity, and hence on the ripening rate, texture, flavour and quality of cheese.

16.3.1   Control of water activity (a_w) and contribution to preservation

The preservative action of NaCl in cheese is mainly due to its effect on the water activity (a_w) of the medium: $a_w = p/p_0$, where p and p_0 are the vapour pressure of the water in cheese and of pure water, respectively. If the cheese is at equilibrium with its gaseous atmosphere, then $a_w = \text{ERH}/100$, where ERH is the equilibrium relative humidity. On exposure to air of a given RH, water in the food, which does not contribute to its vapour pressure, may be considered as not being free, and not available for microbial growth. Typical values of the minimum water activity for the growth of various pathogenic microorganisms in foods and of the a_w of some cheese varieties are shown in Table 16.2. It is apparent that the a_w of most cheese varieties is not low enough to prevent the growth of yeasts and moulds and many bacteria, but together with low pH and low temperature, it is quite effective in controlling microbial growth in cheese.

The a_w of foods depends on its moisture content and the concentration of low molecular mass solutes in the moisture phase (Russell and Gould, 1991). Salt increases the osmotic pressure of the aqueous phase of foods, causing

Table 16.2 Typical water activity (a_w) values of some cheeses and the minimum water activity required for growth of some pathogens

Cheese	Typical a_w	Pathogen	Minimum a_w
Quark	0.99	*Shigella* spp	0.96
Cottage cheese	0.99	*Yersinia enterocolitica*	0.96
Camembert	0.98	*Pseudomonas spp*	0.95
Emmental	0.97	*Escherichia coli*	0.95
Gorgonzola	0.97	*Clostridium botulinum*	0.94
Edam	0.96	*Salmonella* spp	0.94
Cheddar	0.95	*Listeria monocytogenes*	0.92
Gouda	0.95	*Micrococcus* spp	0.87
Parmesan	0.92	*Staphylococcus aureus* (aerobic)	0.86
		Most yeasts and moulds	0.80
		Osmophilic yeasts and moulds	0.55

[a] Compiled from data of Rüegg and Blanc (1981) and Russell and Gould (1991).

dehydration of bacterial cells, killing them or, at least, preventing their growth. In cheese, solutes comprise mainly soluble salts, lactate, low molecular weight peptides, free amino acids, and native whey proteins. The a_w of young cheese is determined to a large extent by the concentration of NaCl in the aqueous phase (Marcos, 1993): $a_w = 1 - 0.033 [NaCl_m] = 1 - 0.00565 [NaCl]$, where $[NaCl_m]$ is the molality of NaCl, i.e., moles NaCl per litre of H_2O and $[NaCl]$ is the concentration of NaCl as g/100 g cheese moisture. As cheese matures, the contribution of low molecular weight peptides and free amino acids to the depression of a_w increases.

16.3.2 Control of microbial growth

Cheese is a fermented milk product, where the pH is reduced from 6.6 in milk to a typical value of ≤ 5.3 in fresh curd due to the conversion of the lactose in milk to lactic acid by the added starter culture. Together with the desired pH, water activity and redox potential, salt has a major influence on the cheese microbiology, inhibiting the growth of pathogens and controlling the populations of starter bacteria and adventitious, non-starter lactic acid bacteria (NSLAB) in the final cheese. The latter bacteria, which can vary with respect to species and strain, are generally harmless but are undesirable, especially in high numbers (e.g. $>10^7$ cfu/g), as they can contribute to unpredictable flavour and to the racemization of L(+)-lactate to D(-)-lactate and the formation of insoluble Ca-DL-lactate crystals which appear as white specks in the cheese (Kubantseva *et al.*, 2004).

The growth of mesophillic (e.g. *Lc. lactis ssp.* strains *cremoris* and *lactis*) and thermophilic starter culture lactic acid bacteria in cheese curd is markedly influenced by the concentration of NaCl, as reflected by changes in population density, lactose utilization and/or by their ability to reduce curd pH during cheese manufacture (Irvine and Price, 1961; Lawrence and Gilles, 1969; O'Connor,

1970, 1973a,b). Starter activity is stimulated by 2% (w/w) salt-in-moisture (S/M), not affected by 4% (w/w) S/M and strongly inhibited at 5% w/w. Hence, increasing the S/M content of Cheddar cheese from 4.8 to 5.5% (w/w) results in a sharp increase in the pH of Cheddar and a concomitant decrease in the grading scores (O'Connor, 1974; Thomas and Pearce, 1981). Although lactic acid production can be uncoupled from cell growth, it is likely that acid production at low salt levels is accompanied by high cell numbers which tend to lead to bitterness (Lowrie and Lawrence, 1972). Not surprisingly, bitterness in Cheddar cheese is markedly influenced by S/M level over a very narrow range: *Lc. lactis ssp. cremoris* HP generally yielded bitter cheese at S/M levels <4.3%, w/w, but rarely at >4.9%, w/w (Lawrence and Gilles, 1969; Lindsay *et al.*, 1982).

The above studies show that inhibition of starter occurs within quite a narrow pH range (Fig. 16.2), emphasizing the importance of precise control of S/M

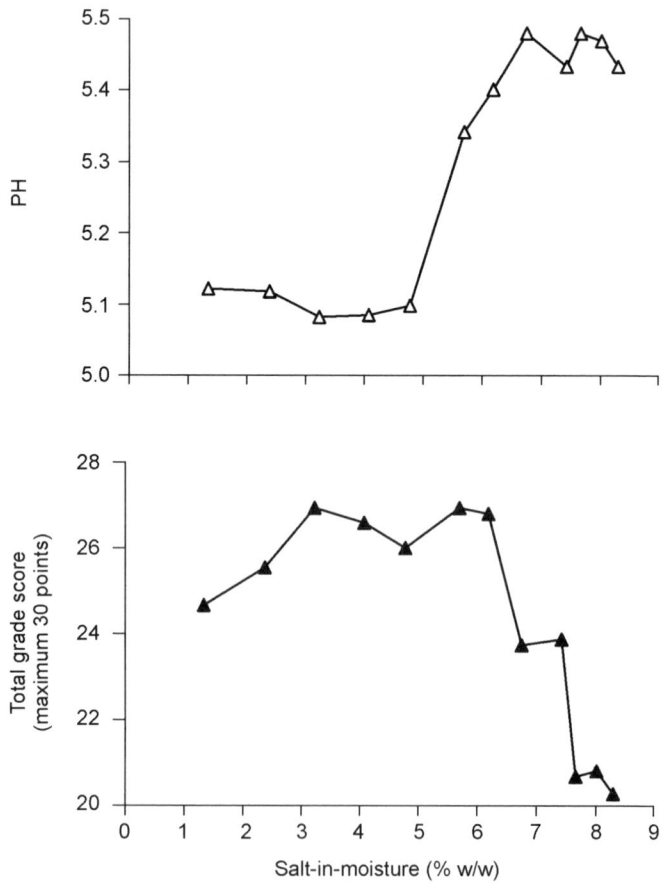

Fig. 16.2 Relationship between the salt-in-moisture (S/M) level and the pH (△) at eight weeks, and between the S/M and the total grade score (maximum 30) (▲) of Cheddar cheeses made from the same vats but salted at different levels (drawn from data of O'Connor, 1974).

level. However, the sensitivity of starter cultures to salt varies. At pH 5.3, *Lactococcus lactis ssp. lactis* strains are generally more salt-tolerant than strains of *Lc. lactis ssp. cremoris* but there is also considerable variation in salt sensitivity between strains of *Lc. lactis ssp. cremoris* (Martley and Lawrence, 1972; Turner and Thomas, 1980). Therefore, the influence of NaCl concentration on post-salting acid production in Cheddar cheese obviously depends on the starter used and a general value for S/M inhibition of mesophilic Lactococci cannot be definitely stated.

Streptococcus salivarius ssp. thermophilus, a starter bacterium frequently used in high cook-temperature cheeses such as Romano and Mozzarella, is considerably less salt-tolerant than *Lc. lactis ssp. lactis* (Rüegg and Blanc, 1981); its critical NaCl concentration is 0.4 M (2.34%, w/w), corresponding to an a_w of 0.984, compared with 1.1 M NaCl ($a_w = 0.965$) for *Lc. lactis ssp. lactis*. *Lb. delbrueckii ssp. helveticus* and *Lb. lactis ssp. lactis* were also less salt tolerant, being inhibited by 0.95 M and 0.90 M NaCl, respectively.

Propionibacteria are used as part of the starter culture in the manufacture of Emmental-type cheeses in which they metabolize lactic acid to produce CO_2, the gas, which is responsible for the characteristic eyes, and propionic acid which is thought to contribute to the sweet nutty flavour. The data of Rüegg and Blanc (1981) show that *P. shermanii* was the most salt tolerant of the starter species investigated: its critical NaCl concentration was 1.15 M (~6.7%, w/w; $a_w = 0.955$). However, as for Lactococci, salt sensitivity is strain dependent (Rollman and Sjosstrom, 1946): in a lactate medium, 6%, w/w, NaCl was required to inhibit the growth of a fast-growing strain of *Propionibacteria* at pH 7.0 and 3%, w/w, at pH 5.2, whereas a slow-growing strain was more salt-tolerant at pH 5.2 than at pH 7.0. Boyaval *et al.* (1999) showed that the growth of *P. freudenreichii ssp. shermanii* CIP 103027 in various media was increasingly inhibited by salt in the range (0 to 11%, w/w) and that inhibition was due to the osmotic effect rather than NaCl *per se*. The level of NaCl required to cause complete inhibition depended on the medium, an effect attributed to the presence of different types and levels of osmo-protective compounds in different media. In milk, ~1.5 and 5.9% (w/w) NaCl were required to double the generation time and inhibit growth completely.

It is generally agreed that NSLAB are more salt resistant than starter bacteria (Turner and Thomas, 1980; Thomas and Pearce, 1981; Lane *et al.*, 1997). Jordan and Cogan (1993) found that ~90% of NSLAB strains *(Lactobacillus casei, Lb. plantarum* and *Lb. curvatus)* isolated from commercial Cheddar grew in the presence of 6%, w/w, NaCl while 58% grew in the presence of 8%, w/w, NaCl. However, the results of Bechaz *et al.* (1998), which showed significantly higher populations of NSLAB in reduced salt Cheddar (1.0%, w/w) than in the control (1.8%, w/w), suggest that salt level has a major effect on the growth of NSLAB.

Mould growth, which is paramount to the ripening of cheeses such as Roquefort and other Blue cheeses, and Camembert-style cheeses, is also affected by the level of NaCl. Germination of *P. roqueforti* spores is stimulated by 1%, w/w, NaCl but inhibited by >3–6%, w/w, NaCl, depending on strain. However, the growth of germinated spores on malt extract agar or in cheese curd is less dependent on NaCl concentration than is germination and some strains grow in cheese curd containing 10%, w/w, NaCl, although growth is retarded compared to that in curd containing less NaCl (Godinho and Fox, 1981a,b). Interestingly, Blue cheeses are among the most heavily salted varieties, with 3–5%, w/w, NaCl (Stilton <3%, w/w). Growth of *P. camemberti* is also stimulated by low levels of NaCl, and mould growth on Camembert cheese is poor at NaCl levels <0.8% (w/w) (O'Nulain, 1986).

16.3.3   Regulation role of NaCl on enzyme activity in cheese
Coagulant
Rennet-induced gelation of milk is achieved by the addition of acid proteinases (e.g., chymosin, pepsins), also referred to as coagulants or rennets. For most rennet-curd cheeses, apart from those cooked at high temperatures (e.g. >50°C) such as Emmental, ~ 6 to 15% of coagulant added to the cheese milk is retained in the cheese (Guinee and Wilkinson, 1992). For many cheese varieties (internal bacterial-ripened types such as Cheddar, Gouda), the residual coagulant is the major agent responsible for the enzymatic degradation of casein to peptides during maturation, an event that has a major influence on the age-related softening of cheese. Moreover, the peptides are further degraded by the action of bacterial peptidases to produce free amino acids, which contribute directly to flavour development and serve as precursors for other flavour compounds.

In most hard and semi-hard, bacterially-ripened cheeses, α_{s1}-casein undergoes considerable proteolysis but β-casein remains relatively intact until an advanced stage of ripening (Phelan *et al.*, 1973; Kristiansen *et al.*, 1999; Fenelon and Guinee, 2000; Feeney *et al.*, 2001). The early hydrolysis (e.g. within ~1 month of manufacture) of α_{s1}-casein at the Phe_{23}–Phe_{24} peptide bond, by residual chymosin, leads to a marked weakening of the *para*-casein matrix, reductions in fracture stress and firmness, and an increase in softness (Guinee, 2003). The degradation of α_{s1}-casein and the concomitant formation of peptides are paralleled by an increase in the level of cheese N soluble at pH 4.6.

Numerous studies have reported an inverse relationship between the salt content and levels of proteolysis (as measured electrophoretically by degradation of α_{s1}-casein, and/or by the levels of pH 4.6-soluble N or non-protein N) in various cheeses, including Blue, Camembert, Cheddar, Danbo, Ragusano; Romano, Feta and other cheeses (see Guinee and Fox, 2004). The degradation of α_{s1}-casein is retarded by very low levels (e.g. 0.5%, w/w) of salt in Cheddar (Phelan *et al.*, 1973; Thomas and Pearce, 1981; Kelly *et al.*, 1996; Mistry and Kasperson, 1998). For most cheeses, including Cheddar (Kelly *et al.*, 1996) and Mozzarella (Guo *et al.*, 1997), the degree of β-casein degradation is much less than that of α_{s1}-casein and decreases rapidly as salt-in-moisture in Cheddar is increased from 0.37 to 5.4% (w/w).

Model studies in dilute solutions have shown that the inhibitory effect of NaCl on proteolysis of sodium caseinate, α_{s1}-casein and β-casein is pH-dependent, with the extent of inhibition generally decreasing with pH in the range 6.6 to 5.4 (see Upadhyay et al., 2004). At low pH, NaCl also alters the proteolytic specificity of the coagulants, chymosin and pepsins. The inhibitory effect of NaCl on the proteolytic activitity of the *Rhizormucor miehei* and *Cryphonectria parasitica proteinases* (coagulants) on β-casein is less than that on chymosin or pepsins (see Upadhyay et al., 2004).

Milk proteinase
Milk contains several indigenous proteinases, the most significant of which is plasmin. The level of residual plasmin activity in cheese depends on several factors including: the temperature to which the curd is cooked during manufacture, the level of curd washing with water, the pH of the curd at whey drainage, and the pH of the final cheese which affects activity of the residual plasmin. Plasmin appears to make a significant contribution to the maturation of Emmental, Blarney and Romano-type cheeses but contributes less to proteolysis in Cheddar or low-moisture-part-skim Mozzarella (Guinee and Fox, 2004; Upadhyay et al., 2004). The activity of the alkaline milk proteinase in model cheeses, without residual coagulant, was stimulated by concentrations of NaCl up to a maximum at 2% (w/w) but was markedly inhibited by higher concentrations of NaCl (4–8%, w/w) (Noomen, 1978).

Microbial enzymes
For most cheese varieties, microbial enzymes play a major role in the development of the desired flavour (McSweeney, 2004). Starter and non-starter, lactic acid bacteria possess an extensive proteolytic system, comprised of a cell-envelope associated proteinase (CEP) (lactocepin: EC 3.4.21.96), and intracellular peptidases, with the latter being released into the cheese matrix on death and autolysis (Beresford and Williams, 2004; Upadhyay et al., 2004). These peptidases hydrolyse peptides, produced by residual coagulant and/or plasmin, into free amino acids (FAA). Furthermore, intracellular peptidases degrade bitter peptides and thereby reduce the incidence of bitterness, a serious flavour defect in cheese that is generally attributed to the presence of peptides derived from β-casein by the actions of residual coagulant and/or starter cell CEP (see Guinee and Fox, 2004). Starter bacteria also possess intracellular lipases (see Collins et al., 2004) which hydrolyse triglycerides into free fatty acids (FFA). Both FAA and FFA contribute directly to flavour or serve as precursor compounds in the formation of other flavours such as alcohols, esters, aldehydes and ketones.

The CEPs of *Lactococcus*, the major species of starter culture bacteria used in cheese manufacture, may be categorized into two broad groups based on their specificity on caseins, P_I- and P_{III}-type proteinases (see Upadhyay et al., 2004). The different proteinase types may contribute differently to cheese flavour because of differences in specificity and the peptides they produce from casein (Reid and Coolbear, 1999). It has been postulated that starter culture strains possessing P_I- type proteinase (e.g., *Lc. lactis* subsp. *cremoris* strains HP and

WG2) degrade β-casein rapidly and may, thereby, contribute to bitter flavour in cheese. The bitterness of peptides is strongly correlated with hydrophobicity (Guigoz and Solms, 1976; Bumberger and Belitz, 1993). The bitter peptides in cheese appear to arise primarily from β-casein (see Visser *et al.*, 1983a,b; Frister *et al.*, 2000), as might be expected since β-casein is the most hydrophobic casein; however, peptides from α_{s1}- and α_{s2}-caseins, especially those containing proline, probably also contribute to bitterness in cheese (Lee and Warthesen, 1995; Frister *et al.*, 2000). In contrast to P_I-type proteinase, P_{III}-type proteinases hydrolyse β-casein differently and rapidly degrade α_{s1} and κ-caseins; starter cultures (e.g. produced by *Lc. lactis* subsp. *cremoris* strains AM1, SK11) having P_{III}-type proteinase tend to give non-bitter cheese.

The association between low salt level and the occurrence of bitterness in cheese (Lindsay *et al.*, 1982) suggests that the activity of starter proteinase is inhibited by a moderately high level of NaCl. Exterkate and Alting (1993) reported that the activity of P_I-type CEP from *Lc. lactis* subsp. *cremoris* strain HP on α_{s1}-casein (f1-23) at pH 6.5 in buffer systems was inhibited and altered by the presence of NaCl (4%, w/w). In contrast the activity of P_{III}-type proteinase on α_{s1}-casein (f1-23) was slightly stimulated at 4% (w/w) NaCl, but again specificity was altered (Exterkate and Alting, 1993). Reid and Coolbear (1998, 1999) reported that the specificity and activity of both P_I-type (from *Lc. lactis* subsp. *lactis* BN1) and P_{III}-type (from *Lc. lactis* subsp. *cremoris* SK11) CEPs at pH 6.4 on α_{s1}-, β- and κ-caseins in model systems, designed to simulate the cheese environment, were influenced by interactive relationship between a_w (0.95 to 1), NaCl (5%, w/v) and type of humecant (added to control the a_w). Gobetti *et al.* (1999b) reported that increasing the level of NaCl from 2.5 to 7.5% (w/v) inhibited the activity of CEP isolated from *Lc. lactis ssp. lactis* T12, an effect attributed to lowering effect of NaCl on a_w.

The autolysis of starter bacteria and the simultaneous release of intracellular peptidases into the cheese environment have a major impact on secondary proteolysis and flavour development in cheese. Increasing the S/M from 3 to 6%, w/w, reduced the activities of intracellular aminopeptidase, dipeptidyl-aminopeptidase and carboxypeptidase from *Leuconostoc mesenteroides ssp. mesenteroides* strain K_1G_8 but had little effect on the intracellular esterase activity (Vafopoulou-Matrojiannaki, 1999). Gobbetti *et al.* (1999b) found that the effect of NaCl on the activities of a range of intracellular peptidases and lipase activities of starter and NSLAB was both enzyme- and species-specific. Similarly, the lipolytic activities were influenced by interactive effects of pH and NaCl concentration to a degree dependent on bacterial species and strain. These authors concluded that interactions between S/M, pH and a_w were mainly responsible for changes in enzyme activity under conditions simulating cheese-making activity. In a similar study on the effect of salt (0 to 7.5%, w/w), pH (5.5 to 7.0) and temperature (4 to 16°C) on the peptidases of 11 strains of NSLAB bacteria isolated from cheese, Gobetti *et al.* (1999a) reported that the peptidase activities (aminopeptidases N and A, and proline iminopeptidase) of *Lb. casei* subsp. *pseudoplantarum* and *Lb. curvatus* were markedly reduced at low pH and high salt levels. Conversely, the peptidases of *Lb. casei* subsp. *casei* and *Lb. plantarum* were quite insensitive to pH and not very sensitive to NaCl.

16.4   The effects of NaCl on casein hydration

It is generally recognized that the extent of protein hydration has a major influence on the structure, physico-chemical stability and physical properties of protein-based products. The extent of casein aggregation, or hydration, affects the microstructure and nature of attractions between protein molecules in dairy products. Consequently, it has a major influence on several aspects of product quality: rheology, texture and cooking characteristics of cheese; texture and mouth-feel of yoghurt; and re-hydration characteristics of casein in food formulation (see Fox and McSweeney, 2003).

16.4.1   Casein hydration in model systems

Protein hydration is generally considered to be the water that is more-or-less immobilized by a protein (Damodaran, 1997). Casein hydration, and particularly casein micelle hydration, is a very useful tool for studying casein-casein inter-actions, e.g. cheese texture (Creamer, 1985) or in heat-treated milks (Creamer and Matheson, 1980). Korolczuk (1982) determined that a single equation could relate apparent viscosity to casein concentration and protein hydration in a simple fashion. Creamer (1985) showed that casein micelle hydration was affected by pH, NaCl and rennet treatment. A peak in casein micelle hydration was observed at pH 5.2 and while the hydration was reduced following the action of rennet, this peak in hydration was still retained. The addition of NaCl was shown to increase casein micelle hydration at all pH values above 4.5 probably by displacing calcium or calcium phosphate from the casein micelle with a concomitant increase in the number of ionic groups and a consequent increase in the volume of the matrix. The interaction between NaCl, pH and rennet action were central to the texture and subsequent functional behaviour of cheese.

Hermansson (1975) observed that the apparent viscosity of sodium caseinate solutions (>10%, w/w) was significantly increased in the presence of added NaCl. Carr *et al.* (2002) also observed that the apparent viscosity of caseinate solutions increased with increases in salt (NaCl, KCl or NH_4Cl) concentrations over an added ionic strength range of 0 to 1.1 mol L^{-1}. Fay (1996) showed that the apparent viscosity of sodium caseinate (12%, w/w) increased linearly with increasing Na^+ concentration (0–1.4%, w/w) irrespective of whether NaCl, Na_2PO_4 or tri-sodium citrate was utilised as the source of sodium. Where calcium was utilised as the counterion to the polyanionic caseinate, the casein became particulate and the protein less hydrated (Carr *et al.*, 2003). The addition of NaCl to a calcium caseinate might be expected to give a range of products with intermediate viscosity between sodium caseinate and calcium caseinate.

However, this is not the case and there is a peak in the viscosity-mineral composition plot (Carr *et al.*, 2003).

Strange *et al.* (1994) showed that the order of addition of casein and NaCl was critical to the proper functioning of casein in food products. They also observed that, at pH above the isoionic point, the pH of all casein solutions decreased upon the addition of NaCl. They suggested that the Na^+ was bound by casein resulting in the release of H^+. Overbeek and Bungenburg de Jong (1949) ascribed salt effects in protein solutions to the ease with which ionisable groups can dissociate. They saw the effect as charge suppression between the various charge-carrying groups on the protein and consequently the dissociation proceeds more easily at the same pH. More recently, de Kruif and Zhulina (1996) used a salted brush theory to describe casein micelle behaviour. The salted brush is characterized by the fact the charges along the chain are well screened by salt so that neighbouring charges, on either the same or different chains, only weakly interact. The salt concentration can therefore affect the swelling or shrinkage of the caseinate particles.

16.4.2   Casein hydration in natural cheese

Casein hydration in cheese has been measured indirectly by determining the weight of serum expressed (ES) on centrifugation at ~12 500 g at 25°C, and the levels of serum protein (SP). Storage of cheese results in an increase in the water-binding capacity of the protein as reflected by the decrease in the level of expressible serum (ES) and increase in the level of SP (Guo *et al.*, 1997). Increasing the mean level NaCl in brine-salted low-moisture Mozzarella from 0.13% (w/w) (unsalted) to 1.4% (w/w) markedly reduced the level of ES and increased the concentration of SP after 10 d storage at 4°C (Guo *et al.*, 1997). However, the levels of pH 4.6-soluble protein in the sera from both the salted and unsalted cheeses were similar, indicating that the differences in SP were due the solubilization of intact casein by the added NaCl rather than to differences in proteolysis of the casein. Paulson *et al.* (1998) showed that the level of ES in fat-free Mozzarella decreased as the level (%, w/w) of NaCl was increased in the range 0.14 to 0.85% (~0.2 to 1.4% S/M) and changed little as NaCl (%, w/w) was further increased to 2.18%, w/w (3.5%, w/w, S/M).

The above findings for Mozzarella are consistent with those from model systems showing that low concentrations of NaCl (~5–6%, w/w, S/M) enhance the solubilization of casein or *para*-casein (Hardy and Steinberg, 1984; Creamer, 1985; Guinee and Fox, 2004). Such a trend is expected as the protein matrix of Mozzarella, and all rennet-curd cheeses, may be considered as highly concentrated hydrated *para*-casein. The influence of low NaCl concentrations on casein hydration in cheese is also apparent from the higher level of 5% NaCl-soluble N than water-soluble N in Cheddar and other cheese varieties (Reville and Fox, 1978), and from the uptake of water by cheese placed in dilute brines, especially those without calcium (Geurts *et al.*, 1972; Guinee and Fox, 1986).

16.4.3   Casein hydration in pasteurized processed cheese products (PCPs)

The manufacture and factors affecting the functionality and quality of pasteurized processed cheese products (PCPs) has recently been reviewed (Guinee, 2003; Guinee *et al.*, 2004). Production generally involves formulation of the blend, consisting of shredded cheese, and/or optional dairy ingredients and emulsifying salts, with the types and levels of permitted ingredients in the different categories (processed cheeses, cheese food, cheese spread) being determined by legislation. The blend is typically cooked to 75–95°C for 3–5 min under constant shear. The resultant hot homogeneous molten blend is then homogenized, hot packed, cooled and stored at ~8°C.

Emulsifying salts, such as sodium citrates, sodium orthophosphates and polyphosphates are normally used as a processing aid in the manufacture of PCPs. While these salts are not emulsifiers *per se*, they affect, with the aid of heat and shear, a series of concerted physico-chemical changes that mediate the hydration of the insoluble cheese protein (calcium *para*-casein) which in turn facilitates uptake of free moisture (water-binding), emulsification of free fat, structure formation, and the formation of stable homogeneous product. The major changes that occur on addition of emulsifying salts include calcium sequestration, upward adjustment of pH and pH stabilization, *para*-casein hydration and dispersal, emulsification of fat and eventual structure formation. Calcium sequestration involves the exchange of Ca^{2+} of the *para*-casein for the Na^+ of the emulsifying salt, which is inherently driven by the capacity of the emulsifying salt anion to interact with calcium. As already alluded to in Section 16.4.1, replacing calcium with sodium as the counterion to the negatively charged casein, increases the protein hydration and hence, alters the textural properties of the cheese. The level of emulsifying salts added to PCPs is varied with category of PCP, type of ES, composition of blend ingredients, processing conditions (heat and shear) and the desired characteristics in the final PCP; the level is typically ~1.5% (w/w).

16.4.4   Effect of NaCl in table spreads

Table spreads, whether of the high or low-fat type, are usually water-in-fat-plastic solids. Low-fat spreads, butter and margarine are emulsions of the water-in-oil type. The level of NaCl in the aqueous phase can vary but is usually in the range 3 to 6%, w/w, depending on the fat content. The water-in-oil pre-emulsions of fat spreads are always stabilized by cold shearing of the fat phase to a plastic consistency. Before this solidification step, instability of the pre-emulsion can arise due to either phase separation or phase inversion (Mulder and Walstra, 1974). It is intuitively evident that the likelihood of phase inversion increases as the fraction of added disperse phase is increased. The processing of low-fat spreads comprises two steps, namely preparation of an aqueous phase-in-oil emulsion while stirring followed by pumping the emulsion through one or two scraped surface coolers in series at a defined agitation rate and at a defined refrigerant temperature. Patent literature has suggested that the higher the aqueous phase viscosity, the greater stability to inversion (Platt, 1988). Sodium caseinate is often the protein of choice to aid in the stabilisation of the water-in-oil emulsion. While the introduction of NaCl into the aqueous phase was initially for organoleptic reasons, the interaction between NaCl and caseinate is also a significant effect. The viscosity of a caseinate solution is an indicator of the degree of bound water absorbed by the hydrophilic groups as well as the water trapped inside the aggregated molecules (Korolczuk, 1982). Sodium caseinate contributed to the stability of the water-in-oil emulsion through steric and water binding effects (Keogh, 1992).

The same author concluded that NaCl made a significant contribution to the aqueous phase viscosity. The effects of NaCl, $NaHPO_4$, trisodium citrate and pH on the water-binding capacity of caseinate are well documented. These ions in turn strongly influence the interfacial rheology of the system, which reflects the conformational state of the protein and the rigidity of the side-chains (Dalgleish, 1989; Dickinson, 1989). While final emulsion stability and its stability to inversion may be related to the viscosity of the caseinate-based aqueous phase, the interaction between the level of fat-soluble emulsifier and the aqueous caseinate may also be significant (Barfod et al, 1989). Significant reductions in the level of NaCl in the aqueous phase can lead to inversion problems during processing, and alternative methods of increasing the aqueous phase viscosity may have to be approached, i.e. hydrocolloids.

16.5   Effect of NaCl on cheese functionality

Owing to its effects on casein hydration, NaCl level has a major effect on the functionality and quality of protein-based products.

16.5.1   Cheese rheology

The rheology of cheese may be defined as its deformation when subjected to stress, and its rheological properties as those that determine its response to a stress or strain, as applied when compressing, shearing and/or cutting, for example during processing (e.g., portioning, slicing, shredding, grating) and consumption (slicing, spreading, mastication and chewing) (O'Callaghan and Guinee, 2004). The rheological properties of cheese are of importance as they influence its processability (e.g., sliceability), suitability as an ingredient in culinary and formulated/assembled foods, its texture and eating quality (e.g., degrees of hardness/softness, brittleness, degree of mastication required for chewing) (Guinee and Kilcawley, 2004).

Owing to its large impact on the hydration of casein (Section 16.4), which is the main structural component of cheese, NaCl has a major influence on the rheology and texture of cheese. This is readily observed on sensorially contrasting salt-free cheeses with their salted counterparts immediately after salting: while the former are relatively soft, pasty and/or adhesive depending on age, the latter are firmer. In contrast, high salt concentrations lead to shortness and crumbliness, dryness and/or hardness, as observed for high salt cheeses such as Gaziantep, Domiati and Feta. Large strain deformation testing using texture analysers has been extensively used to study the effects of salt concentration, or salt-in-moisture (S/M), on rheological properties such as firmness (σ_{max}, e.g., force required to attain a given compression, or to push a probe to a given depth into cheese), fracture stress (σ_f), and/or fracture strain (ϵ_f). These studies have shown that increases in S/M within the range 0.4 to 12% (w/w) result in increased σ_{max}, σ_f, and sensory hardness of various cheeses including Camembert, Cheddar, Feta, Gaziantep, Mozzarella and Muenster (cf. Guinee and Fox, 2004).

Pagana and Hardy (1986) described the relationship between the σ_{max} of mini Camembert cheeses and S/M content in the range ~2–22% (w/w) at different levels of compression (10 to 70%) as being exponential: $\sigma_{max} = ae^{b.s/m}$, where a and b are parameters of the exponential function, with a dependent on the degree of deformation. The increases in σ_{max} and σ_f with increasing NaCl are in part due to the concomitant changes in composition, e.g. reduction in moisture level and increase in protein, and to the indirect effects of salt on proteolysis, pH, *para*-casein hydration/solubility and conformation (Guinee and Kilcawley, 2004). Euston *et al.* (2002) reported an interactive effect between salt level and pH on the microstructure and rheology of model skim milk cheeses with similar gross composition; increasing the NaCl content from ~1.5 to 3.5% at pH 4.6 increased the degree of swelling of the strands comprising the protein network of the cheese (indicative of a greater degree of hydration), reduced cheese elasticity, and increased fracture strain (reflecting a more plastic behaviour). A similar, but more pronounced, effect was observed on raising the pH of the model cheeses from 4.6 to 5.2 while holding the level of NaCl constant at 1.5% (w/w).

Pagana and Hardy (1986) reported a linear decrease in fracture strain (ϵ_f) of Camembert cheese as the S/M content was increased in the range ~2–22%, w/w; this relationship indicated a positive relationship between brittleness and salt level. A somewhat different relationship between ϵ_f and S/M content in the range 0.2 to 10.5% (w/w) was noted for model Gouda cheeses by Visser (1991): similar to results of Pagana and Hardy (1986), ϵ_f increased monotonically with S/M in the range 0 to ~4.5%,w/w, but then decreased sharply to a value which was about half the maximum at 5.5%, w/w, S/M and thereafter remained relatively constant as the S/M was increased to 11.3%, w/w. The discrepancy between the latter studies may be related to differences in the S/M and pH which for a fixed salt level would affect the degree of casein hydration and, hence, the ratio of viscous-to-elastic behaviour of the protein matrix (Euston *et al.*, 2002).

Similar to σ_f and σ_{max}, the increase in ϵ_f with S/M is probably also attributable to a *salting-in* effect of the *para*-casein coinciding with the concomitant increase in *para*-casein hydration as S/M was increased to ~5%, w/w, and a *salting-out* effect with a concomitant loss in casein hydration at high S/M levels (Geurts *et al.*, 1972; Guinee and Fox, 1986). An increase in casein hydration would impart a more viscous (and less elastic) character to the cheese and a transition from elastic fracture behaviour to plastic fracture behaviour, which would necessitate a higher strain for fracture (see O'Callaghan and Guinee, 2004). Conversely, a lower degree of casein hydration at the higher S/M would favour a more elastic casein matrix and an elastic fracture behaviour, i.e. a shorter, firmer, more brittle cheese.

16.5.2 Cooking properties of natural cheeses

Cheese is used extensively as an ingredient in cooking applications in the food service industry, typical examples including toasted sandwiches, pizza, lasagne, burgers, and other hot culinary dishes. In these applications, the heated cheese (80 to 100°C) is required to have one of more characteristics or attributes depending on the dish. Examples of these attributes, which are technically referred to as heat-induced functional properties (functionality), include: meltability (ability to soften), flowability (ability to spread), flow resistance (ability to melt without spreading), stretchability/stringiness (ability to form strings or sheets when extended), oiling-off (ability to exude sufficient oil to create a surface sheen/gloss, prevent drying out, and retain succulent mouth feel). The development of acceptable cooking characteristics in cheeses, such as Mozzarella and Cheddar, generally requires a storage period at low temperatures (4–10°C) (Kindstedt *et al.*, 2004; Guinee and Kilcawley, 2004), the duration of which depends, *inter alia*, on the cheese type, manufacturing conditions, ripening conditions and the specifications set by the customer. During storage, biochemical and microstructural changes occur which contribute to the development of the desired cooking characteristics. These include increases in proteolysis, casein hydration, fat globule coalescence, and swelling of the protein matrix. Studies on cooking properties have focused mainly on Cheddar and Mozzarella, because in volume terms these cheeses are the most extensively used in cooking.

The flowability of both low-moisture part-skim Mozzarella and non-fat Mozzarella increases fairly linearly with salt level in the range 0.1 to 0.5% (w/w) (S/M ≈ 0.2 to 1.2%, w/w) and changes only slightly as the salt level is increased further to ~2.2% (w/w) (S/M ≈ 3.3%, w/w) (Apostolopoulus *et al.*, 1994; Paulson *et al.*, 1998); in contrast, unsalted Mozzarella has very poor cooking properties as reflected by the absence of flow and stretchability. The increase in flowability with salt level coincides with increases in free oil content and water-binding capacity of the *para*-casein matrix, as reflected by a decrease in the level of expressible serum, and increases in serum protein (Guo *et al.*, 1997) and degree of storage-related swelling of the protein matrix (Paulson *et al.*, 1998).

Zonal variations of S/M level in brine-salted Mozzarella during storage lead to variations in functionality in different parts of a cheese block (Kindstedt, 1990). On melting, cheese from the high salt (~3.04%, w/w, at day 2) surface region had a higher apparent viscosity and lower level of free oil, and was tougher and more chewy than cheese from the low-salt (~0.38%, w/w, at day 2) interior region which was smooth, soft, fluid and gelatinous. These differences gradually disappear as equilibrium of S/M concentration throughout the cheese loaf becomes established. The above trends suggest a salting-in effect of the *para*-casein matrix with a concomitant increase in hydration at low S/M levels (e.g., <1.5%, w/w) and a salting-out effect and decrease in *para*-casein hydration at the high salt levels (especially at >6.3%, w/w, S/M). Consequently, Kindstedt and Guo (1997) concluded that, in addition to proteolysis, the increase in casein hydration, at 3 to 4%, w/w, S/M is a major factor contributing to the development of the desirable cooking properties in low-moisture part-skim Mozzarella during storage.

16.5.3   Effect of salt content on overall quality of natural cheese

In addition to its effects on lactose metabolism and pH control, as discussed in Section 16.3.2, salt also exerts a major influence on the levels of proteolysis and lipolysis, which together contribute to the development of texture and flavour in cheeses during maturation (McSweeney, 2004; Collins *et al.*, 2004; Upadhyay *et al.*, 2004). Consequently, the concentration and distribution of salt in cheese have a major influence on various aspects of quality in many cheese varieties; these effects have recently been reviewed (Guinee and Fox, 2004). The following discussion focuses mainly on the effect of salt on the quality of Cheddar cheese.

Cheddar cheese

Primary proteolysis, as measured by the levels of pH 4.6 soluble N, is considerably more extensive in unsalted than in salted Cheddar cheese and consequently the body of the former is less firm (Thakur *et al.*, 1975; Schroeder *et al.*, 1988). A linear relationship between the extent of degradation of both α_{s1}- and β-caseins in young (1 month) cheese and %, w/w, S/M is apparent from the data of Thomas and Pearce (1981) and Kelly *et al.* (1996). The inhibitory effect of salt on the proteolysis of β-casein in Cheddar cheese (Phelan *et al.*, 1973; Kelly *et al.*, 1996) appears to be particularly important in reducing the incidence of bitterness, the occurrence of which is greatly increased at low S/M <4.9% (Lawrence and Gilles, 1969). The effectiveness of NaCl in preventing bitterness is very likely due to its effect on ionic strength; the aggregation of β-casein at elevated ionic strength results in cleavage sites, which are located in the hydrophobic region of the molecule, becoming inaccessible to chymosin (see Guinee and Fox, 2004). In contrast to primary proteolysis, the level of secondary proteolysis, as measured by the level of 5% (w/v) phosphotungstic acid-soluble N, tended to be higher in salted (2.7–5.7%, w/w, S/M) than in unsalted Cheddar at 12 and 24 wk (Kelly *et al.*, 1996).

Little information is available on the influence of salt on lipolysis in Cheddar and other cheeses. Thakur *et al.* (1975) found higher concentrations of all individual fatty acids, except linoleic and linolenic (at certain ages) in unsalted Cheddar (~0.1% NaCl) than in the corresponding control cheese (~1.6%, w/w,

NaCl). However, Lindsay *et al.* (1982) found little difference between the levels of free fatty acids in Cheddar cheeses with low (3.5%, w/w) or intermediate (4.2%, w/w) S/M levels except for myristic and palmitic acids, which were considerably higher in the higher-salt cheese.

Five major studies have considered the effects of composition (including level of salt or S/M) on the quality/grading scores of mature Cheddar cheese. These involved the analysis of 300 commercial Scottish Cheddar cheeses (O'Connor, 1971); 24 commercial Cheddars salted at different rates (O'Connor, 1973a,b); 12 commercial Cheddar cheeses salted at different rates (O'Connor, 1974); an unspecified number of experimental and commercial New Zealand Cheddars (Gilles and Lawrence, 1973); 123 commercial Irish Cheddars (Fox, 1975); 486 experimental New Zealand Cheddar chesses (Pearce and Gilles, 1979); and ~10,000 commercial New Zealand Cheddar cheeses (Lelievere and Gilles, 1982). These studies have identified four *Key Compositional Parameters* (KCPs), namely the levels of S/M, moisture-in-non-fat substances (MNFS), pH, and fat-in-dry matter (FDM), whose impact on quality are inter-dependent. Additionally, two key process parameters (KPPs) were identified as having a large impact on quality and in determining the ranges of the four KCPs which are necessary to give good quality (Gilles and Lawrence, 1973; Lawrence *et al.*, 1984). The KPPs are the rate and extent of acid production in the cheese vat prior to whey drainage, and they determine the proportions of the colloidal calcium and phosphate of milk that are retained in, and the buffering capacity of, the cheese.

While these studies agree that the 4 KCPs are major determinants of Cheddar cheese quality, they disagree on the relative importance of these parameters. However, they concur that defined levels of S/M are critical for quality:

- Cheddar grade deteriorates rapidly at S/M levels <3.0% (w/w) and >6% (w/w) (Fig. 16.2),
- highest grades are achieved with S/M values in the range of ~4.7–5.7% (w/w), which is equivalent to a salt content of ~1.7–2.1% for a cheese with 37.5% moisture,
- the effect of salt level on quality is very dependent on the values of the other three KCPs and the two KPPs.

The importance of salt content with respect to overall quality is scarcely surprising because of its effects on water activity, microbial growth, lactose metabolism, pH, enzymatic activity, lipolysis, proteolysis and casein hydration (as discussed in Sections 16.3–16.4). In addition to 'unclean' flavours, bitterness has been reported consistently as a flavour defect in low-salt cheeses (Lindsay *et al.*, 1982; Kindstedt and Kosikowski, 1986). Texture defects associated with low salt levels include a soft, weak, pasty body, and suggest excessive proteolysis; at high salt levels, the cheese body becomes excessively firm, probably as a consequence of the lower proteolysis and a lower degree of casein hydration.

Other cheese varieties
Compared to Cheddar cheese, relatively little information is available on the relationship between the salt content and grading scores/overall quality of other cheeses; however, some information is available on the effect of salt on proteolysis and lipolysis.

Proteolysis in various cheeses including Blue (Godinho and Fox, 1982; Hewedi and Fox, 1984), Camembert (O'Nulain, 1986), Feta-type (Pappas *et al.*, 1996), Danbo-type (Kristiansen *et al.*, 1999), and Domiati (El-Sissi and Neamat-Allah, 1996) is reduced as the concentration of NaCl is increased. The negative effect of salt on proteolysis is associated with its effect on ionic strength, which affects casein hydration and conformation, and the concomitant reductions in the level of MNFS and a_w (Creamer, 1971; Lawrence and Gilles, 1980; Rüegg and Blanc, 1981; van den Berg and Bruin, 1981).

Likewise, higher salt levels reduce the degree of lipolysis (concentrations of total and individual FFA and/or free fat acidity) in Cheddar (Thakur *et al.*, 1975), Feta (Pappas *et al.*, 1996) and Gaziantep (Kaya *et al.*, 1999) cheeses. However, an opposite trend has been reported for Idiazabal cheese (Najera *et al.*, 1994) and Mahon cheese (apart from caprylic and stearic acids) (Sánchez *et al.*, 2001). Inter-study discrepancies are probably due in part to variations in the ranges of S/M being investigated. Hence, lipolysis in Blue cheese was maximum at 4–6%, w/w, NaCl, and lower at higher and lower salt levels (Godinho and Fox, 1981c); however, the concentrations of methyl ketones were relatively independent of salt concentration. Similarly, the effect of salt on lipolysis in Picante, a hard Portuguese cheese, varied depending on NaCl level: it generally increased with S/M in the range of ~9 to 16% (w/w) and decreased to varying degrees on further increasing the S/M level to 20–23% (w/w) (Freitas and Malcata, 1996). Undoubtedly, other factors such as type of milk, pasteurization of milk, type of coagulant and starter culture, and others (e.g., homogenization of milk) influence the effect of salt on degree of lipolysis (Freitas and Malcata, 1996).

16.6   Approaches to reduce the salt content of cheese products and table spreads

16.6.1   Maintaining salt at the lowest level required for good quality natural cheese

Probably, one of the most effective ways of reducing the salt content of any variety is to consistently maintain the salt content at the minimum level required for optimal quality of that variety. However, inter-batch variation in salt content is quite common in cheese, for many reasons including variations in cheese composition and dimensions, salting conditions and manufacturing technology (cf. Guinee and Fox, 2004). Minimization of salt content to optimum levels is discussed separately for Cheddar and brine-salted varieties.

Cheddar cheese

Recent studies indicate significant variation between the salt content of retailed Cheddar cheeses on the Irish and UK markets: reduced-fat Cheddar with fat <18% (w/w), 1.5–2.6% NaCl (Fenelon *et al.*, 2000); reduced-fat Cheddar with 19–24% fat, 1.2–2.6% NaCl (Fenelon *et al.*, 2000); and full-fat Cheddar, 1.6–2.4% NaCl (Fenelon *et al.*, 2000), and 1.3–2.2% NaCl (Guinee *et al.*, 2000). Similarly, large variations in salt (from ~0.89–1.42%, w/w) and moisture (32–36%, w/w) have been found among leading brands of commercial Cheddar cheeses in the USA (Schroeder *et al.*, 1988). Indeed, variations in the content of salt, and S/M, can occur between cheeses from different vats of milk, between blocks of Cheddar cheeses (typically ~ 20 kg weight) from the same cheese vat (typically with a milk volume equivalent to 120 blocks of cheese), and even within the same block owing to poor distribution of the applied salt throughout the curd mass and to possible variations in curd composition (especially moisture, and acidity) at the time of salting (Guinee and Fox, 2004). Such variation is surprising since the method of salt application (i.e. direct mixing of dry salt with curd chips) used in Cheddar cheese manufacture would appear to be particularly amenable to ensuring accurate control of salt concentration with respect to both level and uniformity (Bennett and Johnston, 2004; Guinee and Fox, 2004). However, it is possible to achieve much tighter control of salt content in full-fat Cheddar as indicated by a recent survey of Irish Cheddar cheese manufacturers: e.g. 1.72 ± 0.12% NaCl, in commercial Cheddar from an Irish plant sampled monthly over the manufacturing season in 2003 (Kelly and Guinee, unpublished results).

Based on the relationship between composition and grade, as discussed in Section 16.5.3, it is clear that for Cheddar cheese, at least, two approaches can be used to reduce NaCl level without compromising on quality:

(i)   eliminating higher-than-necessary NaCl levels (i.e., S/M > 5.7; equivalent to NaCl levels of > 2.23, 2.17, 2.11, 2.05%, w/w, for cheeses of 39, 38, 37 and 36%, w/w, moisture). In addition to the beneficial effect of reducing the sodium content of cheese, this approach is desirable, as it should improve grade scores of Cheddar.

(ii) targeting the lower end (say, 4.7 to 5.0% S/M) of the range prescribed for first grade quality, i.e. eliminating S/M levels > 5% (equivalent to NaCl levels of > 1.95, 1.90, 1.85, 1.80%, w/w, for cheeses of 39, 38, 37 and 36%, w/w, moisture). The reduction of salt to the lower end of the range prescribed for premium quality may be particularly amenable to quick ripened, mild-flavoured Cheddar cheeses.

The implementation of both approaches, especially (ii), necessitates a high degree of process control to ensure that the mean salt concentration and range of salt in cheese is consistently kept within a narrow window of tolerance, with low intra- or inter-block variation. This in turn requires a comprehensive understanding of:

(i) the interactive relationship between the four KCPs and cheese quality,

(ii) the control of cheese composition by manipulation of the cheese making process,

(iii) the need to standardize basic aspects of cheese manufacture (e.g. casein or protein level, casein-to-fat ratio of cheese milk, pH of the milk and/or curd at different stages of manufacture such as set, whey drainage and milling; the addition of starter culture and rennet based on weight of casein in the cheese rather than on the basis of milk volume which is more usual, firmness of gel at cut, vat emptying time, and curd handling ex-vat). Ideally, the curd should be washed, if necessary, to standardize the level of lactose, and ultimately the final level of lactic acid, in the moisture phase of the cheese.

(iv) the factors affecting uptake and distribution of salt in curd.

Familiarity with these intricacies is particularly relevant when making more than one Cheddar recipe in the same plant, and when a seasonal milk supply showing large variations in levels of fat and protein, and rennet coagulation characteristics is used for manufacture.

For any given Cheddar cheese making protocol, the factors that determine the salt content of the final cheese are the level of salt added to the curd, factors affecting its absorbance (uptake) and the uniformity of its distribution throughout the curd mass; these factors have been recently reviewed (Guinee and Fox, 2004). The content of salt, or S/M, in Cheddar cheese increases with the level of salt added to the curd, but at a diminishing rate owing to higher losses in the whey exuding from the curd chips at high rates of salting. For a given level of salt addition, the quantity of salt absorbed by the curd, and hence the level of salt in the final cheese, increases with reductions in curd moisture in the range 45 to 37% (w/w), curd acidity, size of curd chip, depth of salted curd bed, and curd temperature in the range 41 to 24°C, and with increases with curd pH, mellowing time (time between salting and pressing), and degree of mixing salt and curd.

The uniform distribution of the applied salt in the curd is primarily dependent on:

- the flowability of the salt (as affected by changes in temperature and humidity of storage area) and its ability to be accurately metered, conveyed and distributed on the curd bed
- design and maintenance (edge) of mill which determines the size distribution of milled curd chips before salt addition
- plant throughput/capacity which will affect the cleanness of chip cut, size distribution of chips, and level of curd slivers
- design and maintenance of the salting system which affects the weight of salt conveyed, its distribution across the curd mass, and degree of mixing of salt and curd
- design of system for conveyance and filling of salted curd into the block formers, which affects the chip particle distribution within each block former;

since small chips will have a higher salt level than larger ones, salting systems which do not preferentially lead to filling of separate block formers based on a predominance of a particular size range are most desirable to ensure uniform salt level and minimize inter-block variation in cheeses from one vat.

The significant variations in the salt and moisture levels of commercial Cheddar, as discussed above, are unlikely to be the result of differences in salting rate, which, from the authors' experience of commercial Cheddar plants, tends to be fixed within a given plant and not varied with recipe. Most likely, it ensues from intra-vat and inter-vat differences in: curd composition (especially moisture, pH), measurement and distribution of the applied salt, variations in size distribution of curd particles filled into individual block formers. Factors contributing to variation in curd composition at salting, include *inter alia*: seasonal variations in milk pH and protein level which affect gel firmness; non-standardization of cheese making procedure, and inter-vat variations in starter culture activity and curd pH. Moreover, seasonal variations in curd yield and throughput affect milling efficiency and size distribution of curd chips, which in turn influence the degree of salt absorbed, level of compaction of curd chips on the salting belt and the accuracy of the relationship between the height (as measured by sensing fork) and weight of curd being conveyed.

Undoubtedly, a more consistent salt level and distribution could be obtained by:

- the production of curd with more consistent composition (e.g., pH and moisture),
  - e.g., *via* standardization of milk protein level (e.g., by ultrafiltration of cheese milk or supplementation with micellar casein powders) and protein-to-fat ratio;
  - a more defined cheesemaking process where stages/operations are regulated (e.g., levels of rennet and starter in proportion to casein level; pH at set, drain and milling; firmness at cut);
  - the use of cheese vats with dual curd-whey outlets with faster curd/whey emptying, especially from larger vats (e.g., 30,000 L);
  - washing to constant lactose level in the curd;
- more consistent curd throughput by standardization of milk protein level
- more accurate determination of the curd flow and the quantity of salt required by the use of load cells on the salting belt;
- more efficient mixing of salt and curd by the use of curd/salt mixers or tumblers operating in conjunction with the salting belt (Sutherland, 2002; Bennett and Johnston, 2004);
- optimisation of salt flow;
- a narrow curd chip size distribution by optimizing mill performance;
- the use of a curd conveying/filling system that ensures a similar size distribution of salted chips in all block formers.

Brine-salted cheeses

In cheeses which are salted by immersion in brine and/or by surface application of dry salt there is a decreasing salt gradient from the surface to the centre and a decreasing moisture gradient in the opposite direction at the completion of salting (Guinee and Fox, 2004). These gradients disappear slowly due to the slow diffusion of salt from the rind inwards, and in contrast to Cheddar cheese, equilibrium of S/M is practically reached at some stage of ripening. The time required for S/M equilibration is dependent on cheese type, composition, size and shape of cheese and curing conditions: e.g., 7–10 d for Camembert (0.25 kg, flat disc), 4–6 weeks for Edam (2.5 kg sphere), 7–9 weeks for Gouda (10 kg wheel) and >4 months for Emmental (60–130 kg wheel) (Sutherland, 2002).

While the effects of alteration of salt level on the biochemistry and texture/functionality of these cheeses have been extensively investigated (Sections 16.3–16.5), the authors are unaware of published studies on the relationship between salt content and grading score or quality of commercial brine-salted/surface dry-salted cheeses. This is probably in part because of the large zonal variations in salt and moisture in such cheeses immediately after salting and for some time afterwards, which would make the relationship of a mean salt content and quality less meaningful, at least for young cheese. Hence, establishing a minimum range of salt, or S/M, level required to maintain optimum quality for these cheeses would first necessitate the establishment of such relationships in mature cheeses where equilibrium of salt-in-moisture has been established. Then, the manufacturing process would be optimized to ensure that salt level was maintained consistently within the prescribed range, and a plan devised to ensure representative sampling of cheese and salt level within a cheese loaf.

The factors affecting salt uptake and the mean salt content of brine-salted cheeses include: curd composition (moisture, lactose/lactate, and pH); curd dimensions and shape, which determine the ratios of surface area to volume and curved surface to planar surface; brine concentration and temperature; salting time; initial S/M level of curd and pre-salting, where combinations of dry-salting and brine salting are used (Sutherland, 2002; Guinee and Fox, 2004).

16.6.2 Reducing salt to a content below the minimum level generally considered necessary for good quality in a particular natural cheese variety

Lindsay *et al.* (1982) found that reducing salt content from 1.75 (4.9% S.M) to 1.25% (3.5% S/M) did not significantly affect the taste/texture attributes and overall preference scores for Cheddar cheeses (~35%, w/w, moisture) over a 9-month ageing period at 3°C. Similarly, Schroeder *et al.* (1988) reported no detectable differences in the flavour, texture or overall desirability between cheeses with salt levels of 1.4 (~4.1% S/M; 35% moisture) and 1.1% (3.1% S/M; ~36% moisture) at various times over a 7-month storage period at 4.5°C. While further reduction of salt content reduced consumer scores, cheeses with 0.7% (2% S/M) received acceptable overall desirability ratings. At salt levels <0.7%, cheeses became excessively adhesive (stickiness), acid, bitter, soft, lacked saltiness and had unpleasant aftertaste; the latter changes coincided with large increases in proteolysis.

Because of its relatively large serving size (~112 g compared to ~66 g for other cheeses), Cottage cheese has been viewed as a potentially high source of dietary sodium (Marsh *et al.*, 1980). Wyatt (1983) found that a 35% reduction in NaCl (from 1 to 0.65%, w/w) level in Cottage cheese did not influence consumer response; however, reduction by ≥50% resulted in significantly lower scores than the control (1%, w/w, NaCl).

The above studies suggest scope to reduce the salt content of Cheddar and Cottage cheeses from their typical mean values by ~20–30%, without significantly impairing quality. However, in light of the effect of salt level and the interactive effects of the four KCPs on biochemistry, microbiology and grading score/quality of Cheddar cheese, as discussed above (Section 16.6.1), the potential of such an approach, at least in the case of Cheddar, may be most suited to cheeses which are mild-flavoured and short-ripened, or to cheeses with a relatively low levels of MNFS and FDM. Additionally, the quality of commercial reduced-sodium Cheddar, and indeed other rennet-curd cheeses, can be influenced by factors such as pitching pH, the type (proteolytic activity) and amount of residual coagulant in the cheese, types and counts of starter and non-starter bacteria, and ripening temperature. Thus, process modifications such as bactofugation of the milk and reduction of the cheese moisture levels, were deemed necessary to prevent butyric acid fermentation by *Clostridium tyrobutyricum* and its undesirable effects (van den Berg *et al.*, 2004) in Gouda cheese in which the salt level was reduced by ~20% (from 4 to 3% salt in dry matter) (van den Berg *et al.*, 1986). It is also noteworthy, that the ripening temperatures in the studies of Lindsay *et al.* (1982) and Schroeder *et al.* (1988) are quite low, i.e. 3 to 4.5°C compared to typical values of ~8°C. While low ripening temperature would reduce the potential to the development of off-flavours (e.g., bitterness associated with more extensive degradation of β-casein, undesirable flavours due to high counts of NSLAB (Beresford and Williams, 2004)) or other defects in reduced-salt cheese, it would lengthen the storage time required for maturation.

16.6.3   Substitution of NaCl in natural cheese by other salts such as KCl

Studies on the taste of salts (Mooster, 1980; Murphy *et al.*, 1981) indicate that certain cations elicit saltiness while their anions tend to inhibit this perception and contribute their own taste (Karahadian and Lindsay, 1984). Compared to sodium, other cations (potassium, magnesium, calcium and ammonium) educe more bitterness and less saltiness. Compared to chloride, other anions (phosphates and citrates) depress saltiness and contribute more directly to flavour, with phosphate imparting a chemical metallic type flavour (Mooster, 1980). Owing to the varying effects of different anions and cations on saltiness, the partial substitution of NaCl by an alternative salt with a non-sodium cation offers potential as a means of reducing sodium in cheese. Consequently, KCl, $MgCl_2$ or $CaCl_2$ have been investigated as potential substitutes for NaCl in the production of low-sodium cheeses.

Partial substitution of NaCl with KCl
Substitution of NaCl with a KCl/NaCl (1:1) mixture, added at a level which gave an ionic strength similar to the control cheese (1.4% NaCl), gave low-sodium Cheddar cheese (~0.44%, w/w, sodium compared to 1.25%, w/w, in the control) which at 16 weeks was not significantly different from the control in terms of proteolysis, texture flavour and acceptability (Fitzgerald and Buckley, 1985). Similarly, Reddy and Marth (1995) found that the partial substitution of NaCl with KCl did not affect the types of starter and NSLAB bacteria in Cheddar cheeses. In contrast, Lindsay et al. (1982) reported that low-salt (1.25%, w/w – based on chloride salts of Na and K) and medium-salt (1.5%, w/w) Cheddar cheeses prepared using 1:1 molar mixtures of NaCl/KCl had higher levels of free fatty acids (especially C16 and C18 congeners), tended to have a lower perceived saltiness, and were more bitter and astringent than the corresponding cheeses salted using NaCl only; indeed, for the medium salt Cheddar cheeses, the use of a 1:1 NaCl/KCl blend significantly reduced the overall preference score awarded to the cheeses at ripening times >3 months because of bitterness and low intensity of saltiness.

Martens et al. (1976) reported the successful manufacture of low-sodium Gouda cheese using mixtures of NaCl and KCl in curd manufacture and brining. While the Na and K levels (mg %) in the control cheese were ~830–650 and 120, those of the reduced-sodium cheese were 200 and ≥200, respectively. However, reduction of salt in dry matter in Gouda cheese by ~20% was found to increase the risk of butyric acid fermentation (van den Berg et al., 1986).

Katsiari et al. (1997) reported that the sodium content of Feta cheeses was successfully reduced by 50%, by partial replacement of NaCl with NaCl: KCl mixtures (3:1 or 1:1, w/w) during the combined surface dry salting/brine-salting procedure, without affecting gross composition, water activity, lipolysis (Katsiari et al., 2000a), proteolyis (Katsiari et al., 2000b), or sensory or textural properties. Similarly, a 50% reduction in the sodium content of Kefalograviera cheese, by substituting KCl for NaCl, had no effect on lipolysis or proteolysis (Katsiari et al., 2001a,b).

Substitution of NaCl (1.26%, w/w) with a 1:1 commercial mixture of NaCl:KCl (each at 0.63%, w/w) did not significantly affect the flavour scores of Cottage cheese (Demott et al. (1984). Yet, a subsequent study on Cottage cheese by the same authors (Demott et al., 1986) reported a significant reduction in flavour scores when the sodium content was reduced from 0.63 to 0.33% (w/w) by substitution of NaCl with a commercial NaCl/KCl mixture.

The above results show that partial substitution of NaCl with a 50:50 (%, w/w) mixture of NaCl and KCl generally does not markedly alter biochemical, textural, microbiological characteristics of cheeses. Yet, some inter-study discrepancies exist in relation to the effect of a 50:50 (%, w/w) NaCl:KCl as a substitute for NaCl alone on the grading and sensory quality of cheese. Hence, the use of NaCl: KCl mixtures with weight ratios > 50:50 (e.g. 70:30 or 60:40) would probably prove more attractive to some cheese manufacturers; such NaCl:KCl blends also offer significant reductions in sodium level, e.g. from ~0.79% (w/w) Na in a standard Cheddar cheese (salted using NaCl only; salt content, 1.9%, w/w) to 0.54 or 0.47% w/w Na, respectively.

Total substitution of NaCl with KCl
In contrast to the above, Cheddar cheese salted with KCl alone is extremely bitter, an effect, which may be associated with the concomitant increase in concentrations of K^+ ion (Mooster, 1980; Murphy *et al.*, 1981) and free fatty acids in the cheese (Fitzgerald and Buckley, 1985). Similarly, the use of $MgCl_2$ or $CaCl_2$ alone, or in 1:1 mixtures with NaCl, are not suitable NaCl substitutes in Cheddar (Fitzgerald and Buckley, 1985) or Gruyère (Lefier *et al.*, 1987) cheeses, because of associated texture (crumbly, soft, greasy) and flavour (metallic, bitter) defects. These defects in Cheddar coincided with increased levels of proteolysis. Nevertheless, some patents for the production of low-sodium cheeses, based on the replacement of NaCl with $MgCl_2$, have been filed (Lefier *et al.*, 1987, 1990; Brocard and Meunier, 1992).

16.6.4   Increasing the protein content of cheesemilk by supplementation with ultrafiltered whole milk retentate (UFMR)

Kosikowski (1983) reported a significant improvement in the quality of low-sodium Cheddar cheeses (~1.0%, w/w, NaCl; ~420 mg Na/100 g) from protein-enriched milks (4.4 to 6.3% protein) prepared by supplementing whole milk with increasing levels of UFMR (4.5:1 concentration factor, CF; ~14.5%, w/w, fat and 12.4%, w/w, protein) to give concentration ratios (ratio of protein of retentate supplemented milk to the protein of the control milk) ranging from 1.1:1 to 1.9:1. Increasing the concentration ratio (CR) in this range coincided with a slight reduction in moisture level (from ~37 to 36%, w/w) and marked increases in Ca and P (from ~0.49 to 0.65%, w/w, and 0.40 to 0.49%, w/w, respectively). Noteworthy, is the atypically low contents of Ca (~0.49%, w/w) and P (~0.4%, w/w) in the control; the Ca and P of Cheddar are usually much higher, e.g. typically ~0.72%, w/w, and 0.52%, w/w (USDA, 1976; Guinee *et al.*, 2000). The grading scores for quality (flavour and body/texture charac-teristics) after 2 and 4 months ripening at 10°C increased, more or less, pro-gressively on increasing the CR from 1.1:1 to 1.9:1, with CRs of 1.5:1 to 1.9:1 giving very good/excellent quality cheeses with typical Cheddar flavour and without any loss in sensory saltiness. In contrast, the control low-sodium cheeses prepared from whole milk (3.4% protein) received poor grading scores because of the flavour (acid, bitter, lack typical flavour) and texture (weak, pasty, short) defects. The positive effects of protein enrichment were attributed to the higher levels of calcium and phosphate, which increased buffering capacity and, thereby, prevented low pH in the cheese and associated defects such as excessive proteolysis and bitterness (Kindstedt and Kosikowski, 1986).

Further studies using this approach were undertaken by Lindsay et al. (1985) who investigated the following on the quality of reduced-sodium Cheddar cheese (~0.81%, w/w) ripened at 3°C: cheese was made from either non-supplemented milk (control; C), or milks supplemented with: UFMR (CF, 2:1) giving a concentration ratio of 1.9:1 (2.1UFMR1.9:1); UFMR (CF, 4.5:1) giving a concentration ratio of 1.9:1 (4.5UFMR1.9:1), as in the study of Kosikowski (1983); or a combination of UFMR (CF, 4.5:1) and reverse osmosis whole milk retentate (ROMR; CF, 1.33:1) giving a concentration ratio of 1.6:1 (UFMR+ROMR1.6:1). While the Ca content of the control cheese (~0.62%, w/w) was lower than that of the retentate-supplemented cheeses (~0.67%, w/w), it was much higher than that of the control cheese in the study of Kosikowski (1983). At 3 months, the 4.5UFMR1.9:1 cheese received the lowest overall preference scores mainly because of the low intensities of saltiness and tartness; the other cheeses did not differ significantly. However, at 5 months, the preference ratings for all cheeses were similar, indicating that retentate supplementation did not affect quality of reduced-sodium cheese. The discrepancy between the proceeding studies on the impact of UFMR supplementation on quality may be due to a number of factors, inter alia, the differences in Ca content, ripening temperature, and others.

Lindsay et al. (1985) evaluated the effect of substitution of 2-month-old full-salt Cheddar cheese with 2-month-old low-sodium Cheddar manufactured from retentate supplemented milks (C, 2.1UFMR1.9:1, 4.5UFMR1.9:1, UFMR+ROMR1.6:1) as described above, on the quality of processed cheese slices. The resultant low-sodium processed cheeses (0.32 to 0.34%, w/w Na) were softer than the processed cheeses made from the full-salt Cheddar (~0.73% Na); however, that made using the 4.5UFMR1.9:1 Cheddar cheese received significantly lower preference scores than those made using either low-sodium cheeses C, 2.1UFMR1.9:1, or UFMR+ROMR1.6:1, all of which received similar scores.

16.6.5   The addition of flavour-enhancing substances to natural cheese

This approach is based on the flavour accentuation to compensate for the reduced saltiness. The addition of autolyzed yeast extract to low-sodium Cottage cheese had little effect on flavour, but significantly increased the count of psychrotrophic bacteria that developed during storage at 4°C (Demott et al., 1986).

16.6.6   Other methods of reducing sodium in natural cheese

Other methods of manufacturing low-salt cheese include alterations of the cheese making procedure to exploit the preservation effects of heat and reduce the risk of spoilage when the level of salt added to the curd is reduced. Examples include: (i) the patented manufacture of cheese without added salt by washing curd at low temperature (~20°C; to remove lactose) and heating curd in the cheese mould to a core temperature of 85°C (Drews, 1991); (ii) high intensity centrifugation of milk and re-incorporation of bacteria rich portion after sterilization followed by a normal cheese manufacture apart from a shorter brining time (Wessanen BV, 1983).

16.6.7   Low-sodium processed cheese products (PCPs)

PCPs contain a relatively high level of Na because of the addition of sodium phosphate/citrate emulsifying salts and NaCl (or other ingredients containing NaCl) in their formulation. While the level in PCPs depends on many factors (e.g., category, level and type of added ES, level of Na in added ingredients), reported values for sodium include: 1 to 1.5%, (w/w) in retail PCPs available on the Irish market; 1.25% (w/w) in experimental processed Cheddar cheese (Karahadian and Lindsay, 1984), 1.2–1.5% (w/w) (USDA, 1976). In contrast, the values for natural cheeses are generally lower, e.g. ~0.7 and 0.26% in Cheddar and Swiss cheeses, respectively (USDA, 1976). The proportions of total sodium contributed by the different ingredients vary with PCP category, formulation (e.g., types, and sodium content of cheeses and optional ingredients) and blend of ES. For a typical PCP containing 1.0% (w/w) sodium and formulated from Cheddar cheese (80%, w/w), butter (0.5%), emulsifying salts (0.5% trisodium citrate dihydrate, 0.5% disodium hydrogen orthophosphate dihydrate, and 0.5% trisodium orthophosphate), NaCl (0.3%) and water (19.5%), the respective contributions of cheese, ES and added NaCl to total sodium are ~61, 27% and 12%. For a given level of added ES, the contribution to sodium is obviously dependent on the blend of ES. The various approaches used to reduce sodium in PCPs are discussed below.

The use of reduced-sodium cheese and/or the use of potassium phosphate or citrate emulsifying salts
Karahadian and Lindsay (1984) produced acceptable low-sodium processed cheese (75% reduction, 0.34, w/w, Na in product) by using reduced-sodium Cheddar cheese and/or various combinations of potassium-based emulsifying salts (citrates, phosphates). A similar process for reduced-sodium processed cheese, based on the selective use of Na-, K- and Ca-based phosphates, was patented by Henson (1999).

The addition of flavour-enhancing substances
The incorporation of autolyzed yeast extract in reduced-sodium processed cheeses at a level of 0.4% (w/w) has been reported to cause off flavours (burnt, scorched, meaty, brothy) and reduce preference scores (Karahadian and Lindsay, 1984). The addition of delta-gluconolactone, a delayed acidulant, increased the perception of saltiness in reduced-sodium processed cheeses (570 *versus* 1250 mg Na/100 g cheese) but did not significantly affect flavour (Karahadian

and Lindsay, 1984). Other materials, including ammonium citrates and phosphates (as substitutes for sodium citrates and phosphates), glycinamide hydrochloride, monosodium glutamate and/or 5′-ribonucleotides increased saltiness in the reduced-sodium processed cheeses, but led to pronounced metallic/bitter flavours.

Emulsifying-salt free processed cheese products
The omission of emulsifying salt from a processed cheese substantially reduces the sodium level, e.g. by 20–40% depending on level and type added. However, the application of heat and shear to the cheese blend while processing in the absence of emulsifying salts normally results in a heterogeneous, gummy mass with extensive oiling-off and water separation on cooling. These defects arise from shearing of the natural membranes of the fat globules trapped within the calcium *para*-casein network of the cheese and partial destruction of the network itself. Free moisture and de-emulsified liquefied milk fat seep through the looser, more-open matrix. However, such problems can be overcome by careful blending of different cheeses and optimization of process conditions/shear so as to obtain a stable oil-in-water emulsion (McAuliffe and O'Mullane, 1989; Guinee, 1991). A typical formulation for emulsifying salt-free PCPs (Table 16.3) may include:

(i) some cheeses in the blend with a low calcium content, a casein- rather than a calcium *para*-casein-network, and a low protein-to-water ratio (e.g., fresh acid curd cheeses such as Quark, Cream cheese and Fromage frais);

Table 16.3 Formulations for processed cheese food with and without emulsifying salts

Ingredients	% Inclusion	
	With	Without
Cheddar cheese		
Medium-aged	42	–
Mature	18	31.5*
Extra mature	–	13.5*
Quark	–	10.0
Cream cheese	–	16.0
Skim milk powder	5	5
Sweet whey powder	5	5
WPC 35	4	5
Butter oil	3	4
Skim milk ultrafiltrate (25% protein)	–	3
Emulsifying salts		
Disodium orthophosphate	1.5	–
Disodium pyrophosphate	0.5	–
Trisodium citrate	1.0	–
Water	20	7
Total	100	100

(ii) dairy ingredients, which possess protein with a good water-binding and emulsifying capacity such as evaporated skim milk or milk ultrafiltrates. The latter provide casein which is in a natural hydrated state;

(iii) a higher proportion of mature/extra mature rennet curd cheese (Cheddar) in the blend. The mature cheese helps to achieve an equally intense desirable cheese flavour, when using the milder-flavoured acid-curd cheeses, and to obtain a more-stable finished product (as the protein in extra mature cheddar cheese is more water-soluble than that of a less mature equivalent).

Other ingredients such as hydrocolloids may also enhance texture and mouth feel properties of emulsifying-salt free PCPs. Saltiness could be enhanced by addition of ingredients such as monosodium glutamate, autolyzed yeast extracts, 'high cured cheese', cheese powders, enzyme-modified cheese, cheese pastes and/or acidulants.

16.6.8   Reducing the sodium content in table spreads

The original function of salt in butter and dairy spreads was that of a preservative. This functional need for salt in these products is no longer required, as a consequence of improved standards of hygiene in the manufacturing process. The improved standards pertaining to the food cold-chain as well as domestic refrigeration are also significant. For organoleptic reasons some consumers may desire a slight salty taste. Microbial deterioration of low-fat spreads is strongly correlated with water droplet size. Ideally, droplets should be less than 20 μm diameter in low-fat spreads. Marked reduction in salt levels may lead to unacceptable decreases in aqueous phase viscosity where caseinate is used as the protein of choice thus leading to a destabilization or coarsening of the water-in-oil emulsion. However, mixtures of protein (caseinate, buttermilk powder, skim milk powder) and hydrocolloids together with calcium-containing salts have been routinely used to increase the viscosity or water-binding capacity of the aqueous phase. Roberts *et al.* (2000) have outlined the textural properties of low-fat aqueous phase formulations using ternary mixtures of caseinate, alginate and starch in combination with increasing concentrations of calcium.

Reducing the pH of the aqueous phase also inhibits microbial growth (lactic spreads) and allows for marked reductions in salt level, however, formulations may have to be altered to allow for the change in functionality of proteins and hydrocolloids at acidic pH.

16.7   Conclusions

The contribution of cheese to dietary sodium depends on consumption rate and type of cheese, and is estimated to be 10–20% of the RDA in the adult human at highest per capita cheese consumption rate of ~70 g/day. The salt content of cheese ranges from ~0.7 to 6% (w/w), depending on the variety and characteristics. Why these levels of salt, what is their basis, and can they be reduced?

In early cheese making (ca. 3000–4000 BC), salt would have been added deliberately to cheese curd for its preservation effects, with level of addition depending on availability, region, season and climatic conditions (temperature, humidity). This early preservation method of curd, along with other parameters (e.g., type of milk, microbial flora; and make procedure), led to the evolution of regional cheese varieties differing in salt content and other characteristics. With time, these early 'farm-house' cheeses came to be produced in factories (17th/ 18th centuries) where the manufacturing principle was retained but the manufacturing procedure was refined and standardized to give today's well-known cheese varieties such as Cheddar, Mozzarella, Gouda, etc. Salt has two major functions in cheese: it acts as a preservative and it contributes directly to flavour. Together with the desired pH, water activity and redox potential, salt assists preservation of cheese by minimization of spoilage and preventing the growth of pathogens in cheese. In addition to the above functions, salt level has a major effect on cheese composition, microbial growth, enzymatic activities, and on biochemical changes, such as glycolysis, proteolysis, lipolysis, and *para*-casein hydration, that occur during ripening. Consequently, the salt level markedly influences cheese flavor and aroma, rheology and texture properties, cooking performance and, hence, overall quality.

Various technological approaches offer strong potential to reduce the level of sodium in natural cheese: reducing added NaCl to the minimum level required for optimum quality, partial substitution of NaCl with KCl, maximizing the retention of calcium phosphate in the curd by supplementation of cheese milk with reverse osmosis/ultrafiltered milk retentate, and combinations of the foregoing. For pasteurized processed cheeses, useful approaches to salt reduction include: reducing both the level of NaCl and sodium-based emulsifying salts (e.g., by the substitution of orthophosphates with phosphates) added to formulations to the minimum required for acceptable quality; inclusion of low-sodium natural cheeses in the formulation; development of emulsifying-salt free products relying on the natural hydration/emulsification and texturizing characteristics of milk proteins, acid-curd cheeses, and/or hydrocolloids; partial substitution of sodium phosphate and citrates salt with their potassium salts. The effective realization of any of these approaches, in particular their combined effects, will necessitate a high degree of process control (e.g., standardization of cheese making procedure: milk protein level by UF, rennet-to-casein ratio, etc.) and more studies relating quality (grading scores, consumer acceptability) of commercially manufactured cheese to salt level. Undoubtedly, technological developments that afford manufacturers greater process control, especially in relation to salt distribution in natural cheese, will be useful. Perhaps further exploitation of the other preservation mechanisms (pH reduction, dehydration, and temperature control) operating in cheese and their interactive synergy with NaCl may afford further potential for reducing sodium level.

Compared to cheese, table spreads are a recent invention (1980s), the impetus for development being the supply of a low-fat substitute for butter in a market where there is growing emphasis on reducing the intake of dietary fat. There is the capacity to reduce the sodium levels in table spreads provided that taste is acceptable.

For both cheese (natural and processed), table spreads, and, indeed, other food products, our long-term ability to reduce sodium will be further enhanced by developments in sensory research investigating: product factors (e.g., structure, rheology, texture) affecting the release of salty flavour during mastication (Guinard *et al.*, 1998; Neyraud *et al.*, 2003) and how the perception of taste (e.g., saltiness) during mastication is affected by the presence of other taste and/or odour compounds (Schifferstein and Verlegh, 1996; Stevenson *et al.*, 1999; Djordjevic *et al.*, 2004; Kettenmann *et al.*, 2005; Shepherd, 2005). Such research may enable the enhancement of saltiness in reduced-sodium products via the exploitation of synergistic effects of various non-salt tastes and odours on the perception of saltiness.

16.8   References

ABERNETHY J D (1979), 'Sodium and potassium in high blood pressure', *Food Technol.*, **33**, 57–59, 64.

ABOU-EL- NOUR A M (1998), 'Effect of sodium chloride, a mixture of sodium chloride and potassium chloride on the curd characteristics', *Egypt. J. Dairy Sci.* **26**, 193–202.

ANON. (1980), 'Dietary salt' *Food Technol.* **34**, 85–91.

ANON. (1983), 'Low-sodium cheese potential – Is it worth more than a grain of salt?', *Dairy Record* **84** (3), 41–42.

ANON. (1993), 'KCl spells quality in low-salt cheese', *Dairy Foods* **94** (3), 72.

ANON. (1995), 'Cheese manufacturers see markets for "healthy" cheeses having most potential', *Cheese Reporter* 120 (15) 16.

APOSTOLOPOULOS C, BINES V E and MARSHALL R J (1994), 'Effect of post-cheddaring manufacturing parameters on the metability and free oil of *Mozzarella*', *J. Soc. Dairy Technol*, **47**, 84–87.

BARFOD N M. KROG N and BUCHEIM W (1989), 'Lipid-protein-emulsifier-water interactions in whippable emulsions' in Kinsella J E and Soucie W G, *Food Proteins*, Champaign, IL, AOCS, 144–158.

BEARD T C, BLIZZARD L, O'BRIEN D J and DWYER T (1997), 'Association between blood pressure and dietary factors in the dietary and nutritional survey of British adults', *Arch. Internal Med.* **157**, 234–238.

BECHAZ S R, HICKEY M W, LIMSOWTIN G K Y and MORGAN A G (1998), 'Low-salt Cheddar: a microbial investigation', *Aus. J. Dairy Technol.* **53**, 128.

BEILIN L J (1999), 'Lifestyle and hypertension – An overview' *Clin. Exper. Hypertension* **21**, 749–762.

BENNETT R J and JOHNSTON K A (2004), 'General aspects of cheese technology', in Fox P F, McSweeney P L H, Cogan T M and Guinee T P, *Cheese Chemistry, Physics and Microbiology,. Major Cheese Groups*, 3rd edn, Vol. 2, Amsterdam, Elsevier Academic Press, 3–50.

BERESFORD T and WILLIAMS A (2004), 'The microbiology of cheese ripening', in Fox P F, McSweeney P L H, Cogan T M and Guinee T P, *Cheese Chemistry, Physics and Microbiology, General Aspect*, 3rd edn, Vol. 1, Amsterdam, Elsevier Academic Press, 278–317.

BOYAVAL P, DEBORDE C, CORRE C, BLANCO C and BÉGUÉ É (1999), 'Stress and osmo-protection in propionibacteria', *Lait* **79**, 59–69.

BROCARD C and MEUNIER C (1992), 'Process for the manufacture of low-sodium cheese enriched with magnesium', French Patent Application, FR 2676 890 A1.

BUMBERGER E and BELITZ H D (1993), 'Bitter taste of enzymic hydrolysates of casein. 1. Isolation, structural and sensorial analysis of peptides from tryptic hydrolysates of β-casein', *Z. Lebensm. Unters. Forsch.* **197** 14–19.

CAPPUCCIO F P, KALAITZIDIS R, DUNECLIFT S and EASTWOD J B (2000), 'Unravelling the links between calcium secretion, salt intake, hypertension, kidney stones and bone metabolism', *J. Nephrol.* **13**, 169–177.

CARR A J, MUNRO P A and CAMPANELLA O H (2002), 'Effect of added monovalent or divalent cations on the rheology of sodium caseinate solutions', *Int Dairy J*, **12**, 487–492.

CARR A J, SOUTHWARD C R and CREAMER L K (2003), 'Protein hydration and viscosity of dairy fluids', in Fox P F and McSweeney P L H, *Advanced Dairy Chemistry, Volume 1, Proteins*, New York, Kluwer Academic/Plenum, 1289–1318.

COLLINS Y F, McSWEENEY P L H and WILKINSON M G (2004), 'Lipolysis and catabolism of fatty acids in cheese', in Fox P F, McSweeney P L H, Cogan T M and Guinee T P, *Cheese Chemistry, Physics and Microbiology, General Aspect*, 3rd edn, Vol. 1, Amsterdam, Elsevier Academic Press, 373–389.

CREAMER L K (1971), 'Beta-casein hydrolysis in Cheddar cheese ripening', *NZ J. Dairy Sci. Technol.* **6**, 91.

CREAMER L K (1985), 'Water absorption by renneted casein micelles', *Milchwissenschaft*, **40**, 589–591.

CREAMER L K and MATHESON A R (1980), 'Effect of heat treatment on the proteins of pasteurized skim milk', *NZ J Dairy Sci Technol*, **15**, 37–49.

CUTLER J A (1999), 'The effects of reducing sodium and increasing potassium intake for control of hypertension and improving health', *Clin. Exper. Hypertension* **21**, 768–783.

DALGLEISH D G (1989), 'Protein-stabilised emulsions and their properties' in Hardman T M, *Water and Food Quality*, London, Elsevier Applied Science, 211–250.

DAMODARAN S (1997), 'Food proteins; an overview', in Damodaran S and Paraf A, *Food Proteins and Their Applications*, New York, Marcel Dekker, Inc., 1–24.

DE KRUIF C G and ZHULINA E B (1996), 'κ-casein as a polyelectrolyte brush on the surface of casein micelles', *Colloids and Surfaces A: Physicochemical and Engineering Aspects*, **117**, 151–159.

DEMOTT B J, HITCHCOCK J J and SANDERS O G (1984), 'Sodium concentration of selected dairy products and acceptability of a sodium substitute in Cottage cheese', *J. Dairy Sci.* **67**, 1539–1543.

DEMOTT B J, HITCHCOCK J P and DAVIDSON P M (1986), 'Use of sodium substitutes in Cottage cheese and buttermilk', *J. Food Prot.* **49**, 117–120.

DENTON D, WEISINGER R, MUNDY N I, WICKINGS E J, DIXSON A, MOISSON P, PINGARD A M, SHADE R , CAREY D, ARDAILLOU R , PAILLARD F, CHAPMAN J , THILLET J and MICHEL J B (1995) 'The effect of increased salt intake on blood pressure of chimpanzees', *Nature Med* 1(10). Magnesium Online Library, http://www.mgwater.com/saltbp.shtml.

DICKINSON E (1989), 'Surface and emulsifying properties of caseins' *J Dairy Res*, **56**, 471–477.

DJORDJEVIC J, ZATORRE R J and JONES-GOTMAN M (2004), 'Odor-induced changes in taste perception', *Exper. Brain Res.* **159**, 405–408.

DREWS M (1991), 'Käse ohne Kochsalzzusatz und das Verfahre seiner Herstellung', German Patent, DE 3923615A1.

EL-SISSI M G M and NEAMAT ALLAH A A (1996), 'Effect of salt levels on ripening acceleration of Domiati cheese', *Egypt. J. Dairy Sci.* **24**, 265–275.

EUSTON S R, PISKA I, WIUM H and QVIST K B (2002), 'Controlling the structure and rheological properties of model cheeses systems', *Aus. J. Dairy Technol.* **57**, 145–152.

EXTERKATE F A and ALTING A C (1993), 'The conversion of the α_{s1}-casein-(1-23)-fragment by the free and bound form of the cell-envelope proteinase of *Lactococcus lactis* subsp. *cremoris* under conditions prevailing in cheese', *Systematic Applied Microbiol.* **16**, 1–8.

FAY J (1996), M Sc Thesis, National University of Ireland.

FEENY E P, FOX P F and GUINEE T P (2001), 'Effect of ripening temperature on the quality of low moisture Mozzarella cheese: 1. Composition and proteolysis', *Lait* **81**, 463–474.

FELDMAN R D and SCHMIDT N D (1999), 'Moderate dietary salt restriction increases vascular and systemic insulin resistance', *Am. J. Hypertension* **12**, 643–647.

FENELON M A and GUINEE T P (2000), 'Primary proteolysis and textural changes during ripening in Cheddar cheeses manufactured to different fat contents', *Int. Dairy J.* **10**, 151–158.

FENELON M A, GUINEE T P, DELAHUNTY C, MURRAY J and CROWE F (2000), 'Composition and sensory attributes of retail Cheddar cheeses with different fat contents', *J. Food Comp. Analysis* **13**, 13–26.

FITZGERALD E and BUCKLEY J (1985), 'Effect of total and partial substitution of sodium chloride on the quality of Cheddar cheese', *J. Dairy Sci.* **68**, 3127–3134.

FOX P F (1975), 'Influence of cheese composition on quality', *Irish J. Agric. Res.* **14**, 33–42.

FOX P F and McSWEENEY P L H (2003), *Advanced Dairy Chemistry, Volume 1. Proteins*. New York, Kluwer Academic/Plenum Publishers.

FREITAS A C and MALCATA F X (1996), 'Influence of milk type, coagulant, salting procedure and ripening time on the final characteristics of *Picante* cheese', *Int. Dairy J.* **6**, 1099–1116.

FRISTER H, MICHAELIS M, SCHWERDTFEGER T, FOLKENBERG D M and SORENSEN N K (2000), 'Evaluation of bitterness in Cheddar cheese', *Milchwissenschaft* **55**, 691–695.

GEURTS T J, WALSTRA P and MULDER H (1972), 'Brine composition and the prevention of the defect "soft rind" in cheese', *Neth. Milk Dairy J.* **26**, 168–179.

GILES J and LAWRENCE R C (1973), 'The assessment of Cheddar cheese quality by compositional analysis', *NZ J. Dairy Sci. Technol.* **8**, 148–151.

GOBBETTI M, LANCIOTTI R, DE ANGELIS M, CORBO M R, MASSINI R and FOX P F (1999a), 'Study of the effects of temperature, pH and NaCl on the peptidase activities on non-starter lactic acid bacteria (NSLAB) by quadratic response surface methodology', *Int. Dairy J.* **9**, 865–875.

GOBBETTI M, LANCIOTTI R, DE ANGELIS M, CORBO M R, MASSINI R and FOX P F (1999b), 'Study of the effects of temperature, pH, NaCl and a_w on the proteolytic and lipolytic activities of cheese-related lactic bacteria by quadratic response surface methodology', *Enz. Microb. Technol.* **25**, 795–809.

GODINHO M and FOX P F (1981a), 'Effect of NaCl on the germination and growth of *Penicillium roqueforti*', *Milchwissenschaft* **36**, 205–208.

GODINHO M and FOX P F (1981b), 'Ripening of Blue cheese: salt diffusion rates and mould growth', *Milchwissenschaft* **36**, 329–333.

GODINHO M and FOX P F (1981c), 'Ripening of Blue cheese. Influence of salting rate on lipolysis and carbonyl formation', *Milchwissenschaft* **36**, 476–478.

GODINHO M and FOX P F (1982), 'Ripening of Blue cheese', Influence of salting rate on proteolysis', *Milchwissenschaft* **37**, 72–75.

GRUFFERTY M B and FOX P F (1985), 'Effect of added NaCl on some physicochemical properties of milk', *Irish J. Food Sci. Technol.* **9**, 1–9.

GUIGOZ Y and SOLMS J (1976), 'Bitter peptides, occurrence and structure', *Chemical Senses and Flavor*, **2**, 71-84.

GUINARD J X., ZOUMAS-MORSE C and WALCHAK C (1998), 'Relation between parotid saliva flow and composition and the perception of gustatory and trigeminal stimuli in foods', *Physiol. & Behav.* **63**, 109–118.

GUINEE T P (1991), 'Natural stabilization of food', in Keogh M K, *First Food Ingredients Symposium*, Fermoy, Ireland, National Dairy Products Research Centre Moorepark, 74–87.

GUINEE T P (2003), 'Role of protein in cheese and cheese products', in Fox P F, McSweeney P L H, *Advanced Dairy Chemistry, Vol. 1 Proteins*, New York, Kluwer Academic/Plenum Press, 1083–1174.

GUINEE T P and FOX P F (1986), 'Transport of sodium chloride and water in Romano-type cheese slices during brining', *Food Chem.* **19**, 49–64.

GUINEE T P and FOX P F (2004), 'Salt in cheese: Physical, chemical and biological aspects', in Fox P F, McSweeney P L H, Cogan T M and Guinee T P, *Cheese Chemistry, Physics and Microbiology, General Aspect*, 3rd edn, Vol. 1, Amsterdam, Elsevier Academic Press, 207–259.

GUINEE T P and KILCAWLEY K N (2004), 'Cheese as an ingredient', in Fox P F, McSweeney P L H, Cogan T M and Guinee T P, *Cheese Chemistry, Physics and Microbiology, Major Cheese Groups*, 3rd edn, Vol. 2, Amsterdam, Elsevier Academic Press, 395–428.

GUINEE T P and WILKINSON M G (1992), 'Rennet coagulation and coagulants in cheese manufacture', *J. Society Diary Technol.* **45**, 94–104.

GUINEE T P, HARRINGTON D, CORCORAN M O, MULHOLLAND E O and MULLINS C (2000), 'The composition and functional properties of commercial Mozzarella, Cheddar and analogue pizza cheese', *Int. J. Dairy Technol.* **53**, 51–56.

GUINEE T P, CARIĆ M and KALÁB M (2004), 'Pasteurized processed cheese and substitute/ imitation cheese products', in Fox P F, McSweeney P L H, Cogan T M and Guinee T P, *Cheese Chemistry, Physics and Microbiology, Major Cheese Groups*, 3rd edn, Vol. 2, Amsterdam, Elsevier Academic Press, 349–394.

GUO M R, GILMORE J K A and KINDSTEDT P S (1997), 'Effect of sodium chloride on the serum phase of Mozzarella cheese', *J. Dairy Sci.* **80**, 3092–3098.

HARDY J and STEINBERG M P (1984), 'Interaction between sodium chloride and paracasein as determined by water sorption', *J. Food Sci.*, **49**, 127–131, 136.

HENSON L S (1999), 'Reduced sodium process cheese and method for making it', US Patent No. 5871797.

HERMANSSON A M (1975), 'Functional properties of proteins for foods-flow properties', *J Textue Stud*, **5**, 425–439.

HEWEDI M and FOX P F (1984), Ripening of Blue cheese. Characterization of proteolysis. *Milchwissenschaft*, **39**, 198–201.

INSTITUTE OF FOOD TECHNOLOGISTS' EXPERTS PANEL ON FOOD SAFETY AND NUTRITION AND

THE COMMITTEE ON PUBLIC INFORMATION (1980), 'Scientific status summary – dietary salt', *Food Technol.* **34** (1), 85–91.

IRVINE D M and PRICE W V (1961), 'Influence of salt on the development of acid by lactic starters in skim milk and in curd submerged in brine', *J. Dairy Sci.* **44**, 243–248.

JORDAN K N and COGAN T M (1993), 'Identification and growth of non-starter lactic acid bacteria in Irish Cheddar cheese', *Irish J. Agric. Food Res.* **32**, 47–55.

KAPLAN N M (2000), 'The dietary guideline for sodium: should we shake it up? No', *Amer. J. Clin. Nutr.* **71**, 1020–1026.

KARAHADIAN C and LINDSAY R (1984), 'Flavour and textural properties of reduced-sodium process American cheeses', *J. Dairy Sci.* **67**, 1892–1904.

KATSIARI M C, VOUTSINAS L P, ALICHANIDIS E and ROUSSIS I G (1997), 'Reduction of sodium content in Feta cheese by partial substitution of NaCl by KCl', *Int. Dairy J.* **7**, 465–472.

KATSIARI M C, VOUTSINAS L P, ALICHANIDIS E and ROUSSIS I G (2000a), 'Lipolysis in reduced sodium Feta cheese made by partial substitution of NaCl by KCl', *Int. Dairy J.* **10**, 369–373.

KATSIARI M C, ALICHANIDIS E, VOUTSINAS L P and ROUSSIS I G (2000b), 'Proteolysis in reduced sodium Feta cheese made by partial substitution of NaCl by KCl', *Int. Dairy J.* **10**, 635–646.

KATSIARI M C, VOUTSINAS L P, ALICHANIDIS E and ROUSSIS I G (2001a), 'Lipolysis in reduced sodium Kefalograviera cheese made by partial replacement of NaCl with KCl', *Food Chem.* **72**, 193–197.

KATSIARI M C, ALICHANIDIS E, VOUTSINAS L P and ROUSSIS I G (2001b), 'Proteolysis in reduced sodium Kefalograviera cheese made by partial replacement of NaCl with KCl', *Food Chem.* **73**, 31–43.

KAYA S, KAYA A and ONER M D (1999), 'The effect of salt concentration on rancidity in Gaziantep cheese', *J. Sci. Food Agric.* **79**, 213–219.

KELLY M, FOX P F and McSWEENEY P L H (1996), 'Effect of salt-in-moisture on proteolysis in Cheddar-type cheese', *Milchwissenschaft* **51**, 498–501.

KEOGH M K (1992), 'The stability to inversion of a concentrated water-in-oil emulsion'. Ph D Thesis, National University of Ireland.

KETTENMANN B, MUELLER C, WILLE C and KOBAL G (2005), 'Odor and taste interaction on brain responses in humans', *Chem. Senses* **30** (Suppl. 1), i234–i235.

KINDSTEDT P S (1990), 'Physico chemical aspects of Mozzarella cheese', in, *2nd Cheese Symposium*, T.M. Cogan, ed., Dairy Products Research Centre, Moorepark, Fermoy, Co. Cork, Ireland, pp. 95–106.

KINDSTEDT P S and GUO M R (1997), 'Recent developments in the science and technology of Pizza cheese', *Aust. J. Dairy Technol.* **52**, 41–43.

KINDSTEDT P S and KOSIKOWSKI F V (1986), 'Chemical and biochemical advantages of ultrafiltration retentates in the manufacture of low sodium chloride Cheddar cheese', *J. Dairy Sci.* **69** (Suppl. 2), 78 (1 page).

KINDSTEDT P S, CARIĆ M and MILANOVIĆ S (2004), 'Pasta-Filata cheeses', in Fox P F, McSweeney P L H, Cogan T M and Guinee T P, *Cheese Chemistry, Physics and Microbiology, Major Cheese Groups*, 3rd edn, Vol. 2, Amsterdam, Elsevier Academic Press, 251–277.

KORHONEN M H, LITMANEN H, RAURAMAA R, VÄISÄNEN S B, NISKANEN L and USITUPA M I (1999), 'Adherence to the salt restriction diet among people with mildly elevated blood pressure', *Europ. J. Clin. Nitr.* **53**, 880–885.

KORHONEN M H, JARVINEN-R M K, SARKKINEN E S and UUSITUPA M I J (2000), 'Effects of a salt-restricted diet on the intake of other nutrients', *Amer. J. Clin. Nutr.* **72**, 414–420.

KOROLCZUK J (1982), 'Hydration and viscosity of casein solutions', *Milchwissenschaft*, **37**, 274–276.

KOSIKOWSKI F V (1983), 'Low sodium Cheddar cheeses through whole milk retentate supplementation', *J. Dairy Sci.* **66**, 2494–2500.

KRISTIANSEN K R, DEDING A S, JENSEN D F, ARDÖ Y and QVIST K B (1999), 'Influence of salt content on ripening of semi-hard round-eyed cheese of Danbo-type', *Milchwissenschaft* **54**, 19–23.

KUBANTSEVA N, HARETEL R W and SWEARINGEN P A (2004), 'Factors affecting solubility of calcium lactate in aqueous solutions', *J. Dairy Sci.* **87**, 863–867.

LANE C N, FOX P F, WALSH E M, FOLKERTSMA B and McSWEENEY P L H (1997), 'Effect of compositional and environmental factors on the growth of indigenous non-starter lactic acid bacteria in Cheddar cheese', *Lait* **77**, 561–573.

LAWRENCE R C and GILLES J (1969), 'The formation of bitterness in cheese: A critical evaluation', *NZ J. Dairy Sci. Technol.* **4**, 189–196.

LAWRENCE R C and GILLES J (1980). 'The assessment of potential quality of young Cheddar cheese', *NZ J. Dairy Sci. Technol.* **15**, 1–12.

LAWRENCE R C, HEAP H A and GILLES J (1984), 'A controlled approach to cheese technology', *J. Dairy Sci.* **67**, 1632–1645.

LEE K and WARTHESEN J J (1995), 'Characteristics of bitter peptides isolated from molecular weight fractions of Cheddar cheese treated with peptidase extracts', *J. Dairy Sci.* **78**, (Suppl. 1) 103 (1 page).

LEFIER D, GRAPPIN R, GROSCLAUDE G and CURTAT G (1987), 'Sensory properties and nutritional quality of low-sodium Gruyère cheese', *Lait* **67**, 451–464.

LEFIER D, DUBOZ G and GRAPPIN R (1990), 'Low-sodium cheeses and their manufacture', French Patent Application FR 2 648 318 A1.

LELIEVRE J and GILLES J (1982), 'The relationship between the grade (product value) and composition of young commercial Cheddar cheese', *NZ J. Dairy Sci. Technol.* **17**, 69–75.

LINDSAY R C HARGETT S M and BUSH S C (1982), 'Effect of sodium/potassium (1:1) chloride and low sodium chloride concentrations on quality of Cheddar cheese', *J. Dairy Sci.* **65**, 360–370.

LINDSAY R C, KARAHADIAN C and AMUDSON C H (1985), 'Low sodium cheese: an overview and properties of Cheddar cheese made with UF and RO retentate supplemented milk', in *Proc. IDF Seminar, Atlanta, GA, USA*, 8–9 October 1985, 55–76.

LOWRIE R J and LAWRENCE R C (1972), 'Cheddar cheese flavour. IV. A new hypothesis to account for the development of bitterness', *NZ J. Dairy Sci. Technol.* **7**, 51–53.

MARCOS A (1993), 'Water activity in cheese in relation to composition, stability and safety', in, *Cheese: Chemistry, Physics and Microbiology Volume 1. General Aspects*, 2nd edn, P. F. Fox, ed., Chapman and Hall, London, 439–469.

MARSH A C, KLIPPSTEIN R N and KAPLAN S D (1980), 'The sodium content of your food', *USDA Home Garden Bulletin* No. 233.

MARTENS R, VAN DEN POORTEN R and NAUDTS M (1976), 'Production, composition and properties of low-sodium Gouda cheese', *Revue de l'Agriculture* **29**, 681–698 (cited from *Dairy Sci. Abstr.* 1977, **39**, 70).

MARTLEY F G and LAWRENCE R. C (1972), 'The effect of cell numbers in streptococcal chains on plate-counting', *NZ J. Dairy Sci. Technol.* **7**, 38–44.

McAULIFFE J P and O'MULLANE T A (1989), 'Preparation of cheese from natural ingredients', UK Patent Application GB2237178A.

McCARRON D A (1997), 'Role of adequate calcium intake in the prevention and management of salt-sensitive hypertension', *Amer. J. Clin. Nutr.* **65** (Suppl. 2), 712S–716S.

McCARRON D A and REUSSER M (2000), 'The dietary guideline for sodium: should we shake it up? Yes!', *Amer. J. Clin. Nutr.* **71**, 1020–1026.

McSWEENEY P L H (2004), 'Biochemistry of cheese ripening: introduction and overview', in Fox P F, McSweeney P L H, Cogan T M and Guinee T P, *Cheese Chemistry, Physics and Microbiology, General Aspects*, 3rd edn, Vol. 1, Amsterdam, Elsevier Academic Press, 347–360.

MIDGLEY J P, MATTHEW A G, GREENWOOD C M and LOGAN A G (1996), 'Effect of reduced dietary sodium on blood pressure: a meta-analysis of randomized controlled trials', *J. Amer. Med. Assoc.* **275**, 1590–1597.

MISTRY V V and KASPERSON K M (1998), 'Influence of salt on the quality of reduced fat Cheddar cheese', *J. Dairy Sci.* **81**, 1214–1221.

MOOSTER G (1980), 'Membrane transitions in taste receptor cell activation by sodium salts', in Kare M R, Fregly M J and Bernard R A, *Biological and Behavioural Aspects of Salt Intake*, New York, Academic Press Inc., 275–287.

MORRIS C E and DILLON P M (1992), 'Worldwide new product analysis', *Food Eng.* **64**, 103–106.

MOSES C (1980), '*Sodium in Medicine and Health: A Monograph*', Alexandria, VA, USA, Salt Institute, 1–126.

MULDER H and WALSTRA P (1974), 'Isolation of milk fat', in Mulder H and Walstra P, *The Milk Fat Globule*, Wageningen, Pudoc, 228–243.

MURPHY C, CARDELLO A V and BRAND J G (1981), 'Tastes of fifteen halides following water and NaCl: anion and cation effects', *Physiol. Behav.* **26**, 1083–1095.

NAJERA A I, BARRON L J R and BARCINA Y (1994), 'Changes in free fatty acids during ripening of Idiazabal cheese: influence of brining time and smoking', *J. Dairy Res.* **61**, 281–288.

NARHINEN M, NISSIEN A, PENTTILA P L, SOMONEN O, CERNERUD, L and PUSKA P (1998), 'Salt content labelling of foods in supermarkets in Finland', *Agric. Food Sci. Finland* **7**, 447–453.

NEYRAUD E, PRINZ J and DRANSFIELD E (2003), 'NaCl and sugar release, salivation and taste during mastication of salted chewing gum', *Physiol. & Behav.* **79**, 731–737.

NOOMEN A (1978), 'Activity of proteolytic enzymes in simulated soft cheeses (Meshanger type). 1. Activity of milk protease', *Neth. Milk Dairy J.* **32**, 26.

O'CALLAGHAN D J and GUINEE T P (2004), 'Rheology and texture of cheese', in Fox P F, McSweeney P L H, Cogan T M and Guinee T P, *Cheese Chemistry, Physics and Microbiology, General Aspects*, 3rd edn, Vol. 1, Amsterdam, Elsevier Academic Press, 511–540.

O'CONNOR C B (1970), 'Sugars in Cheddar cheese aged 9 weeks made with various rates of salt addition', *IFST Proc.* **3**, 116–117.

O'CONNOR C B (1971), 'Composition and quality of some commercialized Cheddar cheese', *Irish Agric. Creamery Rev.* **24** (6), 5–6.

O'CONNOR C B (1973a), 'The quality and composition of Cheddar cheese. Effect of various rates of salt addition', *Irish Agric. Creamery Rev.* **26** (10), 5–7.

O'CONNOR C B (1973b), 'The quality and composition of Cheddar cheese: Effect of various rates of salt addition', *Irish Agric.Creamery Rev.* **26** (11), 19–22.

O'CONNOR C B (1974), 'The quality and composition of Cheddar cheese. Effect of various rates of salt addition. Part III', *Irish Agric. Creamery Rev.* **27** (1), 11–13.

O'SHAUGHNESSY K M and KARET F E (2004), 'Salt handling and hypertension', *J. Clin. Invest.* **113**, 1075–1081.

O'NULAIN, M (1986), '*Manufacture of Modified Camembert and Cheddar Cheeses*', M.Sc. Thesis, National University of Ireland, Cork.

OVERBEEK J T G and BUNGENBURG DE JONG H G (1949), 'Sols of macromolecular colloids with electrolytic nature', in Kruyt H R, *Colloid Science: Volume II: Reversible Systems*, Amsterdam, Elsevier Publishing Co.

PAGANA M M and HARDY J (1986), 'Effect of salting on some rheological properties of fresh Camembert cheese as measured by uniaxial compression', *Milchwissenschaft* **41**, 210–213.

PAPPAS C P, KONDYLI E, VOUTSINAS L P and MALLATOU H (1996), 'Effects of salting method and storage time on composition and quality of Feta cheese', *J. Soc. Dairy Technol.* **49**, 113–118.

PAULSON B M, McMAHON D J and OBERG C J (1998), 'Influence of sodium chloride on appearance, functionality and protein arrangements in nonfat Mozzarella cheese', *J. Dairy Sci.* **81**, 2053–2064.

PEARCE K N and GILLES J (1979), 'Composition and grade of Cheddar cheese manufactured over three seasons', *N.Z. J. Dairy Sci. Technol.* **14**, 63–71.

PEARSE M J and MACKINLAY A G (1989), 'Biochemical aspects of syneresis: A review', *J. Dairy Sci.* **72**, 1401–1407.

PHELAN J A, GUINEY J and FOX P F (1973), 'Proteolysis of β-casein in Cheddar cheese', *J. Dairy Res.* **40**, 105–112.

PLATT B L (1988), 'Low fat spread' European Patent No 0 256 712.

REDDY K A and MARTH E H (1995), 'Lactic acid bacteria in Cheddar cheese made with sodium chloride, potassium chloride or mixtures of the two salts', *J. Food Prot.* **58**, 62–69.

REID J R and COOLBEAR T (1998), 'Altered specificity of lactococcal proteinase PI (Lactocepin I) in humecant systems reflecting the water activity and salt content of Cheddar cheese', *Appl. Environ. Microbiol.* **64**, 588–593.

REID J R and COOLBEAR T (1999), 'Specificity of *Lactococcus lactis* subsp. *cremoris* SK11 proteinase, lactocepin III, in low-water activity, high salt concentration humecant systems and its stability compared with that of lactocepin I', *Appl. Environ. Microbiol.* **65**, 2947–2953.

REVILLE W J and FOX P F (1978), 'Soluble protein in Cheddar cheese: a comparison of analytical methods', *Irish J. Fd. Sci. Technol.* **2**, 67–76.

ROBERTS S A, KASAPIS S and DE SANTOS LOPEZ I (2000), 'Textural properties of a model aqueous phase in low fat products. Part 2: Alginate/caseinate blends and three-component systems', *Int J Food Sci Technol*, **35**, 227–234.

ROLLMAN N O and SJOSTROM G (1946), 'The behaviour of some strains of propionic acid in bacteria the presence of salt, salt-petre and heat', *Svenska Mejeritidningen*, **38**, 199–201, 209–212 (cited from *Dairy* Sci. *Abstr.*, 1948–50, **11**, 33).

RÜEGG M and BLANC B (1981), 'Influence of water activity on the manufacture and aging of cheese', in, Rockland L B and Stewart G F, *Water Activity, Influences on Food Quality*, New York, Academic Press, 791–811.

RUSSELL N J and GOULD G W (1991), *Food Preservatives*, Glasgow, Blackie Academic & Professional.

SANCHEZ E S, SIMAL S, FEMENIA A, BENEDITO J and ROSSELLO C (2001), 'Effect of acoustic brining on lipolysis and sensory characteristics of Mahon cheese', *J. Food Sci.* **66**, 892–896.

SCHIFFERSTEIN H N J and VERLEGH P W J (1996), 'The role of congruency and pleasantness in odor-induced taste enhancement' *Acta Psychologica* **94**, 87–105.

SCHROEDER C L, BODYFELT F W, WYATT C J and McDANIEL M R (1988), 'Reduction of sodium chloride in Cheddar cheese: Effect on sensory, microbiological, and chemical properties', *J. Dairy Sci.* **71**, 2010–2020.

SCHULZ-COLLINS D and SENGE B (2004), Acid-and acid/rennet-curd cheeses. Part A: Quark, Cream cheese and related varieties, in Fox P F, McSweeney P L H, Cogan T M and Guinee T P, *Cheese Chemistry, Physics and Microbiology, Major Cheese Groups*, 3rd edn, Vol. 2, Amsterdam, Elsevier Academic Press, 301–328.

SHEPHERD G M (2005), 'Outline of a theory of olfactory processing and its relevance to humans. *Chem. Senses* **30** (Suppl. 1) i3–i5.

STEVENSON R J, PRESCOTT J and BOAKES R A (1999), 'Confusing tastes and smells: how odours can influence the perception of sweet and sour tastes', *Chemical Senses* **24**, 627–635.

STRANGE E D, VAN HEKKEN D L and HOLSINGER V H (2004), 'Effect of sodium chloride on the solubility of caseins', *J. Dairy Sci.* **77**, 1216–1222.

SUTHERLAND B J (2002), 'Salting of cheese', in Roginski H, Fuquay J W and Fox P F, *Encyclopedia of Dairy Sciences*, London, Academic Press, 293–300.

THAKUR M K, KIRK J R and HEDRICK T I (1975), 'Changes during ripening of unsalted Cheddar cheese', *J. Dairy Sci.* **58**, 175–180.

THOMAS T D and PEARCE K N (1981), 'Influence of salt on lactose fermentation and proteolysis in Cheddar cheese', *NZ J. Dairy* Sci. *Technol.* **16**, 253–259.

TURNER K W and THOMAS D T (1980), 'Lactose fermentation in Cheddar cheese and the effect of salt', *NZ J. Dairy* Sci. *Technol.* **15**, 265–176.

UPADHYAY V K, McSWEENEY P L H, MAGBOUL A A A and FOX P F (2004), 'Proteolysis in cheese during ripening', in Fox P F, McSweeney P L H, Cogan T M and Guinee T P, *Cheese Chemistry, Physics and Microbiology. General Aspects*, 3rd edn, Vol. 1, Amsterdam, Elsevier Academic Press, 391–433.

USDA (1976), *Agriculture Handbook, No.8-1: Composition of Food and Egg Products, Raw-Processed-Prepared*, L.P. Posati and M.L. Orr, Washington, DC, US Government Printing Office.

VAFOPOULOU-MATROJIANNAKI A (1999), 'Influence of pH and NaCl on proteolytic and esterolytic activity of intracellular extract of *Leuconostoc mesenteroides* subsp. *mesenteroides* strain K_1G_8', *Milchwissenschaft* **54**, 314–316.

VAN DEN BERG C and BRUIN S (1981), Water activity and its estimation in food systems: theoretical aspects, in Rockland L B and Stewart G F, *Water Activity: Influences on Food Quality*, London, Academic Press, 1–61.

VAN DEN BERG G, MEIJER W C, DÜSTERHöFT E-M and SMIT G (2004), 'Gouda and related cheeses', in Fox P F, McSweeney P L H, Cogan T M and Guinee T P, *Cheese Chemistry, Physics and Microbiology, Major Cheese Groups*, 3rd edn, Vol. 2, Amsterdam, Elsevier Academic Press, 103–140.

VAN DEN BERG G, DE VRIES A E and STADHOUDERS J (1986), 'The salt content of Gouda cheese', *Voedingsmiddelentechnologie*, 19(7), 37–39 (cited from *Dairy Sci. Abstr.* 1988, **50**, 210).

VISSER J (1991), 'Factors affecting the rheological and fracture properties of hard and semi-hard cheese', in *Rheological and Fracture Properties of Cheese, IDF Bulletin 268*, Brussels, International Dairy Federation, 49–61.

VISSER S, HUP G, EXTERKATE F A and STADHOUDERS J (1983a), 'Bitter flavour in cheese. 2. Model studies on the formation and degradation of bitter peptides by proteolytic enzymes from calf rennet, starter cells and starter cell fractions', *Neth. Milk Dairy J.* **37**, 169–180.

VISSER S, SLANGEN K J, HUP G and STADHOUDERS J (1983b), 'Bitter flavour in cheese. 3. Comparative gel-chromatographic analysis of hydrophobic peptide fractions from twelve Gouda-type cheeses and identification of bitter peptides isolated from a cheese made with *Streptococcus cremoris* strain HP', *Neth. Milk Dairy J.*, **37**, 181–192.

WESSANEN B V (1983), 'Verfahren zur Herstellung von einem mineralarmen Käse', German Patent, DE 3217310A1.

WYATT C J (1983), 'Acceptability of reduced sodium in breads, cottage cheese, and pickles', *J. Food Sci.* **48**, 1300–1302.

17

Reducing salt in canned foods

T. Robinson, H.J. Heinz Company Limited, UK

17.1   Introduction

The function of canning for food preservation is well-established, originating at the turn of the 19th century as a means to preserve foods in glass jars for troops during the Napoleonic wars. Metal cans first came into commercial production in 1813 in England (Canned Food Alliance, 2006). Heinz canned its first batch of baked beans in tomato sauce in 1895 and at the turn of the century they were imported into Britain. They were manufactured for the first time in the UK in 1928. Canning involves sealing the food product within the can itself, then heating the product for sufficient time to sterilise the food. This process is sufficient to allow the foods to be stored at ambient conditions for long periods of time, for example the current shelf-life of a 415 g can of Heinz Baked Beans is 16 months from the time of production, however, it is not uncommon for canned products to be safely consumed many years after they were originally packaged, providing the can remains sealed. For example, a recent news story related the tale of a man celebrating his 50th wedding anniversary by eating a can of chicken given in a hamper as a wedding present, with no ill effects (BBC, 2006).

Foods preserved by the canning process do not necessarily require the addition of salt for preservation purposes. Despite this, the majority of canned foods have salt added to them as part of their traditional recipe, most often for cultural or taste reasons rather than any preservative effect. In the present climate of concern over the amount of salt in people's diets food manufacturers have been encouraged to avoid adding unnecessary levels of salt to their produce, yet very little attention has been paid to the level of salt that is acceptable for an individual to choose and consume a food and obtain pleasure from the process, not merely through satisfying nutritional or safety requirements.

Heinz has always made it its business to take nutrition seriously. The company founder, Henry J. Heinz was a leading figure in the US campaign for Pure Food Laws and there has been a long commitment to a company policy to produce good food from quality ingredients without the use of artificial colours or preservatives, whenever possible. In addition to these commitments we have long been conscious of the evidence for a link between excessive salt consumption and ill health. That is why, in the early 1980s, Heinz began assessing the levels of salt in over 100 of its products and started to gradually reduce salt content alongside the normal recipe redevelopment programme. Recently, as part of the 'Heinz Good Food Every Day^{TM}' food policy, more ambitious changes have been made to the salt levels in the Heinz canned Quick Serve Meals range (QSM; Baked Beans, Soup, Pasta in Tomato Sauce), which are outlined below.

17.2   Typical levels of salt in products and targets for reduction

Table 17.1 summarises the change in salt content of the main products within the Heinz canned QSM range from 1997 up to 2002, the latter figures representing salt levels prior to recent redevelopment work. It can be seen that some recipes had already achieved significant reductions in salt over this time, reflecting Heinz's awareness of the increasing weight of opinion about the negative health effects of excessive salt intakes and wanting to play a role in continuous improvement by doing good by stealth, without drawing attention to recipe changes.

Table 17.1   Examples of Quick Serve Meal sodium reductions 1997–2002

Heinz Variety		Salt (g per 100 g) 1997	2002	% reduction 1997–2002
Beans	Baked Beans	1.25	1.16	7
	Beans and Pork Sausages	1.53	1.38	10
Pasta	Spaghetti in Tomato Sauce	0.90	0.89	2
	Pasta Shapes in Tomato Sauce	1.00	0.80	20
	Spaghetti Hoops	1.02	0.80	22
Soups	Cream of Tomato	1.02	1.01	1
	Cream of Chicken	0.89	0.89	
	Vegetable	1.18	0.97	18
	Oxtail	1.36	1.29	5
	Cream of Mushroom	0.94	0.86	9
	Mulligatawny	1.12	1.09	3
	Chicken Noodle	1.09	0.97	11
	Country Vegetable	1.14	0.89	22
	Tomato and Lentil	1.06	1.04	1
	Big Soup Beef and Vegetable	1.14	0.98	14
	Big Soup Chicken and Vegetable	1.24	1.10	11
	Pea and Ham	1.03	0.90	12

In 2004 Heinz performed a review of the nutritional content of its product portfolio and set a series of nutrient and ingredient criteria that products should meet in the context of a healthy diet. This review of the Heinz nutrition policy was part of a company-wide commitment to providing 'Heinz Good Food Every Day^{TM}'. The quantities of macronutrients in foods were referenced against dietary reference values for the UK population (Department of Health, 1991) to ensure that no food contributed proportionally too much of any one macronutrient (too little in the case of protein) in comparison with its contribution to daily energy intake. A critical part of the nutrition criteria used in the review was the level of sodium contained in each product. As no energy is contributed by sodium, another method of deciding appropriate sodium levels in products was required.

In 2003 the Scientific Advisory Committee on Nutrition report on Salt and Health recommended a target reduction in average intake of salt by the population to 6 g per day (SACN, 2003). The publication of a draft 'Salt Model' by the Food Standards Agency in September 2003 and subsequently revised in February 2005 gave the food industry a first glimpse of the extent of reductions that were expected of them to achieve the Department of Health's target of reducing the nation's salt intake to 6 g per day by 2010 (Department of Health, 2004). Whilst some of the target figures contained within the salt model were ambitious, it was decided by Heinz to utilise these targets to set criteria for maximum or average salt content of specific food categories, e.g. baked beans, retail soups and pasta in tomato sauce. Priority was given to reducing salt levels in popular products so that the greatest impact could be made upon consumers' intakes of salt and also in products commonly consumed by children.

A pragmatic approach was taken to achieving the targets set out in the salt model, it was accepted that some of the reductions could not be achieved overnight without negatively impacting upon product taste and consumer satisfaction. Therefore, for some products, a framework of interim targets was set, thus enabling a progressive reduction in salt content over time, whilst still striving to achieve reductions as quickly as possible.

17.3   Methods to reduce salt levels while retaining quality and safety

As mentioned above, the sterilisation process that is involved in canned food production means that for most products salt is not required for preservation. The major reason for its inclusion is related to taste and therefore the quality of the product. Along with safety, product quality and taste 'performance' is of primary importance to food manufacturers, since a consumer will only purchase a product on subsequent occasions if they have enjoyed it, choosing another option if the product has not met with their satisfaction. Of overriding concern to manufacturers is the need to ensure that any changes to recipe do not impact negatively upon acceptability, but rather that they result in a product preferred to the previous recipe. From research conducted in the early 1980s (unpublished data) Heinz knew that consumers desired healthier foods, such as options lower in salt, but were not willing to trade-off taste preference to achieve this. The gradual decline in salt content of baked beans, soup and pasta recipes from 1984 to 2002 went largely unnoticed because the step changes were gradual and too small to be detected by the consumer over such a broad period of time. To meet the growing demand for further reductions in salt in recipes, however, a more ambitious approach was needed which would require detailed research and careful management if consumer acceptance was to be maintained.

17.4   Pre-reduction consumer/sensory research

Although Heinz had historically been working to reduce the levels of added sodium in its range of QSM the decision was made to drive some more ambitious reductions to help meet the UK government's goal of reducing the population's daily salt intake to an average of 6 g/day. It was acknowledged that care would need to be taken to establish the optimal level of reduction that would allow a significant amount to be removed from the product, but which would not negatively affect consumer perception and acceptance of the reformulated product. A programme of research was commissioned that included consumer preference and sensory analysis components to determine this optimal level of salt reduction. To this end, a number of objectives were set out at the beginning of research:

- to understand the impact of different levels of salt reduction on overall preference/product liking and purchase intent
- to determine at what level of salt reduction consumers notice the difference between products
- to establish at which level of salt reduction this negatively impacts on appeal and likelihood to buy
- to identify the key drivers of taste and any interactions between the product attributes as a result of salt reduction.

As the impetus for reduction was in the QSM market, key recipes from this category were chosen for research, Baked Beans in Tomato Sauce, Cream of Tomato Soup and Spaghetti in Tomato sauce.

17.4.1   Case Study 1
Consumer acceptance test of Baked Beans
Approximately 160 eligible consumers were recruited to the study, in which they evaluated six different Baked Beans recipes differing only in their salt levels. The order in which consumers were presented with the different recipes was randomised and counterbalanced to ensure that no effect of order of presentation of recipe was observed. The Baked Beans were heated and served

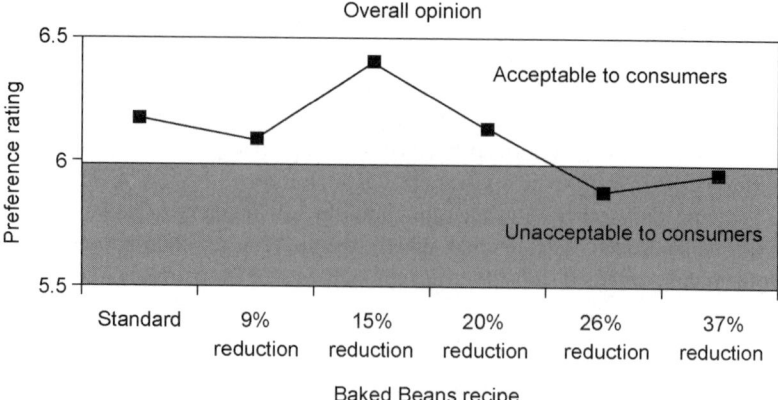

Fig. 17.1 Consumer preference ratings of Baked Beans salt reduction recipes.

according to the manufacturer's instructions. Each of the six Baked Beans recipes were presented to the consumer across a 45-minute period and questions completed, including consumers' perceptions of the appearance, smell, colour, taste, aftertaste, overall opinion of the Baked Beans and their likelihood to buy that particular recipe.

The six Baked Beans recipes tested were as follows: the then current standard recipe, salt reduction recipes of 9%, 15%, 20%, 26% and 37%. Appearance, smell and colour were largely unaffected by reducing the salt content of the recipe. Taste and aftertaste were rated as significantly lower for the two highest salt reduction recipes. The overall opinion of the consumers tested was that the 15% salt reduction recipe was the most preferred of the recipes, followed by current standard and 20% reduction recipes (Fig. 17.1). This was also reflected by consumers' reported likelihood to buy the 15% reduction recipe above the then current standard recipe.

Overall, the consumer scores for all of the products were similar, however there was a clear delineation of acceptability to the consumer, with the 15% reduction recipe showing no decrement in preference when compared with the then current standard recipe and the two products with the most extreme salt reductions (26% and 37%) being substantially less appealing to the consumer.

Sensory profiling of Baked Beans recipes

A second series of tests was conducted on the same Baked Beans recipes to objectively define and quantify any differences between them. Trained sensory panellists assessed the six recipes using quantitative descriptive analysis in a repeated-measures, controlled randomised manner. Various aspects of the Baked Beans' appearance, aroma, flavour, mouth-feel and aftertaste were described and quantified. The tests concluded that compared to other flavour characteristics such as haricot beans or sweetness, the recognisable salt levels in the recipes were low, although saltiness remained a discriminatory characteristic and a positive driver for product liking. Drivers for liking, such as beans' aftertaste, decreased, whilst drivers for not liking, such as sweetness, increased with increasing salt reduction, indicating that attention should be given to preventing any significant changes in such factors.

Results
In addition to Baked Beans, Cream of Tomato Soup and Spaghetti in Tomato Sauce recipes all underwent similar testing, with the addition of a child population in the consumer testing of Spaghetti in Tomato Sauce. For all of the products tested, saltiness *per se* did not score highly as a key flavour characteristic compared with others such as bean or tomato flavour, yet salt level retained a key interaction with other distinguishable flavour components that act as positive drivers for liking, such as bean flavour (beans), creaminess (soup) and tomato flavour and acidity (spaghetti) and prevented negative drivers such as sweetness (beans) and oiliness (soup). All of these interactions should be considered when planning any salt reduction activity to ensure that essential motivating flavour components are not reduced as a consequence.

Implementation of results
As a result of the recommendations of the consumer and sensory research it was decided to meet the approximate 30% salt reduction required by the FSA (FSA, 2003) in a two-stage process. This was to avoid any consumer acceptance issues that were likely to occur, as suggested by consumer testing of the 26–37% salt reduction recipes. In the first instance, 15% of the added salt in Baked Beans was removed from the recipe. This was followed, a number of months later to allow for consumers' tastes to adjust, by an additional 14% reduction to make the Baked Beans fall within the target average sodium level of 350 mg per 100 g set in the FSA Salt Model (2003).

The research also suggested that a 20% reduction could be made to the salt content of Spaghetti in Tomato Sauce without any negative impact upon consumer acceptance and this was implemented immediately in a single recipe change. Mention should also be made here of the range of Heinz Pasta Shapes in Tomato Sauce, which are distinct and different from the Spaghetti in Tomato Sauce recipe used in the consumer and sensory testing. Pasta shapes are predominantly eaten by children, usually from between the ages of 4–11 years. Because of this it was decided that special attention should be made to salt reductions in this food category. Pasta shapes are often eaten by young children at the time that they are weaned off conventional 'toddler foods' and begin to move onto eating an adult-style diet. Unfortunately these diets do not take into consideration a child's requirement for lower amounts of sodium and can, therefore, be contributing to excessive salt intake in these children. An internal benchmarking exercise showed that there was a substantial step-change in saltiness from our own range of toddler foods, which contain no added salt, to that of Heinz Pasta Shapes in Tomato Sauce. It was reasoned that a child's taste, migrating from weaning/toddler food would accept a much lower salt content than was in the current recipe. Therefore, in addition to the initial 20% reduction in salt as happened to Spaghetti, Heinz Pasta Shapes in Tomato Sauce had a further 35% of its salt content removed. This resulted in a sodium level of 168 mg per 100 g, approximately half of the target average sodium level suggested by the FSA (2003) for the category.

Research into potential salt reductions in Cream of Tomato soup suggested that consumers would only be willing to move to an 8% salt reduction, possibly to a 12% reduction, but that any greater reduction would not be well received by consumers. This degree of salt reduction was, however, not in the vicinity of what was required by the FSA Salt Model (2003). The decision was made to not simply remove salt from the soup recipe, therefore, but to address salt reduction through a total review of all Heinz soup recipes themselves, as detailed below.

17.5   Salt reduction in canned soups

The FSA targets for salt reduction in soups (55% reduction) were relatively more ambitious than those for Baked Beans or Pasta in Tomato sauce (36% reduction). Previous salt removal from Heinz soups had been moderate (up to ~20% reduction) and occurred over a number of years, yet had progressed with little or no comment from the consumer. The more ambitious targets set by the FSA and adopted by Heinz could be expected to be noticed by the consumer if implemented purely by means of reducing the salt added to the existing recipe. It was decided, therefore, that a successful campaign of ambitious salt reduction in soups could only be achieved through a complete change in the soup recipes themselves. Heinz began a reappraisal of each of its soup recipes, returning to kitchen-based recipes, reinvesting in core ingredients, removing as many technological ingredients as was possible and, most importantly, substantially reducing the amount of salt added to the recipes. This created a range of soups that were noticeably different from their predecessors, both nutritionally and sensorially, a brave move in a market sector founded on traditional tastes yet one which Heinz was committed to through its policy of providing its consumers with 'Heinz Good Food Every Day^{TM}'. Table 17.2 shows some examples of the recipe changes that occurred in some classic recipes as a result of this process.

Overall, there was a 24% average reduction in salt content of the soup recipes and every recipe was within the first target of Heinz's own step-wise salt reduction plan, bringing the salt content of a 200 ml serving of any Heinz soup to no more than 1.4 g.

Table 17.3 shows the overall changes in QSM salt content following the recent salt reduction activities, along with a summary of the changes that have taken place since 1997. It can be seen that some substantial reductions have been made in this time and this is reflected by the fact that in the first year of producing these new recipes 19% less salt was used as an ingredient at the Heinz canning factory in Wigan.

Table 17.2   Summary table of soup recipe changes

Variety	Recipe change
Traditional	
Chicken	Chicken increased 40%
	Removal of MSG
	Salt reduced 26%
Vegetable	Removal of MSG
	Sugar reduced 9%
	Carrots increased 13%
	Potatoes increased 13%
	Dried peas replaced with frozen peas
	Green beans removed
	Addition of swedes
	Salt reduced 18%
Tomato	Tomatoes increased 12%
	Oil reduced 18%
	Sugar reduced 16%
	Salt reduced 24%
Minestrone	Addition of olive oil
	Removal of MSG
	Dried leeks replaced with frozen leeks
	Green beans and cheese removed
	Addition of onions and celery
	Total vegetables increased 39%
	Salt reduced 17%
Big Soups	
B/S Chicken & Veg	Chicken increased 28%
	Vegetables increased 5%
	Removal of chicken fat
	Removal of MSG
	Salt reduced 26%
B/S Beef & Veg	Beef increased 20%
	Vegetables increased 10%
	Removal of MSG
	Salt reduced 28%
B/S Lamb & Veg	Lamb increased 25%
	Vegetables increased 10%
	Salt reduced 27%

17.6   Collaborative work within the industry – Project Neptune

In parallel with the in-house ongoing activities on salt reduction, in 2002 a group of leading branded soup and sauce manufacturers joined forces to launch an initiative to explore ways in which they could support the Food Standards Agency's objective of reducing the amount of salt in the UK diet. Membership of this group, under the title of 'Project Neptune' included Baxter's, Campbell's

Table 17.3 Summary of salt reductions in QSM 2002–present

Heinz Variety		Salt (g per 100 g) 2002	2003	% reduction 2002–2003	Current values December 2004	% reduction 2003–2004	% reduction 1997–2004
Beans	Baked Beans	1.16	0.99	15	0.85	12	32
	Beans and Pork Sausages	1.38	1.38		1.01	27	34
Pasta	Spaghetti in Tomato Sauce	0.89	0.78	12	0.62	21	31
	Spaghetti with Sausage in TS	1.27	1.27		1.01	20	20
	Pasta Shapes in Tomato Sauce	0.80	0.64	20	0.42	35	59
	Pasta Shapes and Mini Sausages		0.76		0.49	36	61
	Spaghetti Hoops	0.80	0.63	21	0.42	34	59
	Spaghetti Hoops and Hotdogs	1.50	1.01	33	0.84	17	44
Soups	Cream of Tomato	1.01	0.91	10	0.70	23	32
	Cream of Chicken	0.89	0.86	3	0.64	26	28
	Vegetable	0.97	0.84	13	0.69	18	41
	Oxtail	1.29	1.02	21	0.69	32	49
	Cream of Mushroom	0.86	0.86		0.70	19	26
	Mulligatawny	1.09	1.05	3	0.66	37	41
	Big Soup Beef and Vegetable	0.98	0.90	8	0.65	28	43
	Big Soup Chicken and Vegetable	1.10	0.90	18	0.67	26	46
	Pea and Ham	0.90	0.89	1	0.68	24	24

Grocery Products Ltd, Centura Foods, Heinz, Masterfoods, McCormick Foods, Patak's, Premier Foods and Unilever Bestfoods – spanning a range of household name products in these categories, and was administered by the Food and Drink Federation and in open partnership with the FSA.

At the outset of this initiative, Project Neptune agreed to implement the following:

- A 10% reduction in sodium for branded ambient soups and sauces by the end of 2003 on the baseline figures for 2002.
- To make further reductions, of a similar order of magnitude, in 2004 and 2005, where technologically possible, safe and acceptable to consumers, i.e. a 30% reduction over three years.
- A policy whereby all Project Neptune members would ensure that any reformulated variety would be lower in sodium than the recipe it replaced and any new variety would be lower in sodium than its nearest equivalent in the same range.

The recipe redevelopment work ongoing at Heinz at the onset of this collaboration, combined with the more significant reductions outlined above, resulted in Heinz reaching its 30% commitment within two years, although Heinz continued to remain an active partner in Project Neptune for the duration of the collaboration.

To date so far the group has achieved sodium reductions in the wet soups category, taking the 6% reduction achieved in 2003 to a cumulative reduction of 23.7% to the year ending December 2005.

17.7   Example of reduced sodium/sugar Baked Beans

In addition to the salt reduction work occurring in the traditional Heinz Baked Beans recipe, an additional variety of Baked Beans was launched, to provide the consumer with a choice of product that was substantially lower in salt than the (new) standard recipe. Reduced salt reduced sugar baked beans contained 173 mg of sodium per 100 g, half that of the standard recipe. Acceptance of this recipe by consumers was, however, quite low and it remains a niche product bought, in the main, by individuals specifically looking for lower salt offerings.

17.7.1   Consumer response

The pre-reduction consumer and sensory testing was conducted to ensure that any salt reduction activity did not impact negatively upon consumer satisfaction with the products. The two-stage salt reduction in Baked Beans was subtle enough that very little negative response was monitored by our Careline. Indeed, consumer comments about the changes only occurred after the salt reduction programme had been publicised in the media, suggesting that it was the principle of changing such a well-established recipe that upset a few consumers more than the change itself. Therefore it may be suggested that taking steps to characterise the sensory profile of foods and how these profiles affect consumer satisfaction with foods can provide an effective means of optimising the extent of reduction, which does not impact upon consumer acceptability.

17.8   Conclusions

For the campaign to reduce salt in the nation's diet to be successful it requires effort from many players. It is hoped that the work that Heinz has conducted will have contributed to a successful campaign. Care will always need to be taken by manufacturers to achieve the balance between salt reduction to benefit health and a food's taste and consumers' acceptance of that food. If consumers choose to migrate from a lower salt option of a particular food to a different brand or different type of food that has a higher salt level then this does not benefit consumers' health and is also not appealing to the manufacturer of the lower salt food. Further careful research is required to investigate the acceptance and motivation to consume lower-salt foods in a marketplace where variety allows the customer to let their diet be determined by their tastebuds.

17.9   References

BBC (2006). Husband eats 50-year-old Chicken. BBC Online. Available from http:// news.bbc.co.uk/1/hi/england/manchester/4693520.stm. Accessed 30 May 2006.
CANNED FOOD ALLIANCE (2006). History of Canning Process. Available from http:// www.mealtime.org/default.aspx?id=328. Accessed 30 May 2006.
DEPARTMENT OF HEALTH (1991) Dietary Reference Values for Food Energy and Nutrients for the United Kingdom. Norwich: TSO.
DEPARTMENT OF HEALTH (2004) Choosing Health: Making healthy choices easier. London: HMSO Available from http://www.dh.gov.uk/assetRoot/04/12/07/92/ 04120792.pdf. Accessed 30 May 2006.
FOOD STANDARDS AGENCY (2003) UK Salt Intakes: Modelling Salt Reductions. Available from http://www.food.gov.uk/multimedia/spreadsheets/saltmodel.xls. Accessed 30 May 2006.
FOOD STANDARDS AGENCY (2005) UK Salt Intakes: Modelling Salt Reductions. Available from http://www.food.gov.uk/multimedia/pdfs/saltmodelfeb05.pdf. Accessed 30 May 2006.
SCIENTIFIC ADVISORY COMMITTEE ON NUTRITION (2003) Salt and Health. London: HMSO.

Index